T0076124

ENSNARED BETWEEN HITLER AND STALIN

Ensnared between Hitler and Stalin

Refugee Scientists in the USSR

DAVID ZIMMERMAN

UNIVERSITY OF TORONTO PRESS
Toronto Buffalo London

ISBN 978-1-4875-4365-5 (cloth)   ISBN 978-1-4875-4366-2 (EPUB)
ISBN 978-1-4875-4367-9 (PDF)

Library and Archives Canada Cataloguing in Publication

Title: Ensnared between Hitler and Stalin : refugee scientists in the USSR /
    David Zimmerman.
Names: Zimmerman, David, 1959–, author.
Description: Includes bibliographical references and index.
Identifiers: Canadiana (print) 20220416648 | Canadiana (ebook) 20220416737 |
    ISBN 9781487543655 (cloth) | ISBN 9781487543662 (EPUB) |
    ISBN 9781487543679 (PDF)
Subjects: LCSH: Scientists – Soviet Union – Biography. | LCSH: Scientists – Germany –
    Biography. | LCSH: Refugees – Soviet Union – Biography. | LCSH: Scientists –
    Soviet Union – History – 20th century. | LCSH: Scientists – Germany – History –
    20th century. | LCSH: Refugees – Soviet Union – History – 20th century. |
    LCSH: Forced migration – Soviet Union – History – 20th century. | LCSH: Forced
    migration – Germany – History – 20th century.
Classification: LCC Q141 .Z56 2023 | DDC 509.2/247 – dc23

We wish to acknowledge the land on which the University of Toronto Press operates.
This land is the traditional territory of the Wendat, the Anishnaabeg, the Haudenosaunee,
the Métis, and the Mississaugas of the Credit First Nation.

This book has been published with the help of a grant from the University of Victoria's
Book and Creative Works Subvention Fund.

University of Toronto Press acknowledges the financial support of the Government of
Canada, the Canada Council for the Arts, and the Ontario Arts Council, an agency of
the Government of Ontario, for its publishing activities.

Contents

List of Images   ix

Acknowledgments   xi

Introduction   3

  1 Scholars and Scientists   14

  2 German Scientists in the Soviet Union before Hitler   37

  3 Scientists and Communists   51

  4 Scientists in Flight to the Soviet Union   68

  5 Living in Stalin's Soviet Union   91

  6 Refugee Scholarship in the Soviet Union   114

  7 The Great Terror   134

  8 Into Stalin's Frying Pan   162

  9 From Stalin's Frying Pan into Hitler's Fire   181

10 From the Great Terror to the Shoah   201

11 Survival and Triumph   225

12 The Ensnared in the Cold War   249

13 The Long Ordeal   283

Conclusion: The Ensnared and History   291

Notes   299

Bibliography   325

Index   345

Images

1   The memorial plaque to Fritz Noether   4
2   Heinrich Luftschitz   20
3   Portrait of Herman Muntz   40
4   Alex Weissberg and Eva Striker in Kharkiv   42
5   Fritz Noether with his wife Regina and sister Emmy in
    Germany   62
6   Ena and Kurt Zinneman   66
7   Fritz and Charlotte Houtermans with their first child,
    Giovanna   72
8   Hermann Borchardt's German passport   80
9   The Schein family – Hildegard, Ed, and Marcel   87
10 Konrad Weisselberg and (Anna) Galia Mykalo   101
11 Hans Hellmann with Hans Jr.   126
12 One of the prison recipes written on cigarette paper   171
13 Siegfried Gilde's questionnaire for the registration of health
    professionals in occupied Warsaw   188
14 Helmuth Simons with the Swiss-Jewish-Communist artist Alis
    Guggenheim   218
15 Ernst Simonson recording a ballistocardiogram   240
16 David Rousset and Alexander Weissberg   274
17 Galia and Alex Weisselberg in Kharkiv   287

Acknowledgments

Instrumental in writing this book has been the support of the families of the Ensnared, including George Sadowsky; Art Kharlamov, Mica Nava, and Kiffer Weisselberg; Edgar Schein; Bill Chicurel (Siegfried Gilde's great-nephew); Monica Noether, Evelyn Noether Stokvis, and Margret Stevens; Hans Hellmann Jr, and Petra Netter; and Pamela Zinnemann-Hope. These descendants of the Ensnared willingly shared their personal and family stories, private archives, and photographs. Professor Mikhail Shifman, whose great work in chronicling the events at the Ukrainian Physical Technical Institute, has been instrumental in making this study possible. Professor Shifman also provided invaluable photographs from his personal collection. Shifman is first and foremost among a large number of historians who have written biographical studies of the scientists included in this book.

My colleague Serhy Yelechyk provided invaluable guidance and assistance in locating the NKVD files of Kurt and Ena Zinnemann. Professor Yelechyk and Professor Perry Biddlescombe provided insightful comments on the manuscript. Commodore Jan Drent (Royal Canadian Navy, retired) did a masterful job translating documents and identifying hard-to-find secondary literature. I am deeply grateful to Dr Michal Palacz and Professor Paul Weindling at Oxford Brookes University for their willingness to share their findings on Siegfried Gilde's time at the Czyste Hospital in the Warsaw Ghetto. Dr Palacz and Paula Larson did a superb job in conducting follow-up research for me at the Bodleian Library, Oxford University. Dr Christopher Hesse, Freie Universität Berlin Institut für Publizistik- und Kommunikationswissenschaft, enthusiastically shared his team's research on Hermann Borchardt. Much thanks to Pamela Zinnemann-Hope and Ward Wood Publishing for permission to use the poetry from *On Cigarette Papers*.

Ari Sigal, Reference Librarian at Catawba Valley Community College in Hickory, North Carolina, assisted me with the Gilde-Marx Collection for Holocaust and Genocide Studies and in putting me in contact with Siegfried Gilde's relatives. Dan Flanagan, Archives Assistant, J.W. England Library, University of the Sciences, Paul Doty, Interim Special Collections and Archives Librarian, St Lawrence University, and Ralph A. Pugh, University Archivist and Adjunct Professor of History, Illinois Institute of Technology, all aided my research with their timely answers to my often obscure queries. Professor Eliot Fried, Okinawa Institute of Science and Technology Graduate University, graciously gave his time to explain the significance of Michael Sadowsky's mathematical research. The author wishes to thank Dr Milton Leittenberg for information on Pavlosky and on Soviet biochemical warfare research. Janet Kinrade Dethick assisted with information on the Castello Guglielmi internment camp. My warmest thanks to the staff of the Bodleian Library, Department of Special Collections, and the New York Public Library, Manuscripts and Archives Division, where the bulk of the primary research for this book was undertaken. Göran David Gerby allowed me to publish the portrait of Herman Muntz. The work was only made possible by the tireless work of the UTP editorial team, including Stephen Shapiro and Janice Evans, Stephanie Mazza in marketing, and Matthew Kudelka, whose superb copy editing made sense of my sometimes tangled prose.

This book was supported in part by funding from the Social Sciences and Humanities Research Council of Canada.

Finally, to Wendy Muscat-Tyler, my always zealous editor, cheerleader, and love of my life, who endured my hundreds of eureka moments as I made yet another new discovery, thank you with all my heart. Without you I could never have persevered through this never-ending labour of love.

ENSNARED BETWEEN HITLER AND STALIN

Introduction

In a small cemetery in the German Black Forest town of Gengenbach there is a memorial. The plaque reads:

> In Memoriam
> Noether, Prof.
> Prof. Dr. Fritz Alexander Noether
> 7 Oct. 1884 Erlangen – 10. Sept. 1941 Orel
> Iron Cross 1914–1918
> Victim of two dictatorships
> 1934 Expelled from Germany due to race
> 1938 Accused and sentenced by the Soviets
> 1941 Executed 1988 declared innocent.[1]

The memorial to Professor Fritz Noether was erected by his sons Herman and Gottfried a short time after they received official notification from the Soviet embassy in Washington, D.C., in May 1989, that their father had been executed some forty-eight years earlier in a Soviet NVKD (secret police) prison. Herman and Gottfreid had last seen their father on 22 November 1937, the day he was arrested in Tomsk, Siberia. The only confirmed sighting of Fritz before that letter from the embassy was received occurred in late 1939, when the Austrian physicist Alexander Weissberg caught a glimpse of the German mathematician in a Moscow prison cell. Herman and Gottfreid erected the plaque to their father just in front of the tombstone of their mother Regina, who died in a nearby mental hospital in 1935, having been sent there after suffering a mental collapse during her exile in Tomsk. This memorial was the closest their sons could get to reuniting their parents. Fritz's final resting place remains unknown.

Image 1. The memorial plaque to Fritz Noether (picture by Evelyn Noether Stokvis, 2015).

Fritz Noether was one of a small group of German and Central European scientists and academics who fled Hitler's Germany in the 1930s, seeking refuge in Stalin's Soviet Union. The account that follows will examine the lives of these thirty-six scholars and their families, all of whom found themselves ensnared in the horrific persecution and violence unleashed by Hitler and Stalin. The story of the scholars who sought refuge in the Soviet Union, which has never been fully told, is markedly different from the accounts of those who were fortunate enough to find employment in Great Britain or the United States.

The mass exodus of academics from Germany began in the spring of 1933, when the German Reichstag passed the Law for the Restoration of the Professional Civil Service, thus depriving most Jewish university faculty of their positions. The Nazis' early targeting of university faculty was deliberate, one of several acts designed to remove any potential opposition to their intellectual leadership. On the surface it might seem a straightforward task to identify academic refugees, but this is not the case. Many historians writing about refugee academics have not defined precisely who those academics were, or how many. The best estimate, which this study will use, is between 1,500 and 2,000 displaced academics, a figure derived from the prewar annual reports of the Society for the

Protection of Science and Learning, the British academic rescue organization; however, some recent studies have put the figure as high as 5,000.
Without a comprehensive list of refugee academics, it is impossible to be
more precise, but in large measure, the discrepancy in numbers arises
from the chronological scope of the different studies. There were three
distinct waves of refugee scholars between 1933 and 1941. The first comprised those scholars who lost their positions in Germany between 1933
and 1935. The second began with the passage of the Nuremberg Laws in
1935, which led to the dismissal of even more university faculty as racial
criteria became more restrictive and enforcement was tightened. German scholars were soon joined by those fleeing Austria and Czechoslovakia after the Nazis annexed those countries. As fascism and antisemitism
increased throughout much of Europe, other scholars were displaced
from countries like Spain and Italy. Included in this second wave were all
those who managed to escape the Soviet Union.

In the third wave were those who found themselves driven into exile
from all the countries the Germans and Italians had conquered in the
first two years of the war. Many of the scholars in this exodus had already
fled one or more countries before they were again forced to escape the
Germans and their allies. As we shall see, not all of those who survived
their first escape from the Nazis would survive their second encounter
with them. The discrepancy in numbers is largely a function of whether
these wartime "twice exiled" scholars are included in the count.

In selecting the individuals to include in this study, I have applied the
criteria established by the British Society for the Protection of Science
and Learning (before 1935, the Academic Assistance Council [SPSL or
AAC]); the American Emergency Committee in Aid of Displaced Foreign (before 1938, German) Scholars (AEC); and other international
academic assistance organizations. Academic refugees were generally
defined as individuals who had been displaced from research and teaching positions at universities, colleges, and research institutes. A few exceptions were made for non-academics with doctorates involved in research
in industry or government. Professionals, including medical doctors,
lawyers, and engineers, were also excluded unless they were also university instructors and researchers. Moreover, students were not under the
purview of the academic assistance organizations. Professionals and students were aided by other relief agencies. Similarly excluded were artists,
including musicians, actors, and writers, unless they also met the criteria
for being academic refugees. Though the criteria were relatively simple,
it was no straightforward task to determine who should be included in
this study. Some of the scientists included here would not have fit the criteria for receiving aid from the SPSL or the AEC. At least three had been

rejected by those assistance organizations when they applied to them before or after migrating to the Soviet Union. Every one of the scholars included in this study, however, held a position at a university or research centre, either before or after migrating to the Soviet Union.[2]

Academic refugees were a very small subgroup within the much broader community of those fleeing Hitler and his political and racial oppression in Central Europe. University teachers enjoyed certain distinct advantages over most other would-be migrants. In many disciplines, especially in the sciences and social sciences, there existed an international community linked by publications, shared research interests, and education. Scholars in specific fields knew one another personally. Much more than most people, they travelled widely for study, postdoctoral work, research, visiting lectureships, and conferences. These international activities were fostered by the Carnegie and Rockefeller Foundations, which provided funding for transnational scholarly exchanges. Germany's research excellence and the prestige of its graduate programs had encouraged many American and British students to study there, especially before 1914.[3] As we shall see, these relationships played an important role in the recruitment of scholars by the Soviet Union and in their subsequent efforts to re-establish themselves as exiles.

This study begins well before Hitler came to power in early 1933. Like all Germans and Central Europeans, those who eventually migrated to the Soviet Union had experienced the chaos that descended on their part of the world after the First World War, the traumatic political and social changes following the 1918 armistice, the emergence of the Soviet Union, and the Great Depression. Many of those displaced in the 1930s had already experienced the dreadful existence of refugee life, whether they were fleeing the czar's armies, antisemitic mobs, or the victorious Red Army during the Russian Revolution.[4] Until the Nazis came to power, German universities, despite the deeply entrenched academic antisemitism, were meccas of learning, with almost unmatched opportunities, even for Jews.[5] Academics, however, were among the first victims of Nazi persecution; scholars were being dismissed from German universities as early as March 1933. The vast majority of displaced scholars were Jewish or had sufficient Jewish heritage to be labelled non-Aryan by the German government. A much smaller group were fired for ideological reasons, such as membership in a trade union, or for espousing socialist or communist ideas.

The initial goal of most displaced faculty was to continue their careers, either in other countries or in other settings, such as industrial research facilities. Many scholars were able to flee Germany and re-establish themselves elsewhere with the help of academic refugee organizations.

Eventually, of course, leaving Germany and German-controlled territory became a matter of survival. While the survival rate of academics was perhaps higher than that of other groups, many scholars would later find themselves unable to escape and would perish in the Holocaust. Ultimately, the fate of scholars depended on many factors: the time when those scholars tried to leave Germany, their academic discipline and their status within it, their personal network in the international academic community, and sometimes, simple luck. In the early years of Hitler's regime, when opportunities to escape were relatively more numerous and more likely to lead to a university research position, individuals made decisions to leave Germany without realizing they were probably saving their own lives and the lives of their families. Until well into 1934, the general consensus both within and outside the country was that these dismissals were merely temporary and that the German government would soon reverse its destructive policies.

The first choice for most refugee academics was to relocate to a university research position in the United States, Great Britain, or Western Europe. Other seemingly attractive opportunities existed in the Soviet Union. Many intellectuals in Central Europe held a utopian view of Soviet achievements, which strengthened their desire to migrate there.[6] Even after his expulsion from the Soviet Union in 1938, the physicist Martin Ruhemann maintained that the Soviet approach to science was superior to that of other countries because barriers between Soviet scientists and workers had been broken down for the common good. This "penetration of the scientific approach into the cornfield and the factory, the bathroom and the kitchen is ... essential for development of a society in which each shall work according to his ability and receive according to his needs."[7] Due to the prevalence of this exuberant view of Soviet science among intellectuals in Western Europe in the 1930s, the few historians who have written about the migration of scientists to the Soviet Union have assumed that "only Communists and fellow-travelling intellectuals and artists were permitted to live in the USSR."[8] This may have been true for most of the intellectuals who fled Germany for the Soviet Union after 1933, but scientists fit into a distinct category of immigrants – they were foreign specialists, who could be granted a work permit without affiliation in any communist party provided that they were sponsored by a Soviet research centre.

Certainly, many of the scientists who sought refugee in the Soviet Union were active supporters if not members of Communist parties in Germany or Austria. Before emigrating, a number of these party members went so far as to spy for the Soviets.[9] A careful examination of the migration trajectories of these scholars, however, reveals far more complex motivations. Ideology was hardly the only driver, and most of the academic migrants were not Communists. Also, many scientists did not specifically choose the Soviet Union; rather, they fled there only after exhausting every other avenue of escape. In this book I will examine the full gamut of reasons why these scholars fled to the Soviet Union by tracing their experiences after escaping from Germany.

Once in the Soviet Union, the scientists and their families experienced life in a strange new land, and their adjustment was made more difficult by an alien language and customs. Initially, the professional opportunities for research and teaching persuaded even the most cynical among them that they had made the right decision to migrate there. Prior to the revolution, Russia scientists had been part of the broader European scientific community that reshaped the understanding of the physical world in the late nineteenth and early twentieth centuries. The destruction of the First World War and then the Russian Civil War, which ended with a Bolshevik victory, saw a dramatic decline in science in the early years of the new Soviet state. Many of the best and brightest Russian scientists and university faculty had fled the country; even so, enough scientists remained behind to form a nucleus for rebuilding. They were supported by an unprecedented amount of direct state investment in science, which was considered integral to the Soviet economic development. Under Stalin's Five-Year Plans, the Soviet state had made major investments in university and research institutes and appeared open to the participation of the international scientific community.[10] With massive state support, world-class research centres – such as the Ukrainian Physical Technical Institute (UFTI) in Kharkiv – had sprung up seemingly out of nowhere.[11] These centres boasted some of the best-equipped research facilities in the world and were staffed by brilliant Soviet colleagues. They also seemed to enjoy almost inexhaustible funds to acquire the latest scientific equipment, even if the expertise to operate and maintain it was often lacking. Perhaps, just perhaps, some thought, the Soviet Union would live up to their idealistic expectations.

For the scholars and their families there were signs that the utopian vision of Soviet life as propounded by Western intellectuals was not completely off the mark. In most cities, refugee scholars noted that tremendous efforts were under way to improve the lot of the average Russian. Primary education was being made available to all, public health care was free and was being rapidly expanded, and huge building projects were being launched – state-of-the-art factories and multistorey apartment complexes in addition to modern research facilities. Women seemed to have far more opportunities to pursue academic research and could continue to work even after they had married and started families. Workers were apparently running institutions collectively, based on socialist principles. A rich cultural and social life promoted communist revolutionary values.[12]

Yet even in the best of times, life in the Soviet Union was difficult. For most of the foreign-expert scholars, those hardships were only partly lessened by the privileges they were offered. Anxiety born of insecurity was the sole constant. Even in the relatively calm times between the end of the Great Famine and the start of the purges, they feared the secret police, public denunciation, informers, arbitrary arrest, and expulsion. Setting aside why they had migrated to the Soviet Union, their decisions to do so ultimately proved to be disastrous. Almost all of the foreign scholars, just like so many Soviet citizens, became victims of Stalin's purges, which culminated in the Great Terror of 1937–38. In the end, all but one of the refugee scholars in the Soviet Union would either flee or be expelled, arrested, or killed.

The scholars who migrated to the Soviet Union were just a drop in the proverbial bucket compared to the thousands of Germans and Central Europeans who migrated there in the years leading up to Stalin's purges. There are no reliable estimates of foreigners in the Soviet Union, but, prior to the purges, the country appeared to be a safe haven for Communists from all over the world. Yet during the purges, international Communist Party members were specifically targeted by Stalin. By the end of April 1938, one German party member in Moscow had recorded the arrest of 842 anti-fascist countrymen as a result of the purges, though "in reality there was much more."[13] So many members of the German Communist Party (Kommunistische Partei Deutschlands; KPD) in the Soviet Union perished as a result of the purges that their survival rate was lower than that of their comrades who remained behind in Nazi Germany. The scholars who went to the Soviet Union were quite a distinct group from other refugees and foreign migrants there, since most were not Communist Party members. Only scholars who *were* party members or had become Soviet citizens were arrested. All the others fled or were expelled.

Those scientists who were arrested, but not summarily executed by the NKVD, endured lengthy stretches in Soviet prisons under horrific conditions. They were subjected to mental and physical torture designed to extract false confessions. When not being interrogated, the prisoners lived in unheated cells, so overcrowded that some remained standing all night because there was not enough room for everyone to lie down. The meagre, poor-quality food they received ensured slow starvation. In 1940, three of the remaining imprisoned scholars were handed over to the Gestapo in a prisoner exchange. Remarkably, two of these three survived the Second World War. Afterwards, the two survivors wrote early eyewitness accounts of the purges and the Great Terror. In addition to these memoirs, this study uses eyewitness accounts of life in Soviet prisons, hitherto unrecorded.

Those academics who were able to leave the Soviet Union found themselves forced to seek refuge elsewhere, along with thousands of other refugees trying to escape the Holocaust.[14] Returning to Western and Central Europe, the refugee scholars found themselves competing for visas and employment with hundreds of others, many of whom had been forced to flee Austria, Czechoslovakia, Spain, and other parts of the increasingly troubled continent. Some would re-establish their scholarly careers; many others would merely find safe havens. These were the lucky ones. A few would find themselves thrust back under the control of the Nazis and their allies.

Eight scholars, and in some cases their families as well, would be incarcerated in concentration camps, ghettos, or other parts of the elaborate apparatus set up by the Nazis to exterminate Jews. Six of those scholars survived the Holocaust, and their stories are told here, many of them for the first time. Three scholars, and at least one entire family, were murdered by the Nazis. One scholar and his family avoided arrest, living openly in Vichy France, their Jewish heritage hidden by false documents. The experiences of refugee scholars in the Soviet Union had an intense effect on their psyches.

Some of the scientists who survived both the Great Terror and the Holocaust had an impact on the Cold War that belied their small numbers. Most of this group, exhausted by decades of political turmoil, chose to focus on their academic careers rather than on politics. Remarkably, three of the scholars who had fled the Soviet Union would voluntarily reside in Communist East Germany. Indeed, they were among the most politically powerful academics in that country. These diehard Communists were among the many European intellectuals who refused to believe the "rumours" about Stalin's purges and "held out hope that the Soviet Union would fulfill its revolutionary ambitions." While many intellectuals clung to their belief in a Communist utopia, there was "a small but determined core of American and European intellectuals," many of them ex-Communists, who worked "to disrupt and discredit Soviet-backed front groups. Most were at least sympathetic to Communism or its variants during the 1930s. They feared the resurgence of the Communist idea as well as a cultural void in war-ravaged Western Europe."[15] Some of the scientists who had experienced the Soviet system at its worst became among the fiercest critics of the Soviet Union, of Stalin, and of Communism during the early Cold War. This small group of anti-Communist critics would have a profound influence on the attitudes of influential scientists, scholars, and intellectuals. For a few families, like the Noethers, their terrible experiences in Germany and the Soviet Union would plague their lives for decades, even after both dictators were dead and even after the Soviet Union collapsed in the early 1990s.

This group biography of these thirty-six scholars will show all aspects of the stories of academic refugees far better than studies that focus on the lucky few who fled straight to the United States or Great Britain. Some of the thirty-six would perish after being ensnared in the web of hatred and murder spun by Hitler and Stalin. Only a few, as a result of their own actions, professional and personal connections, the aid of academic rescue organizations, and *fortuna*, would go on to continue their scientific careers after the Second World War.

Much of the literature focuses on scholars who succeeded in re-establishing their careers, especially in the United States or Great Britain. These accounts provide a wealth of information on the contributions those individuals made to shaping, or even founding, particular fields of research in their host countries.[16] A passage from the recent book *Ark of Civilization: Refugee Scholars and Oxford University, 1930–45* illustrates the limitations of these studies:

> As this chapter in this volume makes clear, a very specific type of person was being rescued from the Nazis – only those people (predominantly men) who had made, and who would continue to make, a significant and internationally important contribution to scholarship. Oxford could not rescue everyone, but set itself to rescue, from the storm of Nazism, the greatest representatives of their disciplines. There were, of course, a fleet of such "arks" across the country, but of them all, Oxford was the flagship.[17]

This book will show that the type of narrative described above is the antithesis of the experiences of the vast majority of refugee scholars; it is an elitist view, and a misleading one. The notion that Oxford, or any other university, recruited the best and brightest, who were chosen through some sort of equitable procedure, is at best a gross oversimplification of what actually happened. Many of the scholars who were forced to migrate to the Soviet Union or other less desirable countries were leaders in their fields. Various factors determined the fate of displaced scholars.

This book's examination of these displaced scholars, who were linked together by their ill-fated decision to seek refuge in the Soviet Union, calls for a better understanding of the experiences of *all* the scholars in flight during the dreadful decade of the 1930s. American historian

Laurel Leff recently labelled most of the literature on academic refugees the "Triumphalist Tale." Quite independently, I labelled the writings on the subject the "Triumphant Narrative."[18] Whichever term is used, it can be said that even the most privileged refugees of the tempestuous 1930s and 1940s were likely to have suffered tremendous hardships. For many, the hardships proved fatal, if not to their very survival, then to their careers. This book, like Leff's recent book *Well Worth Saving* and Karin Otto's study of displaced scholars and the German Research Foundation, calls for a more general reappraisal of the academic refugee experience.[19] Ultimately, this need is more than merely academic.

Ensnared also challenges some of the assumptions previously made about the role of academic refugee assistance organizations, especially the AEC, the SPSL, and the French Committee for the Reception and the Organization of the Work of Foreign Scientists (Comité français pour l'accueil et l'organisation du travail des savants étrangers). Past scholarship about these organizations tends to highlight their successes in awarding fellowships to hundreds of displaced university faculty, allowing them to find academic positions in the United States, Britain, and elsewhere.[20] This study does not dismiss that positive view of the rescue organizations, but by examining a group whose members did not receive their support when they first fled Germany, it offers a more broad-based interpretation of their work.[21]

Notes on Sources

Twenty-four of the thirty-six ensnared scholars have individual case files in the records of the AEC and/or the SPSL. These files hold a wealth of information tracing their lives, including their application forms, curricula vitae, correspondence, evaluations, letters of reference, and copies of their publications. While the files of the two agencies contain some duplicate information, they often hold very different documents, including letters, job and grant applications, and immigration records. After the war, when the AEC was shut down, the SPSL attempted to trace missing academic refugees. This information is contained in the individual scholars' files. In addition, in the 1950s, as part of Lord Beveridge's history of the SPSL, his staff conducted a survey of the experiences of the displaced scholars. The collections of the AEC and SPSL must be used in conjunction with each other, something that many historians have failed to do.

Thirty of the scholars or their families have left behind diaries, memoirs, and oral histories and/or have been the subject of biographies. German doctoral dissertations contain a biography of the candidate, and these are invaluable in tracing the early lives of many of these scholars. Of particular note is the work of Misha Shifman, professor of theoretical physics at the University of Minnesota. Shifman's work in finding, translating, and publishing primary and secondary literature pertaining to events at the Ukrainian Physical Technical Institute has been instrumental to that part of this study.

A number of small collections of documents pertaining to these refugee scholars were utilized for this study; many of them have never before been used by historians. These include the private papers of the families of Kurt Zinnemann, Michael Sadowsky, and Konrad Weisselberg (Mica Nava), as well as Martin Ruhemann's correspondence with Sir Francis Simon at the Royal Society. A small collection of letters by Hans Hellmann was unavailable, but these were extensively quoted in his biography. Of particular note is Helmuth Simons's account of his life in France, found in the collection of the Wiener Library.

The dearth of Soviet documents from this period is reflected in the other literature on various aspects of foreign scholars in the Soviet Union, primarily in regard to the UFTI. This lack of documents is in large measure explained by the brief period of time the Ensnared were in the Soviet Union and by the tumult caused by both the purges and the German invasion during the Second World War. The principal Soviet sources used in this and other studies are the transcripts of NKVD interrogations. These files must be used with caution, since all confessions were extracted under some form of torture. Many transcripts consist of clearly fabricated confessions, with fantastic and elaborate stories of intricate conspiracies against the Soviet Union and Stalin himself. A few secret police files, however, may contain relatively reliable accounts provided by the prisoners. Summaries or the complete files of all but one of those arrested have been used. The NKVD files for Kurt and Ena Zinnemann are used here for the first time. Only the file of Siegfried Gilde was unavailable, although the file was viewed by German researchers just after the fall of the Soviet Union.

1 Scholars and Scientists

Before we begin to examine the flight of this group of scholars to the Soviet Union, we must consider their lives before Hitler came to power in 1933. Most previous studies of refugee scholars have tended to consider them as a homogeneous group, linked together simply by the fact that they were all university faculty who had been dismissed by the Nazis. They are further linked by their shared experiences in seeking positions elsewhere as refugees. The homogeneity of this group is emphasized by studies that focus on displaced academics from one specific discipline.[1] This study reveals that the academics who migrated to the Soviet Union in the 1930s were far more heterogeneous, besides differing in marked ways from those academics who found haven in the United States, Great Britain, and other countries. Of course, most of those who found refuge in the Soviet Union were displaced German scholars; that said, there were three other distinct groups of scientists who did not fit this traditional categorization.

Groups of Displaced Scholars in the Soviet Union

The academic migration from Germany and Europe to the Soviet Union commenced before 1933. Those who voluntarily left the region did so, in part, for economic and ideological reasons. The largest number of voluntary migrants, however, departed just before the Hitler came to power, sensing the growing danger to those of Jewish heritage and those whose political views were anathema to the Nazis. Eight scholars were voluntary migrants to the Soviet Union and thus were not initially refugees.

The largest group of displaced scholars were the seventeen who had been dismissed from academic positions in Germany in the months after Hitler came to power. Joining these refugees were two scientists, the physicist Werner Romberg and the linguist Wolfgang Steinitz, who had completed their doctorates in Germany just as the Nazis came to power and had been barred for racial and political reasons from seeking employment there. As well, Edgar Lederer, an Austrian-Jewish biochemist, left his postdoctoral fellowship in Heidelberg in September 1933.

Before leaving Germany, three of the refugees held non-academic positions there. Hans Baerwald, a physicist and electrical engineer, worked in the central laboratory of Siemens and Halske. The biologist Helmuth Simons had worked in a number of industrial research laboratories and as a technical correspondent for a trade journal before joining the German patent office. Like his academic colleagues, he was dismissed under laws that barred "non-Aryans" from the civil service. Hermann Borchardt held a PhD in philosophy and was a gymnasium (high school) "professor" until his termination for racial reasons. All three would be appointed to university research and teaching positions in the Soviet Union. Finally, the physicist Charlotte Houtermans, who had ended her academic career after the birth of her first child, would accompany her husband Fritz to the Soviet Union. As we shall see, her exile from Germany would lead her to resume her scientific career.

The final group who migrated to the Soviet Union were four Central European Jewish scientists, the physicists Guido Beck, Marcel Schein, and Laszlo Tisza, and the chemist Konrad Weisselberg; these scientists were outside of Germany in 1933, when they found themselves unemployed or unable to find academic positions. Beck was laid off from Prague's German University after the Czech government withdrew funding for his professorship. Schein, a Swiss-trained Czech, was perhaps the only academic in history to be made a refugee after being cast out by Switzerland. Tisza was searching for his first position after completing his PhD in Hungary when he was arrested for being a member of the illegal Communist Party. When he was released from prison after a year, he had to leave Hungary. Weisselberg, after completing his PhD at the University of Vienna in 1930, was unable to find an academic position in Austria; he was working in industry prior to his migration to the Soviet Union. Unable to find academic work in Germany after the Nazis came to power, the four turned to the Soviet Union as the country where they hoped to continue their careers.

These thirty-six scientists and scholars found what they hoped was a permanent refuge in the Soviet Union after fleeing Nazi persecution; this number compares favourably to the number of permanently placed refugee scholars in major Western countries such as France, and was exceeded only by the United States, Great Britain, Palestine, and Turkey. What marks the Soviet Union as distinctly different from other countries is that displaced scholars were allowed to immigrate only if they had

Table 1. The Ensnared

Name	Academic field	Reason for dismissal / heritage	Position when dismissed	Date of birth
Baerwald, Hans	Physics/electrical engineering	Jewish	Central Laboratory Siemens & Halske AG. Wernerwerk, Berlin Siemensstadt	1904
Beck, Guido	Physics	Jewish, but laid off in Prague	Lecturer, German University, Prague	1903
Bergman, Stefan	Mathematics	Jewish	Lecturer Privatdoz Institute of Mathematics and Institute of Applied Mathematics, Berlin	1895
Borchardt, Hermann	Philosophy/language	Jewish	Professor, various schools, Berlin	1888
Cohn-Peters, H. Jurgen	Physics	50% Jewish	Researcher, Berlin University	1907
Cohn-Vossen, Stephan	Mathematics	Jewish	Privtdoz with teaching assignments in Geometry University of Cologne	1902
Duschinsky, Fritz	Physics	Jewish	Assistant, Kaiser Wilhem Instutute for Physics, Berlin	1907
Emsheimer, Ernst	Musicology	Voluntary migrant, 1932, Communist	–	1904
Fröhlich, Hermann	Physics	Jewish	Assistant to Prof. Terres at the Technical Chemical Institute, Berlin Technical Hochschule	1907
Gilde, Siegfried	Medicine	Jewish	Hospital researcher, Berlin	1905
Harig, Gerhard	Physics	Political	Assistant, University of Aachen	1902
Hellman, Hans	Chemistry	Wife Jewish	Application to "habilitate" with the scientific contributions above was refused, and he was fired from his post as an assistant professor on 24 December 1933, Hanover	1903

Name	Field	Status	Position	Year
Houtermans, Fritz G.	Physics	Leaves after harassment from Gestapo. SPSL document says for being Jewish but he was only ¼ Jewish	Assistant to Prof. G. Hertz, Technische Hochschule, Berlin	1903
Houtermans, Charlotte	Physics	Leaves with husband	Housewife in 1933	1899
Lange, Fritz (Friedrich)	Physics	Jewish, Communist	Industrial scientist, Berlin	1899
Lederer, Edgar	Biochemistry	Jewish	Postdoctoral fellow, University of Heidelberg	1908
Levy, Saul	Physics	Jewish	Leaves Kaiser-Wilhelm Institute for Physical and Electrical Chemistry, 1932	1897
Luftschitz, Heinrich	Engineering	Jewish	Dresden Technical College	1885
Muntz, Herman (Chaim)	Math	Jewish, pre-Hitler migrant	University of Leningrad, 1929–37	1884
Murawkin (sometimes spelled Muravkin), Herbert	Physics	Jewish, leader Communist Esperanto Organization	University of Berlin	1905
Noether, Fritz	Math	Jewish, compulsory retirement 1934	Professor of mathematics and mechanics, T.H. Breslau	1884
Romberg, Werner	Physics	Socialist Workers Party, anti-Nazi group	Goes directly to Soviet Union after completing doctorate, 1933	1909
Ruhemann, Martin	Physics	Voluntarily resigns in 1932 to go to UFTI, Kharkiv. Jewish	Stuttgart Institute of Physics	1903
Ruhemann, Barbara	Physics	Voluntarily resigns in 1932 to go to Kharkiv. Religious affiliation unknown	Stuttgart Institute of Physics	1905

(Continued)

Table 1. (Continued)

Name	Academic field	Reason for dismissal / heritage	Position when dismissed	Date of birth
Sadowsky, Michael	Math	Married to Jew	Privatdozent at the Technical University in Berlin	1901
Schaxel, Julius	Biology	Ideological dismissal	Professor, University of Jena	1887
Schein, Marcel	Physics	Jewish, but dismissed for Swiss political reasons	Dismissed from Post at University of Zurich	1902
Schlesinger, Lotte	Musicology	Jewish, dismissed 1933	State Academic College of Music, teacher, Berlin	1909
Simons, Helmuth	Biochemistry	Jewish, dismissed 1933	Worked at Patent Office	1893
Simonson, Ernst	Physiology	Jewish, dismissed 1933	Medical Faculty, University of Frankfurt	1898
Steinitz, Wolfgang	Linguistics	Communist, Jewish	Asstitant, Hungarian Institute, Berlin	1905
Tisza, Laszlo	Physics	Jewish, Communist sympathizer	Goes to Soviet Union after being released from Hungarian jail for Communist activities	1907
Wasser, Emanuel	Physics	From Austria in 1932, possibly economic, Jewish	Resigns from University of Vienna to migrate to Soviet Union	1901
Weissberg, Alexander	Physics/engineering	Jewish, Communist	Working in industry before recruitment to UFTI 1931	1901
Weisselberg, Konrad	Chemistry	Jewish, voluntarily leaves Austria	Private industry	1905
Zehden, Walter	Physics	Jewish	After finishing doctorate leaves Kaiser-Wilhelm Institute for Physical and Electrical Chemistry, 1932	1904
Zinneman, Kurt	Medicine	Jewish	Junior Assistant, Hygiene & Bacteriological Institute, Frankfurt, dismissed 1933. At Jewish Hospital, Frankfurt, as surgical assistant, 1933–34	1907

received a permanent post. In Western European countries, the number of refugee academics temporarily placed in positions was far greater than those fortunate to have found stable, long-term employment.

The Background of the Scholars

The vast majority of the scholars were Jewish or of sufficient Jewish heritage to be dismissed on the basis of Nazi racial laws; two were fired for having a Jewish wife; only three were clearly dismissed on political grounds. F.G. "Fritz" Houtermans had a single Jewish grandparent and thus was not subject to the first round of anti-Jewish dismissals; he fled Germany after the Gestapo visited his apartment to question him about his left-wing political activities.[2] Julius Schaxel and Gerhard Harig, neither of them Jewish, both fled the country because of their public association with the KPD or affiliated organizations. In 1933, Harig was arrested by the Gestapo, which held him for several months before releasing him.

Eight of this group had been based in Berlin. Two scholars came from Frankfurt, two from Breslau. Other academics held positions in Munich, Dresden, Aachen, Cologne, Jena, Hanover, and Heidelberg. The large number from the German capital is directly related to the size and diversity of the post-secondary institutions in Berlin, but also to the fact that Jews often found more opportunities open to them in the larger centres. Another reason why so many academic migrants to the Soviet Union were Berliners was that many were members of a closely linked social group of left-wing intellectuals. This group will be studied in detail in chapter 2.

More than half the academics who relocated to the Soviet Union were physical scientists or engineers; 40 per cent of the émigrés were physicists. This composition reflects the Soviet state's focus on science and engineering. There were, however, two medical researchers, a physiologist, four mathematicians, two biologists, a biochemist, two language experts, and two musicologists.

A distinctive feature of the academic refugees in the Soviet Union was their age. Most of these scholars were young, their average age being just thirty-two in 1933. The youngest among them was twenty-four. Only three were older than forty-five. The youth of these scholars explains why the majority displaced from German teaching and research positions were junior academics; research assistants, lecturers, or *Privatdozents* (a position roughly equivalent to sessional instructor). Only two were professors. This unusual demographic profile was a direct consequence of the specific difficulties that younger scholars faced when seeking a coveted permanent position in the United States or Great Britain. In the early

Image 2. Heinrich Luftschitz, "I am, as you can see from the enclosed
photograph not unhealthy or weak, but a healthy man train [*sic*] by sport"
(New York Public Library, Manuscripts and Archives Division).

years of the academic refugee crisis, the Academic Assistance Council
and the AEC discriminated against junior scholars, fearing their employment might generate an antisemitic backlash among their own young
academics and graduate students. As a result, Soviet recruiters found
that the best available candidates were these up-and-coming younger
scholars, who did not yet have a sufficient international reputation to
command a scarce appointment at a Western university.

Of the three older scholars who went to the Soviet Union, only one
migrated there after exhausting all other possibilities. Heinrich Luftschitz was a forty-eight-year-old chemical engineer, specializing in concrete,
who had been dismissed from his positions as director of the Cement
and Mortar Laboratories and Material Testing Office at the University of
Berlin and as a *Privatdozent* at the Technischen Hochschule (Technical

College) in Dresden. His crime, under the Nazis' racial laws, was that he was a Jew. Luftschitz had twenty years' experience and could list sixty scholarly publications, but he was acutely aware that his age would factor against him in his quest for gainful employment. When he first applied to the SPSL, before going to the Soviet Union, he enclosed in his application a carefully crafted photograph of his left profile. He explained, "I am, as you can see from the enclosed photograph not unhealthy or weak, but a healthy man train [*sic*] by sport."[3]

The other older scholars, the two professors, went directly to the Soviet Union without applying elsewhere. Born in 1884, Fritz Noether was the son of the famous mathematician Max Noether. His older sister Emmy was considered by Einstein and others to be "the most important woman in the history" of mathematics. Fritz, too, followed in his father's footsteps; in 1922, he was appointed to the second chair of Higher Mathematics and Mechanics at the Technical University of Breslau. Noether's main expertise was in the mathematics of turbulent flow in liquids; his most famous work was a critical analysis of Werner Heisenberg's dissertation, *On the Stability and Turbulence of Liquid Currents*. Thirty years later, Heisenberg, the father of quantum mechanics, had to admit that "the paper of Noether, which in his time had made the whole theory of instability suspicious, seems to contain some mistake, but the mistake has not yet been found."[4]

Born in 1887, Julius Schaxel had been professor of biology at the University of Jena since 1916. According to Paul Weindling:

> Schaxel sought to establish a theory of development based on successive acts of historical formation when ontogenetic and environmental forces interacted. Far from a naive reductionist, Schaxel tried to combine the psychology of individual perception with a materialist philosophy of life. There was a continual process of historical change in nature. Heredity meant the production of ever-new combinations of genetic material. He was among the few biologists who were critics of racial science, and condemned volkisch demands for racial purity.[5]

Schaxel was one of the leading Communist intellectuals of the Weimar era. This drew him the intense hatred of the generally conservative Weimar-era professoriate. Schaxel was "the later Weimar Republic's most prolific and prominent writer on Marxism and natural science." By 1924, he had ceased active research but "concentrated on popularizing biology in the labour movement."[6]

Given their age, it is no surprise that few of the more junior academics who went to the Soviet Union were yet to be considered leading figures in their fields, although several were thought to be rising stars. One aspect of the refugee experience was the inability of many academics to restore their careers or to reach their full potential as scholars. By focusing on the few who managed to continue as cutting-edge scholars, most studies of displaced scholars have neglected to fully consider the vast damage caused by forced migration. It is impossible to measure lost potential; that said, very few of the scholars who sought refuge in the Soviet Union enjoyed academic success in ways that their early careers seemed to suggest they would. The prolonged chaos caused by their multiple moves in search of safe haven, arising during the most productive period of their lives, often crippled their careers.

Stefan Cohn-Vossen was one of these promising young scholars whose career was cut short by Hitler's rise to power. Born in 1902 in Breslau, Germany (now Wroclaw, Poland), Cohn-Vossen completed his doctorate in mathematics in 1924. He received his habilitation, the second doctorate required in Germany to pursue a permanent university faculty position, at the University of Cologne in 1929. He joined the faculty at Cologne and began to write a series of important articles on geometry, culminating in the publication in 1932 of *Anschauliche Geometrie* (titled in its English edition *Geometry and the Imagination*), a book he co-wrote with the eminent German mathematician David Hilbert. The book was quickly recognized as a brilliant introduction to mathematical geometry. It was widely translated, and republished many times and in many languages. On 29 April 1933, Cohn-Vossen was placed on leave by Cologne, and on 2 September, he was formally dismissed under the terms of the Law for the Restoration of the Civil Service on the grounds of his Jewish heritage. German mathematicians who supported Hitler downplayed Cohn-Vossen's role in *Geometry and the Imagination*, calling him a mere "amanuensis," or merely Hilbert's assistant, who had merely copied and edited the great scholar's manuscript.[7]

Another high flyer was the theoretical physicist Guido Beck. Beck was born in 1903 in Reichenberg (now Liberec), Bohemia, then part of the Austro-Hungarian Empire (after 1919, Czechoslovakia). Beck completed his PhD at the University of Vienna in 1925. Two years later, he wrote a paper showing "that several features of the photoelectric effect, including Einstein's relation, followed from Schrodinger's equation." The equation, only discovered by Schrodinger in 1925, was a groundbreaking mathematical formula for examining quantum mechanics. Beck's work was rediscovered years later and "played a significant role in the discussions of the foundations of quantum physics." Between 1928 and 1930 he wrote a series of pioneering papers in which he tried to "apply the new quantum mechanics to the nucleus." This "bold attempt" was "hampered by its very earliness," since the first paper appeared four years before the discovery of the neutron. In 1929, Beck moved from Vienna to Leipzig, where he spent three years as an assistant to Heisenberg, the father of quantum theory and the uncertainty principle.[8]

Friedrich (Fritz) Houtermans was on the threshold of becoming one of Europe's leading experimental physicists when he fled from the Gestapo in 1933. Houtermans was born in 1903, in Zoppot near the German Baltic city of Danzig, but he grew up in Vienna, where he lived with his mother Elsa. An accomplished scholar, Elsa was the first woman in the Austro-Hungarian capital to earn a PhD in chemistry. Houtermans attended the University of Gottingen in Germany, where he studied under the Nobel laureate James Franck. Houtermans earned his PhD in 1927.

Gottingen was the centre of an international community of physicists, all drawn to be near Franck, theoretical physicist Max Born, and other great scientists based there. Houtermans established several close personal and professional friendships, including with the theoretical physicists Robert Oppenheimer and Linus Pauling (US), Wolfgang Pauli (Germany), and George Gamow (USSR). He also became very close friends with Charlotte Riefenstahl, the one women in a sea of male scientists.

Houtermans was known for his quick wit, his outgoing personality, and his scientific brilliance. According to one story, almost certainly apocryphal, it was Houtermans who first explained that the large contingent of Hungarian physicists were actually from Mars. Afraid that their accent might reveal their true identity, the Martians pretended to be Hungarians, since that language was as incomprehensible to other people as Martian, and the Martians, like Hungarians, were unable to speak other human languages without a thick accent. Unlike many of his contemporaries, Houtermans never hid that his mother was half-Jewish. "He used to counter antisemitic remarks with the retort, 'When your ancestors were still living in trees, mine were already forging checks.'"[9] In the years ahead, Houtermans would live to regret such flippancy in dealing with those who espoused hatred of Jews.

In 1928, Houtermans moved to Berlin to work under Gustav Hertz, who had been the co-winner of the 1925 Nobel Prize with Franck. That same year, Houtermans and Gamow published a paper examining ways to refine the latter's theory of alpha decay and how the "theory could be applied to specific nuclei." This paper put Houtermans at the cutting edge of the new field of quantum mechanics, which was revolutionizing human understanding of the universe. More was soon to follow.

In 1929, Houtermans and visiting British physicist Robert Atkinson published a paper postulating that nuclear reactions were the "source of stellar energy." The two scientists, along with Gamow, coined the term "thermonuclear" to describe the reactions inside suns. The origins of this pioneering work on thermonuclear reactions went back to Gottingen, during a walk with Atkinson, who was visiting the university as part of a two-year Rockefeller fellowship to study in German universities. During this walk, the scientists discussed Gamow's recent theoretical work and some revolutionary experiments on nuclear reactions at Cambridge University. Atkinson hypothesized that this work might explain nuclear reactions within stars. Excited by this hypothesis, Houtermans developed the theoretical framework for the idea. This work was so important that, according to one account, it served as the inspiration for John Cockcroft at the Cavendish Laboratory in Cambridge, to begin work on the first particle accelerator to test the concept of nuclear reactions.[10]

In Berlin, Houtermans collaborated with Max Knoll to build one of the world's first electron microscopes. He reported his initial successes at a small regional conference in February 1932; just one year later, before he could develop the apparatus further, he was forced to flee Berlin. By the time he left Germany, Houtermans had received his Doctor Habilitatus the second doctorate required in Germany to teach at universities, and he had been appointed a full-time *Privatdozent* or lecturer. Before his career was interrupted, Houtermans was well on his way to becoming a professor.[11]

Fritz Lange, another promising experimental physicist at the University of Berlin, would follow Houtermans to the Soviet Union. Born in the German capital in 1899, Lange completed his doctorate at the Physical Institute of the University of Berlin in 1924, under the supervision of Nobel Prize laureate Walter Nernst. Lange's most famous and daring experiment was conducted in 1927 and 1928 with Berlin colleagues Arno Brasch and Curt Urban on Monte Generoso, high in the Swiss Alps. He and his co-workers realized that the key to unlocking the secrets of atomic structure that were being predicted by the latest theoretical work was to accelerate charged particles to bombard the nuclei.[12] At the time, a machine had yet to be invented that could generate a high enough charge for this purpose – between 2 and 3 million volts. So the scientists strung a 600 metre cable between two of the mountain's peaks and then suspended from it a wire net of 400 square metres. The net contained thousands of metal points to capture lightning from the frequent, intense thunderstorms for which the region was famous. The net worked, and large voltage pulses were observed, but they encountered difficulties tapping into the charge so that they could complete their experiment. Moreover, there was "no discharge tube for the acceleration of particle beams which could withstand such voltages." The experiment was abandoned after Urban fell to his death while working on the apparatus. While a failure, this experiment is remembered today as a pioneering effort to create a practical particle accelerator.[13]

Perhaps the most talented of this cadre of young scientists was Hans Hellmann, who by 1933 had made his mark as a pioneer of quantum chemistry. Born in Wilhelmshaven, Germany, in 1903, he received his doctorate in physics at the University of Stuttgart at the age of twenty-five, under the supervision of Professor Erich Regener. In 1929, Hellmann was awarded a postdoctoral fellowship at the Hanover Institute of Technology. Hanover placed a strong emphasis on applied chemistry; it was, however, only 100 kilometres from Gottingen, the centre for quantum physics. Hellmann decided to focus his postdoctoral research on "the quantum mechanical interpretation of chemical bonding and molecular structure." Over the next four years, Hellman made a series of ground-breaking discoveries; he was also one of the first to apply the Dirac equation to atomic structure in order to refute "speculation that spin is only an apparent property of bound electrons." British physicist Paul Dirac had published his equation the previous year. The equation indicates how subatomic particles like electrons behave when they travel at close to the speed of light.[14]

Hellmann went on to make important contributions to the diatomics-in-molecules "technique for computing approximate electronic energies of polyatomic molecules from known information about their constituent diatomic and atomic fragments."[15] In 1933 he published an important paper in *Zeitschrift fur Physik* (Journal of Physics) in which he outlined several key equations that applied quantum mechanics to chemical bonding. Included in the article was what would later be called the Hellmann–Feynman theorem, "a key ingredient of the quantum mechanical treatment of forces acting on nuclei in molecules and solids."[16] The theory is named after Hellmann and Richard Feynman, the American Nobel laureate who independently rediscovered the equation some years later.[17]

Later that year, Hellmann completed another pioneering article, this one co-authored with W. Jost, explaining to the general chemistry community "the basic ideas of quantum chemistry," including how quantum mechanics affected chemical bonding. Hellmann hoped his impressive body of postdoctoral research would enable him to "habilitate" and move into a permanent university research position. He was, however, a supporter of left-wing causes, and in 1929 he had married Viktoria Bernstein, a refugee of Jewish Ukrainian heritage who had converted to Lutheranism. Because of his politics and his mixed marriage, Hellmann was denied his second doctorate, and on 24 December 1933 he was dismissed from his position at Stuttgart.[18]

Edgar Lederer was a promising young biochemist when he headed east in his quest for a permanent scientific position. Lederer, the scion of a prominent Austrian Jewish family, had completed his doctorate at the University of Vienna in 1930, when he was just twenty-two years old. For the next three years he worked as a postdoctoral fellow in the laboratory of Professor Richard Kuhn at the Kaiser Wilhelm Institute for Medicine at Heidelberg. At Heidelberg, Lederer made a "striking and in fact revolutionary contribution" to organic chemistry and biochemistry by perfecting a means to separate and purify pigments using chromatography. This single discovery would have virtually guaranteed Lederer a permanent scientific research position had he not been forced to flee Germany after Hitler came to power.[19]

Many promising scholars fled to the Soviet Union, but many others were simply too young to have established themselves in the academy on completion of their PhDs. Werner Romberg and Laszlo Tisza would be among those fortunate to survive and, eventually, to become leading scholars in their fields. Three Berlin-based scholars were not as fortunate. Experimental physicists Herbert Murawkin and Fritz Duschinsky, and medical scientist Siegfried Gilde, would not survive the Second World War. In 1931, at Friedrich-Wilhelms-Universität in Berlin, Murawkin completed his thesis on mass spectrometry, which included developing a mass spectrometer, which he used to examine glass, metals, and salts.[20] The same year, he published four articles based on his dissertation in *Annals of Physics* (*Annalen der Physik*), a leading journal in the discipline.[21] These were the only articles he would ever publish. Duschinsky was a Czech Jew, who had studied in Prague and Paris before completing his doctorate in 1931 at the University of Berlin, magna cum laude. His supervisor was Professor Peter Prinsgsheim. Even before he completed his degree, Duschinsky was appointed an assistant to the experimental physicist Professor Karl Weissenberg at the Kaiser-Wilhelm Institute for Physics, also in Berlin. He published three scholarly papers in *Journal for Physics* (*Zeitschrift fur Physik*) before fleeing Germany in April 1933.[22] Gilde, a native Berliner, completed his doctorate at Friedrich-Wilhelms-Universität in 1932, in which he examined the comparative effects of two enzymes on the circulatory system. Upon graduation, he married Ruth Gelinek of Katowice, Poland. He would never get the opportunity to publish his results or to establish himself as a research scientist.[23]

The Central European Academic Diaspora before 1933

Surprisingly, only a minority of the refugee academics in this study (fourteen out of thirty-four whose birthplace is known) were born in Germany. Martin Ruhemann was born in Britain to German parents and held dual Anglo-German citizenship. A significant number came from elsewhere in Central Europe. Nine had been born in the old Austro-Hungarian Empire. Four were born in what was, before 1918, Russia, mainly in Poland.

After the dissolution of the Austro-Hungarian and Russian Empires, the citizenship of many Central Europeans was rather ambiguous. Emanuel Wasser and Guido Beck, for instance, were both Austrian citizens because of the time they had spent at the University of Vienna, but the former was born in what became Poland, the latter in Czechoslovakia. Each of the new nations created after the First World War had different rules for citizenship. Wasser, though born in Lvov (Lviv) in what became Poland, lost any rights to citizenship there because he had lived in Vienna after the First World War; this, even though his parents became Polish citizens when the new state was formed. Beck, however, retained his Czech citizenship. Determining nationality was even more complex than this, since under German law, all permanent faculty were classified as civil servants and thus had to be German citizens. At least four of the scholars were dual nationals of Germany and another nation.[24] Scholars from Central Europe, therefore, were an important subgroup among foreign scholars in Soviet universities. Some Central European scientists went to the Soviet Union in search of economic and intellectual opportunities; Jewish scholars throughout Central Europe went there because they had found themselves excluded from the much larger German academic community after 1933.

German racial laws, and, more broadly, antisemitism throughout Central Europe, affected not only German scholars but also German-speaking Jews as a whole. Prior to 1933, Germany had been more open than Austria, Poland, and Hungary to Jewish scholars. In the 1920s, Austrian universities became especially hostile to Jewish students and faculty. Guido Beck, the Czech-Austrian physicist, and the Viennese-born biochemist Edgar Lederer recounted being forced to leave Austria because of incessant harassment against Jews. Konrad Weisselberg, who remained in Austria after finishing his PhD in 1930, believed that his failure to secure an academic appointment was a direct result of his Jewish heritage.

At the University of Vienna in the 1920s, Jewish students had to endure twice-yearly *Judenkrawalle* (Jewish riots) organized by the *deutschnationale* (German nationalist) students. Lederer, who was blond and blue-eyed, was not himself physically accosted, but neither did he ever try to hide his Jewish heritage. As Lederer began his doctoral work, a petition was circulated, signed by fifteen students in the lab, declaring that "we do not want the Jewish student Lederer to come here." When this failed to block Lederer's entry into the program, an anonymous letter was sent to his supervisor claiming that the Jew Lederer was spreading wicked rumours about him. This letter poisoned Lederer's relationship with his supervisor. The chilly climate in the laboratory convinced Lederer to work day and night to complete his thesis as quickly as possible; when he received his degree only two years later, he realized he had no future in Austria.[25] He explained: "As a Jew, there was no way I could have even an assistant post at the University of Vienna, at least what my professor had told me without flourish. It is not known that Austria has always been much more antisemitic than Germany. In Vienna, even Sigmund Freud was never a full professor. And yet again this antisemitism did not reach the proportions encountered in a city like Graz in Styria."[26]

Some of the scholars displaced by the Nazis had already experienced being refugees. Long before the First World War, antisemitic violence was a constant threat to Jews in Central and Eastern Europe. The tumult that engulfed the region after the First World War and the postwar redrawing of national boundaries led to a massive forced movement of both Jewish and non-Jewish populations. The mathematician Michael Sadowsky was born in 1902 in what was then Dorpat, Russia. His parents were of "the greek-catholic faith [sic]," and his father had been a physics professor at "the universities of Dorpat, Petersbourg [sic], and Prague" successively. When the war began, the family returned to Dorpat; then in 1916 the approach of the German army forced the Sadowskys to flee to Viborg, Finland. After the Russian Revolution, Dorpat became Tartu, in newly independent Estonia, and Finland too broke away from what was now the Soviet Union. The Sadowskys were unwilling to migrate to the latter, yet they were unwelcome in Estonia as ethnic Russians. The family was allowed to remain in Finland but was denied citizenship there and thus became stateless. Michael only acquired German citizenship in 1928, after accepting his first academic position at the Technische Hochschule (engineering college) in Berlin.[27]

Konrad Weisselberg's family also experienced multiple periods as internal refugees within the Austro-Hungarian Empire. Weisselberg's father was "an assimilated German-speaking" Jewish timber merchant in the town of Berlad (sometimes spelled Barlad or Birlad) in Bukovina, then the easternmost province of the Austro-Hungarian Empire.[28] The region is now part of Romania. In 1907, when Konrad was just two years old, an antisemitic club, recently formed by leading teachers, priests, and political leaders of the town, incited high school students to riot against the enrolment of Jews. When the students who participated in the riots were expelled, "the others organized and equipped themselves with axes and clubs, and burst into the Jewish quarter destroying and looting. Eighty shops owned by Jewish merchants and craftsmen were damaged during the rampage."[29] The Weisselberg family fled to Chernovtsy, the capital of the region. In 1914, with the Russian army approaching Chernovtsy, the Weisselbergs fled again, this time west to Vienna. Here the family settled, and Konrad completed his education, receiving a doctorate in chemistry from the University of Vienna in 1930.

Martin Ruhemann's father Siegfried was a Prussian-born physicist. In 1881, Siegfried received his PhD from the Friedrich-Wilhelms-Universität in Berlin. Four years later he accepted a position as assistant to the Jacksonian Professor of Experimental Philosophy at Cambridge University in England. There, he overcame the initial hostility of those who viewed him as a German interloper and enjoyed a highly successful thirty-year teaching and research career, highlighted by his election to the Royal Society in 1914. When Martin was born in 1903, Siegfried decided to become a naturalized British subject. This did not save him from the explosion of anti-German hatred that spread throughout Britain during the First World War. After the sinking of the *Lusitania* in 1915, local newspapers attacked the university for paying the salary of a German professor. Siegfried received threatening letters, Martin was shunned by his schoolmates, and the family doctor refused to continue to treat them. Siegfried felt compelled to resign from his position at Cambridge, and the family lived quietly there until they were able to return to Berlin in 1919.[30]

Many of those who avoided becoming refugees before 1933 had their lives dramatically affected by the political, social, and economic turmoil that gripped Germany during, and in the fifteen years following, the First World War. Helmuth Simons was among a generation of scientists who were never able to establish a university or industrial career as a result of the tumult of this period. Simons was twenty-one years old when the First World War broke out. In 1914, he had just completed an undergraduate program in zoology at the University of Freiburg, with physics and chemistry as subsidiary subjects. In the summer just before the war commenced, Simons took a postgraduate course in parasitology and undertook his first research at the Zoological Station at Villefranche, on the Côte d'Azur in southern France. When the war began, he was training laboratory assistants to identify parasites in medical samples and beginning his doctoral research studying nagana, a parasitic disease of cattle, antelope, and other livestock in southern Africa, spread by the tsetse fly. On the basis of his work on parasitical infections, Simons was appointed the malaria expert for the German army's 7th Corps. He also lectured to Army medical personnel on the disease.

In 1919, Simons was finally able to complete his doctorate in zoology and biology at the University of Würzburg. He then undertook postdoctoral research at the medical school in Düsseldorf and at the Hygiene

Institute at Cologne. Financial difficulties forced him to abandon a university research career, and he turned to applied industrial science. He became chief chemist in the industrial laboratory of a firm that specialized in combating insect pests. The firm was in the Rhineland, which was under French occupation in the 1920s. The difficulties of conducting business under the occupation, and the hyperinflation that wrecked the economy of the Weimar Republic, forced the company to close its laboratory in 1924. The following year, Simons was hired to establish a research laboratory at a large flour mill "until this also came to an end in consequence of the financial crisis." For the next six years he kept his family fed by working on the occasional government research contract and as a chemical expert and reporter for a major trade paper for the flour industry. Finally, in 1932, he found what he hoped was stable, long-term employment in the German patent office as an expert adviser on food chemistry. Simons was Jewish, and like all civil servants, he became subject to Nazi racial laws. He was dismissed from the patent office in the summer of 1933.[31]

Like Simons, Hermann Borchardt was unable to pursue an academic career at a German university, despite receiving his doctorate in philosophy at University of Greifswald in 1917. Born in 1888, his parents were Lewis and Bertha (nee Borchardt) Joelson. Hermann's PhD supervisor was the Neo-Kantian philosopher Johannes Rehmke. Rehmke thought so highly of his former student that he immediately recommended him for habilitation. Instead of habilitating in hopes of a university career, Hermann opted for a livelihood teaching high school, and in 1921 he started teaching at the Koellnisches Gymnasium in Berlin.[32]

No clear reason is available why Hermann did not pursue a career in post-secondary education, but there can be little doubt that just as with Simons, the postwar economic situation precluded a university appointment. There is some evidence that antisemitism also played a significant role in Hermann's career choice. Discrimination against Jews in German universities was less than in other countries, like Austria, but it was still prevalent, especially in the humanities. Antisemitism permeated German society. Hermann was unable to secure a permanent state teaching position until 1928, one year after he legally changed him last name from the very Jewish-sounding Joelson to Borchardt, his mother's maiden name. In the official document authorizing the name change, the reason he gave for making the request was so that he could preserve his mother's family name, which would otherwise go extinct. Later on, Borchardt wrote over the document in red ink, explaining that he changed his name because of the increasing antisemitism he faced in his daily life.[33] Despite his efforts to hide his heritage, he too was dismissed for racial reasons in 1933. There are, no doubt, others among the refugee scholars who experienced similar hardships long before Hitler came to power. Unfortunately, there is little extant information about the early lives of many of these scholars, and it is impossible to expand this account further.

Women Refugee Scholars in the Soviet Union

As might be expected by the composition of university faculty in the 1930s, the vast majority of the refugee academics were male, although women scholars, wives, and also children are a significant part of this story. Barbara Ruhemann, nee Zarniko, Martin's wife, was among those scientists who migrated to the Soviet Union before 1933. Born in 1905 in the East Prussian town of Heiligenbeil (now the Russian town of Mamonovo), she was the daughter of a mill owner. It is quite likely that her family was caught up in the tumult of the Russian invasion of the region in 1914. While Heiligenbeil was not occupied by the Russians, her family may have been among the 350,000 refugees who fled west.

In 1925, Barbara finished her teacher training, but eschewing this traditional female profession, she moved to Berlin to study physics and mathematics. Two years later, she won a fellowship from the Academic Exchange Service that allowed her to spend a year at Barnard College, Columbia University, New York. She earned her baccalaureate certificate at Barnard before returning to Germany to complete her doctorate. While in graduate school in Berlin, Barbara met Martin, a fellow graduate student and her future husband. She earned her PhD in low-temperature physics in 1931 from Friedrich-Wilhelms-Universität. Before and immediately after completing her dissertation, she co-authored a number of papers with her supervisor, Professor Franz Simon. Unlike many of her married contemporaries, Barbara did not give up physics to become a housewife; rather, she and Martin were hired to teach physics at the University of Stuttgart.[34]

Charlotte Houtermans, née Riefenstahl, went to the Soviet Union as an accompanying spouse, though she was a physicist in her own right. Charlotte was born in Bielefeld, Germany, in 1899, the daughter of a journalist. After graduating from the gymnasium, she completed her teacher training. After her father's death, however, her mother asked her to help support her and a younger brother. For a number of years, Charlotte taught religion and geography at a private school. When she learned that her mother was putting aside the money she was sending home to fund her brother's education, she decided to apply the money toward her own university education instead.

In 1922, Charlotte began her studies at Gottingen. She learned mathematics from "such geniuses as Richard Courant and David Hilbert, and physics and physical chemistry from Max Born, Gustav Hertz, James Franck, Hertha Sponer, who later ... became the first woman on the physics faculty at Duke University, and Gustav Tammann."[35] Life was harsh for Charlotte and for most of the other students at Gottingen; the hyperinflation that ravaged Weimar Germany in the early 1920s had rendered her savings worthless. She was forced to take jobs teaching just to ward off starvation, and, like many of her fellow students, she could only afford one meal a day. Despite these hardships, life at Gottingen was vibrant and exciting. She explained: "These were the years when quantum mechanics was born, when the spin of the electron was discovered. One sat around the small marble tables at the café *Cron und Lanz*, and somebody would explain matrices or the Schrodinger equation, writing directly on the table." The café's waiters tolerated this, and lent money to the students, whom they addressed as "Herr Doktor," anticipating repayment with interest once their customers had graduated and had begun to receive salaries.

Many of the students Charlotte met would play crucial roles in her life. Among them were George Gamow, Robert Atkinson, the British physicist Patrick Blackett, and the Hungarians Michael Polanyi and Leo Szilard. She also met leading physicists – most importantly for Charlotte, the great Danish scientist, Niels Bohr. Two of Charlotte's classmates stood out for her as potential suitors, for their brilliance, charm, and, perhaps, comparative affluence. Robert Oppenheimer, the great American theoretical physicist and future "father of the Atomic Bomb," courted Charlotte as "best he could in his stiff and excessively polite manner." Oppenheimer, Charlotte later commented, "seemed strange to me, and far away, as far as a distant nebula." She would later confess that "perhaps, I underappreciated him at that time."

Fritz Houtermans, whom Charlotte affectionately called Fisl, was very different. He was a sophisticated and well-dressed Viennese, whose dark "Italian" features and confident, outgoing personality made him irresistible to many women. During the 1926–27 academic year, Charlotte fell under Fisl's spell – she found him "more engaging and more intriguing" with each meeting. She described feelings not uncommon to many who are hopelessly in love:

> What must have puzzled me most was that there seemed to be no ordinary explanation for his personality, no common denominator to shed light on his actions or sayings, his likes or dislikes. He was brusque and tender, aloof and social, attentive and remote. Because I believed he

loved me, I set out to understand him, maybe not rationally, but by trusting and loving him and believing everything he said in his good moments.[36]

By the end of the academic year, Fritz and Charlotte had reached an understanding that they would be married as soon as Fritz had established himself in a secure academic position. Charlotte had already been in contact with Professor Edna Carter, head of the physics department at Vassar College, to teach at the prestigious women's college for a year. After a successful year at Vassar, she taught for one more year at Winthrop College in South Carolina. Charlotte's two years in the United States would later prove invaluable. After Fritz was arrested by the NKVD, she found herself destitute, and her US connection would smooth her subsequent escape from the Soviet Union,

In 1930, Charlotte returned to Germany, settling in Berlin to be near Fritz. Both attended an international physics congress in Odessa at the invitation of Gamow. On a ship across the Black Sea, Fritz proposed to Charlotte. When she accepted, Fritz, in his typical impulsive fashion, declared that he could not wait. At the next port of call, the Georgian town of Batumi, the two were married. The witnesses were fellow physicists Rudolf Peierls and Wolfgang Pauli. Back in Berlin, Charlotte did not resume her academic career, instead working at a series of industrial laboratories. In April 1932, the Houtermans' first child, Giovanna, was born. Giovanna's birth should have marked a happy time for Fritz and Charlotte, with their family started and Fritz's professional career seemingly secure, but the politically active couple were well aware that the political clouds were darkening as the Nazis gathered strength. Just over a year later, the family would be forced to leave Germany.

The musicologist and composer Charlotte "Lotte" Schlesinger was the sole female academic refugee to choose the Soviet Union of her own accord.[37] Born in 1909, the daughter of a violinist and granddaughter of the famous Prague rabbi Moyzis Stark, Schlesinger was a child prodigy. After her parents' divorce, Schlesinger was raised by her uncle Rudolph Schwarzkopf, a journalist and movie producer. Schwartzkopf arranged for Schlesinger to have the very best musical education possible. At the age of eleven, she began studying composition, music theory, and piano. Just four years later, she successfully applied to join the composition class of Franz Schreker at the State Academic College of Music (Staatliche Akademisches Hochschule fur Musik) in Berlin. "A year later, her first public performance as a composer took place within a lecture evening of the composition class."

In 1927, Schlesinger joined the music seminar at College of Music, graduating two years later summa cum laude. In April 1930 she joined the academy's staff as a part-time extraordinary teacher. She composed a number of classical works, including quartets and cantatas, often as part of the college's lecture-performance evenings. In 1929, one of her compositions won the prestigious City of Berlin Beethoven Prize. She was recognized as a promising young composer who used "the form and expressive forces of classical modernism."[38] She would later confess that her years in Berlin had been the "best time" of her life. But her happiness was interrupted by an encounter at a "Bohemian Party" she attended, where she fell madly in love with a "tall, incredibly handsome, just addictive, funny, and very intelligent" young scientist named Fritz Houtermans. Fritz was accompanied by his fiancée, also named Charlotte. Schlesinger became the couple's best friend, always secretly in love with Fritz, never knowing whether Charlotte Houtermans ever suspected her desires. As its sole professional musician, Schlesinger became an "indispensable participant" in the Houtermans's social circle of Communist and socialist scientific intellectuals, a group that included Alexander Weissberg before he left for the Soviet Union. As happened with so many other scholars, Schlesinger's happy time came to an abrupt end when the Nazis came to power. She was dismissed from her position in March 1933 because she was a Jew, and the entire program of music education at the college closed just a few weeks later.[39]

Other Foreign Scholars in the Soviet Union

The thirty-six scholars in this study were selected using a much more catholic definition of academic refugees than was used by the AEC and the SPSL. Undoubtedly, other German and Central European refugee scholars chose to go to the Soviet Union; unfortunately, evidence for them is scanty at best. For instance, Professor Alexander Nathansohn, an ocean biologist, is mentioned as having settled there in just one Academic Assistance Council document, from October 1934. Nathansohn was born in 1878 in Brzezany, East Prussia (now Poland), and did his most important scientific work prior to the First World War. His life was well-documented until 1933, when he left Germany for Turin, Italy, to work as an industrial chemist. There is a no information on his whereabouts between 1933 and 1938, when he reappeared in Turin. If he was in the Soviet Union, nothing indicates where he lived or worked.[40]

Even less is known about Ludwig Rosenfeld, an Austrian engineer who worked for the Moscow State Oil Research Institute. Rosenfeld's name appears on a handwritten addendum to a list of scholars displaced from the Soviet Union found in the files of the SPSL. Despite considerable effort, no further information has come to light on his background, his time in the Soviet Union, or his later years. Other German academic scientists known to have migrated to the Soviet Union include the geologist Friedrich Stammberger, the biologist Georg Schneider, and the anatomist Louis Jacobsohn-Lask.[41]

Another refugee scientist excluded from the study is Boris Davison, who joined the exodus of foreign scholars from the Soviet Union that occurred in 1937 and 1938. Davison was a brilliant mathematical physicist specializing in hydrology (the study of the motion of groundwater). As indicated by his Anglo-Russian name, his grandfather was British, and Davison was a dual British–Soviet citizen. In 1938 the Soviet authorities presented Davison with the choice of renouncing his British citizenship or leaving. He chose to depart, and arrived in London that summer. As an academic refugee, Davison's case was unique; the SPSL was unsure whether, as a British subject, he qualified for assistance. The society, however, did view him as one more victim of the purges, and eventually they did what they could to find him a position. Davison had an impressive scientific career in Britain, the United States, and Canada, which included a brief stint at Los Alamos, where he worked on the atomic bomb program.[42]

There were also a number of foreign scholars and scientists in the Soviet Union from countries outside of Germany and the part of Central Europe directly affected by the Nazis' rise to power. Most of these migrants were Communists or sympathizers. Unlike most of the Ensnared, they were voluntary migrants, and when the purges began, safe havens were available to them in their home countries. Perhaps the best-known of these scholars was the American geneticist Hermann Muller, who would win a Nobel Prize in Medicine in 1946. Muller's story will briefly intersect with that of the Ensnared. With that exception, these scientists have been omitted from this study as well.[43]

Conclusion

This overview of the background of the scholars and their families who fled Hitler for the Soviet Union, with some important variations, could be a survey of almost any group of German and Central European academics who fled or were forced to leave after 1933. What differentiates these academics from other displaced scholars is that they chose to go to the Soviet Union and what happened to them after they did. The story of the Ensnared is not one of triumphant escapes, in which refugees rebuilt their research and teaching careers, as did many of those who found haven in Britain or the United States. Those were the lucky few. The reasons why the Ensnared chose Stalin's Soviet Union, despite ample evidence that it might well become as dangerous a place as Germany, will be explored in the next three chapters.

2 German Scientists in the Soviet Union before Hitler

European scholars had always enjoyed a degree of international mobility. With the inception of the research university in nineteenth-century Germany, and the new type of doctorate degree, a new international professional credential was established. Even without a PhD, a well-published scholar could establish a professional status and reputation that transcended national boundaries.

While not typical for the period, the 1881 appointment of Siegfried Ruhemann, Martin's father, as the assistant to the Jacksonian Professor of Experimental Philosophy at Cambridge University, was not unique. In 1845, August Wilhelm Hofmann, another Prussian, became the first director of the Royal College of Chemistry in London. Although Hofmann returned to Prussia in 1864, he retained enough influence in Britain to recommend Ruhemann to James Dewar, the Jacksonian Professor. This is just one of many examples of the international migration of scientists and other academics, a practice that long predated the forced exodus of so many Central European scholars after Hitler came to power.

By the beginning of the twentieth century a small but growing number of Central European academics had migrated voluntarily to the United States. This migration accelerated after the First World War, when research funding at German universities was greatly reduced along with the number of positions available. American universities and research laboratories were comparatively well-funded and had an interest in importing some of the best minds of Europe. Certainly some American professors objected to the hiring of foreign scholars, and this opposition would increase dramatically after 1929. The period of voluntary migrations came to a sudden end with the Great Depression. More than 1,000 faculty members found themselves laid off by cash-strapped American universities. Among them was Michael Sadowsky, whose life will be discussed in the next chapter.

The decision to go to the United States was a difficult one for any European scholar, for the vast distances between research centres, and the markedly different social and cultural environment, often produced profound feelings of isolation. By European standards, American professors often had a much heavier teaching load, and teaching required a good command of English. Still, before 1930, the difficulties of migrating across the Atlantic paled in comparison to the risks of going east to the Soviet Union.[1]

It is interesting to note that the first great exodus of refugee academics from universities occurred not in 1933, but a decade earlier, as a result of the Russian Civil War. Somewhere between 3,500 and 4,000 university professors, teachers, and scientists fled Russia as a result of the chaos of the period and the victory of the Bolsheviks. That exodus amounted to more than one third of the academic community. There was little organized international assistance for this exodus of displaced scholars. Some Russian scholars were able to continue their careers elsewhere, often in exotic locales like China, South America, and Africa, but many were not so fortunate.[2] Most of the academic aid organizations established in 1933 to assist German scholars received requests from former Russian academics to be included in the relief effort. In 1937 the SPSL estimated that there remained eight hundred displaced Russian scholars still hoping to return to the academy.[3]

Foreign Experts in the Soviet Union before 1930: The Case of Herman Muntz

There was a great need for foreign experts in early Soviet-era rebuilding programs, a need greatly exacerbated by the massive migration of the Russian intellectual and professional elite after the revolution. Many foreign experts came to be part of the great Communist social experiment. There was, however, no open market encouraging non-Russian scholars to move to the Soviet Union, where the free flow of foreigners, even foreign technical and scientific experts, was closely regulated by both the state and the party. At the end of 1929 there were fewer than 5,000 foreign workers in the Soviet Union, because of government restrictions on migrants and also because foreigners often faced atrocious living and working conditions. In the 1920s these hardships were so severe that many of those who were most ideologically committed to socialism returned to their capitalist home countries.[4] Walter Elsasser, in 1929, was the first foreign physicist hired to work at the then brand-new Ukrainian Physical Technical Institute (UFTI); he was compelled to leave the Soviet Union just six months later after "falling gravely ill with jaundice, then endemic in the Ukraine." After recovering in Germany, Elsasser declined to return to finish his contract, unwilling to face the harsh living conditions in Kharkiv.[5]

Herman Muntz, who became a professor of mathematics at Leningrad University in 1929, is the earliest academic migrant included in this study. Born in Łódź, Russia (later Poland) in 1884, Muntz earned his PhD in 1910 at the University of Berlin. In 1912 he made his most significant mathematical discovery, a generalization of the Weierstrass approximation theorem, which states "that every continuous function on a finite interval can be uniformly approximated by algebraic polynomials." This theorem was "of major importance in mathematical analysis and a foundation for approximation theory." According to one account of Muntz's legacy in mathematics, since 1940 more than 150 papers have cited his name in their titles. Muntz's influence has moved well beyond his basic theory: "In addition there are Muntz polynomials, Muntz spaces, Muntz systems, Muntz type problems, Muntz series, Muntz-Jackson Theorems, and Muntz-Laguerre filters. The Muntz Theorem is at the heart of the Tau Method and the Chebyshev-like techniques introduced by Cornelius Lanczos. In other words, Muntz has come the closest a mathematician can get to attaining a little piece of immortality."[6]

Muntz, like so many other Central Europeans, experienced the turmoil that rocked the region in the first quarter of the twentieth century; he was also a victim of some of the more cruel aspects of the German university system. During the First World War, while living in the German state of Hessen, he was considered an enemy alien. Although he was not interned, his travel within Germany was restricted and no university would offer him a teaching position. Despite continuing to publish a steady stream of important papers, he could not find a permanent university teaching position even after the armistice.

Muntz was trapped in a classic Catch-22: he could not get a teaching position because he could not get any university to grant him his habilitation, or second PhD, and he could not get his habilitation because he did not have a teaching position. The problem was not unique to Muntz; universities were under some moral obligation to provide for their habilitants until they found employment. Granting Muntz his habilitation was risky; he had several strikes against him. Unlike many of his contemporaries, he did not hide that he was a Jew – indeed, he preferred to be referred to by his Hebrew name of Chaim – nor was he ashamed of his Polish origins. Moreover, after the war, Muntz experienced some sort of mental crisis or nervous breakdown. Despite his ever-growing list of important papers, and despite working with Albert Einstein for a time

Image 3. Portrait of Herman Muntz, likely painted in the Soviet Union (from the collection of Göran David Gerby).

in the 1920s, he remained unemployable. He eked out a living tutoring, teaching at a series of private schools, and writing paid reviews for journals.

Muntz was also a Zionist, and he wrote several works in support of a Jewish homeland in Palestine. In 1927 he placed high hopes for a research post at the then new Hebrew University of Jerusalem. One again, however, his lack of a habilitation worked against him. Only when he lost this competition did Muntz, in desperation, accept an offer to join the faculty at Leningrad. The Soviets offered Muntz the positon of professor of mathematics and Chair of Differential Equations at Leningrad State University, even without a second doctorate. Realizing that this would be his only opportunity for a university career, he decided to move to the Soviet Union.[7] While not a Communist, he was a brilliant mathematician who had been robbed by circumstances of an opportunity for an academic appointment. There is no information regarding why the Soviets had hired Muntz. Clearly, though, they had found a diamond in the rough at a time when better-established academicians were unlikely to accept any position in the Soviet Union, no matter how promising.

Foreign Experts at the Ukrainian Physical Technical Institute, 1930–1933

In 1930, after a two-year debate, the Sixteenth Congress of the Communist Party decided to admit "'foreign engineers, foremen and skilled workers' in order to utilize 'their experience and knowledge inside of Soviet plants' in connection with the five year plan." A ceiling of 40,000 foreign workers and specialists was established.[8] The Five-Year Plan called for a major expansion of scientific and industrial research capabilities, all of which were to be as advanced as any in the world. So by the early 1930s, new positions were beginning to open at Soviet universities and at stand-alone research institutes – indeed, so many that they could not all be filled from within the country. Thus, the Soviet Union set out to recruit the best and brightest, just as Germany was expelling hundreds who matched that description and Western universities were reeling under the weight of the Great Depression. Between 1930 and 1933 a handful of Central European scientists began migrating east to the Soviet Union. Surprisingly, among this group, ideology was a motivation for only a few.

Of the Central European scientists who went to the Soviet Union before 1933, only one was a devoted Communist who was migrating solely for that reason. Polish-Austrian physicist Alexander Weissberg was a dedicated member of the Austrian and German Communist Parties. He is best remembered today as the author of *The Accused* (aka *Conspiracy of Silence*), which was published just after the Second World War. The book was one of the first exposés of Stalin's purges to be published in the West; it also provides one of best accounts of the experiences of Central European scientists in the Soviet Union in the 1930s.

Weissberg was born in 1901 in Kraków, then part of the Austro-Hungarian Empire. His father, a prosperous Jewish-Polish merchant, moved his family to Vienna. After the First World War, Weissberg studied mathematical physics and engineering at the University of Vienna, graduating in 1926. He held a variety of jobs, including a brief stint as an assistant professor at Berlin Polytechnic, and later, as an adviser to the Argentinian government in the purchase of German machine tools. Weissberg joined a socialist youth organization when he was just seventeen; later he became a member of the Austrian Socialist Party, before joining the Communist Party in 1927. While in Berlin, he became a leading member

Image 4. Alex Weissberg and Eva Striker in Kharkiv (from the collection of Professor Mikhail Shifman).

of a close-knit group of left-wing academics, scientists, artists, journalists, actors, and musicians whose popular intellectual haunt was the Romanisches Café.

Another central figure in this group was Eva Striker, a promising ceramics designer with a Bohemian lifestyle, who had rented a large apartment just five doors from the Romanisches after moving to Berlin from Paris. Striker was a native of Budapest and a member of the renowned Polanyi family. Her uncle, Michael Polanyi, was one of the twentieth century's most accomplished polymaths. Since 1926, he had been a professor at the Kaiser Wilhelm Institute for Physical Chemistry and Electrochemistry in Berlin, and he no doubt introduced many of the city's young physicists to his niece. One of them was Weissberg, who soon became Striker's lover. Weissberg and Striker brought into this Berlin social circle many of those who would later join them in the Soviet Union, including Charlotte and Fritz Houtermans, Charlotte Schlesinger, and

Fritz Lange. Striker's apartment became the scene of many wild parties, often with more than a hundred guests. Weissberg was an intellectual leader of the group, as a self-taught expert on Marxist-Leninist theory. Many other intellectuals came under Weissberg's dynamic spell, including the journalist Arthur Koestler.

In early 1931, Koestler moved to Berlin from Paris to take up a position as science editor for the moderate centrist newspaper *Vossische Zeitung*. Long before, Koestler had attended kindergarten with Striker, at a school run by Eva's mother, Laura. In Paris, Eva and Arthur were reintroduced on a blind date arranged by Eva's mother. The two became lovers, and when Arthur arrived in the German capital, Eva invited him to one of her famous parties. At first Koestler was unimpressed by Eva's new paramour, who was somewhat overweight and gobbled large handfuls of candy while loudly telling hilarious stories. Only after the other guests had left did Weissberg reveal his extraordinary intellectual power. In a discussion on politics and ideology, Koestler later admitted, Weissberg "made dialectical mincemeat of me." Weissberg helped "guide" Koestler to embrace Communism. The two men became both fellow travellers and close friends.[9]

Weissberg later claimed that he migrated to the Soviet Union in March 1931 because of his political beliefs; with his unique combination of scientific, engineering, and industrial expertise, the Soviet Union needed him. Weissberg was recruited by Soviet scientists to join the newly established UFTI (Physical Technical Institute of the Ukraine). Martin Ruhemann, however, provided a different version of the events that led Weissberg to UFTI. According to Ruhemann, Weissberg "had all the trappings of a secret agent, and most people were convinced he was a spy." Ruhemann had his doubts about Weissberg's espionage activities, believing that Weissberg had deliberately cultivated the image of "an international spy." Ruhemann concluded that it was all a front, since "real spies don't act as spies." Ruhemann alleged that rather than being a volunteer recruit to the Soviet Union, Weissburg had engaged in illegal activities for the KPD. As a result of these clandestine activities, Weissburg "became an embarrassment to the Party, who pushed him over to the Soviet Union to get him out of the way."[10] Neither Ruhemann's account nor Weissburg's can be fully trusted. After 1945, the two men had a severe falling out over the former's continued support of Stalin's regime. Ruhemann published his account in an obscure Ukrainian physics journal; it only appeared many years after Weissberg's death. Mikhail Shifman has published two documents found in the Comintern archives indicating that Ruhemann's account is closer to the truth. These documents explain that Weissberg was recruited to do espionage work for the Communist Party but that he proved singularly inept as a spy. On a train trip from Berlin to Vienna "he blabbed that he worked for Russian intelligence to one of his friends." The Berlin police soon issued an arrest warrant for Weissberg. His party handlers, fearing he might prove to be a menace for a number of people who had also engaged in espionage, advised him to move to the Soviet Union. A position was found for him at UFTI.[11]

Shortly after Weissberg moved to Ukraine, Eva Striker followed him, and they were soon married. Striker's departure from the German capital did not end the frequent left-wing social soirées, which now moved to the Houtermans's home. Gatherings there were slightly smaller and less boisterous then those at Striker's, but they provided an opportunity to discuss the latest developments in physics, although most of the conversations concerned politics.[12]

Weissberg worked with several Soviet physicists to develop low-temperature techniques to extract nitrogen from the air for industrial purposes. Soon after arriving in Kharkiv, Weissberg realized that there were not enough skilled Soviet low-temperature physicists to staff the growing industrial research facility. He received permission from Ivan Vassiliyevitch Obremov, director of UFTI, to recruit additional foreign experts. That summer, Weissberg travelled back to Central Europe, where the Czech physicist George Placzek suggested he contact Martin Ruhemann, an assistant at the Stuttgart Institute of Physics, to see if he might be interested in joining him in Ukraine.

Ruhemann arrived at his laboratory late one morning in the summer of 1931 to find a letter with a Munich postmark from Alex Weissberg, offering him a position at UFTI. Ruhemann had never before heard Weissberg's name, and ordinarily he would have disregarded the unusual offer. These were not, however, ordinary times. Martin and Barbara were living with "extreme insecurity" caused by the Great Depression. Both of them were working, but their salaries gave them barely enough to live on. Worse, there was a persistent threat that the university might be forced to reduce their meagre pay or even lay them off. The uncertainty in the Ruhemanns' lives was so great that they had delayed starting a family. Weissberg's offer seemed to promise long-term security, so they decided to visit Weissberg in Munich. This being 1931, Hitler was not factor in the decision of the apolitical couple. They knew "little" about the Soviet Union, although they "admired the Bolsheviks for turning out Kerensky and winning the war of intervention" when the Western powers had invaded postwar Russia. They had seen the film *Battleship Potemkin* in Berlin and had been tremendously impressed, but so had Martin's conservative cousins. Martin summarized the couple's political views: "We thought vaguely that communism was a good thing and would probably have waved red flags if anyone had asked us to do so. But that was about all."[13] Weissberg had a somewhat different view of the couple's political leanings before they immigrated to the USSR. Ruhemann, Weissberg explained, was not a Communist; if he had any political beliefs they were those of an English liberal. Martin's physicist wife Barbara was not yet a Marxist, but according to Weissberg she "was a Prussian, and the sort of stuff fanatics are made of." Weissberg continued: "She was vehement in all her reactions, and her opinions were always formed and her decisions taken for some internal reasons which she never thought it necessary to check against her experience. She was a courageous woman and she devoted herself wholeheartedly to any cause she thought right, but she was just unable to judge objectively."[14]

It is virtually impossible to know whether Ruhemann's or Weissberg's account of the couple's political views before they left for the Soviet Union is accurate. It is hard to believe, however, that Martin was not influenced by the political realities in Germany in the early 1930s. Both of Martin's parents were Jewish; Barbara's heritage is unknown. Francis Simon, Martin and Barbara's doctoral supervisor, saw the potential risks facing Jewish academics. Simon and his wife had visited the Soviet Union in 1930 and, according to his biographer, he had averred that "if conditions in Germany should ever become unbearable, a research career in the Soviet Union in Franz's own field would be a continuing option."[15]

The meeting between the Ruhemanns and Weissberg went well. Weissberg offered a secure well-paying position to Martin, who insisted that Barbara be hired as well. Weissberg quickly agreed to this opportunity to hire a second, highly qualified physicist. Martin and Barbara returned to Stuttgart believing that, for the first time in their marriage, they had a safe and secure future, a feeling that Martin later confessed, in view of subsequent events, was "difficult to understand." Finalizing the paperwork for their employment at UFTI took some time, and by the time the couple departed for the USSR in April 1932, Barbara was already seven months pregnant.[16]

For Martin, the prolonged period before departure was a blessing. Weissberg returned to Ukraine shortly after hiring the couple, and one of the last things he did before leaving Munich was call Martin to explain that at UFTI he would be responsible for developing a gas separation plant, which would use cryogenics to remove nitrogen from the air. Weissberg asked Martin whether this was acceptable. Ruhemann confirmed that it was. He did not tell Weissberg that he actually had no idea how the process worked. Ruhemann spent the next few months in a desperate scramble to become a self-taught expert in gas separation. At first he could find only one article on the process in the Stuttgart physics library.

That article was incomprehensible to the experimental physicist, and Martin first assumed it was because of his lack of theoretical knowledge. When he took the article to his department's theoretician, however, his colleague was unable to shed any light on the subject. Only afterwards did Martin realize that his difficulty in understanding gas separation had nothing to do with its complexity; the problem was that it was a task for chemical engineers, not physicists. At the time, Martin did not even know what chemical engineering was. Gradually, he worked out that by "the use of rather elementary mathematics and a little of the thermodynamics I had learned from Max Planck, it was possible to design an industrial plant and be assured it worked." Thus began Martin's transformation from physicist to chemical engineer.

Other German Academic Migrants before Hitler

Besides Weissberg, only one other voluntary migrant to the Soviet Union was a committed member of the KPD: the musicologist Ernst Emsheimer. Born in 1904 in Frankfurt, Emsheimer studied music theory and history in Vienna before completing his doctorate at the University of Freiburg in 1927. His thesis was a rather conventional study of the seventeenth-century organist and composer Johann Ulrich Steigleder. There is no information on Emsheimer's employment in the five years after completing his doctorate, but he was an active publishing scholar. Applying Marxist ideology, Emsheimer in his postdoctoral work challenged the domination of bourgeois approaches to musicology. His early articles moved away from traditional subjects; he wrote about work songs, jazz culture in Paris, and "The Social Meaning of Music in the Working Day of the Proletariat." Emsheimer and his long-time partner Mia Wilhelmine were active in the KPD, and they viewed the rise of the Nazis with growing alarm. In 1932, several months before Hitler became chancellor, the couple migrated to Leningrad. It is unclear whether they went to the Soviet Union as Communist migrants or as foreign specialists. Whatever the case, within a short time Emsheimer received a contract as a foreign specialist working for the National Academy for the History of Art at the Hermitage's music collection. Ernst and Mia married shortly after they arrived in Leningrad.

Ernst Simonson, a highly regarded physiologist, if not a member of the party, either was highly sympathetic to Communist ideology or would become so during his first three years in the Soviet Union. He had earned his medical degree in 1923 and, one year later, his PhD at the University of Greifswald in Germany. He was a pioneer of industrial physiology, and by the end of the 1920s he had become the Chief of

Industrial Physiology and Hygiene at the Social Hygiene Investigation Office in Frankfurt, as well as a *Privatdozent* (lecturer) at the University of Frankfurt. In the autumn of 1929, he was invited to Kharkiv to found and lead the Physiological Laboratory of the Institution of Labour of the Ukrainian Commissariat of Labour at Kharkiv. The offer was tempting, since Simonson would be leading a large team of up to one hundred researchers, to investigate methods of improving worker safety and comfort as well as productivity. Simonson chose to accept the Soviet offer, but only after he received a commitment from the Prussian state government that they would reinstate him to his positions if he returned to Germany before the end of 1933. It remains unclear, however, whether Simonson actually intended to leave Kharkiv if he and his family found their lives in the Ukraine agreeable.[17]

The difficulty in assessing Simonson's belief in Communism stems from a number of factors related to his abrupt departure from the Soviet Union in 1937 and his desperate search for another place of refuge. As he sought employment in Britain, Canada, and the United States, he never mentioned a past interest in Communism. After arriving in the United States, he seldom discussed his seven years in the Soviet Union. He hid the truth about his time there; indeed, on at least one occasion he lied about his experiences. In 1938, while still looking for employment, he told A.V. Hill, the renowned British physiologist, that he "had never felt comfortable" in the Soviet Union and had never intended to stay there permanently. Only Hitler's coming to power and Simonson's Jewish heritage precluded a return to Germany.[18]

Yet given Simonson's claims that he had never been happy in the Soviet Union, it is difficult to explain a public lecture he gave in 1932. During a brief visit to Germany, he spoke to a Silesian cultural association in Breslau about science in the Soviet Union. In that lecture he described the wonders of Soviet science, which was advancing by leaps and bounds, having been deemed essential for the success of Stalin's Five-Year Plan. Simonson lauded the Soviet state's sponsorship and direction of practical scientific research, which was far superior to policies in Western countries. He also expressed a deep admiration for dialectic materialism and the Communist system. Nothing in his presentation suggested he was considering a return to Germany.[19]

Yet this lecture does not by itself prove that Simonson was a member of the Communist Party. From the late 1920s and throughout the 1930s, there was a genre of literature that lauded Soviet science – more precisely, the Soviets' massive financial support of applied research. The Soviet state's investment in science dwarfed the government funding available to support research in all Western democracies, especially after the

Great Depression crippled governments' budgets. Science in Germany and elsewhere in the West was heavily supported by private philanthropy, most notably the Rockefeller and Carnegie Foundations, but even those resources had been surpassed by the ones made available in the Soviet Union. Some, but by no means all, of those who lauded Soviet science, like British physicist J.D. Bernal, were devout Communists.[20] Others, such as the science journalist J.G. Crowther, who wrote two books praising research in the Soviet Union, respected the Soviet approach to science. Crowther, however, argued that capitalist societies could adopt the Soviet model of state-supported science without upending the entire economic system.[21] It is impossible to know from Simonson's lecture to which of these views he subscribed.

After fleeing the Soviet Union in 1937, Simonson did not include the publication of his talk on Soviet science on his CV, nor did he refer to the fact that he had moved to the Soviet Union three years before Hitler came to power. It is very likely that he did not want his admiration for Communism to become known. It is also doubtful, however, that he was a full-fledged party member, since in 1937 he and his family were able to escape, at a time when most Communists could not. Even if Simonson *was* a Communist, most of the academics who migrated to the Soviet Union before 1933 had done so for professional and economic reasons, not ideological ones.

Emanuel Wasser, an experimental physicist at the University of Vienna, left Austria for the Soviet Union in 1932 for economic reasons, which may have been directly linked to prevailing Austrian antisemitism, which had blocked his further advancement at his institution. Wasser was born in 1901 in Lvov (Lviv), then part of the Austro-Hungarian Empire. In 1920 he entered the University of Vienna. Four years later, on the basis of experimental research on the "photoelectric effect of submicroscopic mercury spheres," he graduated with a doctorate. His dissertation examined the influence of the hydrogen channel radiation light in weak magnetic fields. He spent the following eight years in a series of minor academic positions at the University of Vienna, finally reaching the rank of extraordinary assistant while publishing a number of academic papers "in the region of subatomic particles." In 1932 he accepted an invitation to join a research institute in Leningrad as a "scientific collaborator"; once there, however, he was immediately seconded to become the head of the Photoelectric Laboratory of the Physico-Technical Institute of the Urals, one of the new research centres that had been established by the Soviets in the early 1930s.[22] Wasser was not a displaced scholar, nor was he a Communist; apparently he had simply seized an opportunity for advancement. He only became a refugee academic after his expulsion from the Soviet Union in 1935.

Saul Levy, another experimental physicist, migrated the same year as Wasser. Born in 1897 in Raseiniai, Lithuania, Levy was a bit older than most migrants and, like Wasser, had at best a mediocre career before his move. Levy had begun his undergraduate education at the University of Berlin and finished a doctorate at the German University of Prague in 1927. For the next five years he was employed as an assistant to several professors in Prague and, later, Berlin. In Berlin, he worked at the Kaiser-Wilhelm Institute for Physical and Electrical Chemistry in the laboratory of Professor Rudolf Ladenburg. Ladenburg was a physicist who had achieved significant results in both theory and experimental work. Levy became Ladenburg's main assistant and was, according to Michael Polanyi, "a very able man" as well as "a jolly and agreeable fellow."[23]

Levy assisted Ladenburg and H. Kopfermann in their pioneering experiments investigating the "the dispersion of gaseous neon" contained in a glass tube by the use of electrical discharge. They discovered that when they increased the current, at some point the dispersion of the gas decreased. They correctly deduced that this effect was caused by the increased charge increasing the number of atoms in a higher energy state. Had they continued their experiment systematically, they might have discovered a basic type of laser, decades before Einstein's theory was successfully demonstrated.[24] Working as an assistant, Levy had little opportunity to show his abilities as a scientist. Between 1926 and 1932, he published only six articles, five of which were co-authored with more senior scholars.[25] In 1931 Ladenburg was offered a research position at Princeton University, and a year later a permanent chair of physics at the prestigious American university. With his professor's sudden departure, Levy found himself unemployed. He was fortunate to almost immediately be offered a well-paying and prestigious position working with Professor Grigory Landsberg at the Optical Laboratory of the Physical Institute of Moscow State University.

Levy was not the only German scientist associated with Ladenburg who was hired in 1932 to work at Landsberg's Moscow laboratory. Walter Zehden had only just completed his doctorate, under Ladenburg's supervision, in February, when he left for the Soviet Union. There is no way to know how Levy and Zehden received their positions, but it is likely that Ladenburg, unable to bring his assistant or student to the United States, had found them places where they could escape the Nazis. Ladenburg, an atheist of Jewish heritage, must have been influenced by the political situation in Germany when he decided to migrate to Princeton. Undoubtedly, he felt obligated to assist in the escape of his Jewish assistant and his graduate student.[26] His sense of a moral obligation to help displaced Jewish scientists would continue in the United States, where he involved himself in coordinated efforts to find employment for former colleagues seeking refuge in North America.[27]

Conclusion

The Soviet Union offered prestigious positions with attractive salaries as well as the chance to work in well-funded laboratories. Prior to 1933, those scientists who migrated there tended to be those whose careers had not prospered in Germany or Austria. Even Weissberg, who was, in large measure, motivated by ideology, moved to Kharkiv after a lacklustre scientific career. Those who migrated to the Soviet Union prior to Hitler's ascent to power were not, strictly speaking, academic refugees. The exceptions to this might be Levy and Zheden, both of whom left Berlin before Hitler came to power in large measure because their mentor correctly predicted the impending catastrophe. Those two physicists can also be seen as belonging to a precursor group of academics who fled just before the arrival of the tsunami of racially and politically motivated firings that would sweep through German universities and research centres beginning in the spring of 1933. During the Great Terror, however, the pre-1933 voluntary migrants would either join the refugee scholars fleeing the Soviet Union or find themselves under arrest. The NKVD did not distinguish between voluntary and forced migrant scholars when they commenced their persecution of the ensnared scholars. While the voluntary academic migrants and the refugees would share the same fate, those who came after Hitler's appointment as chancellor had taken a very different route to the Soviet Union. It is to their story that we now turn.

3   Scientists and Communists

When Herbert Strauss reached the conclusion that all of the refugee scholars who fled to the Soviet Union were Communists or fellow-travellers, he based his findings on his study of the much larger community of German and Central European refugee intellectuals – including artists, actors, musicians, journalists, and union leaders – who fled there.[1] Strauss is correct that the vast majority of those who fled Hitler for the Soviet Union were Communists, or fellow-travellers who supported the party's philosophy; but among academic refugees, only a minority were party members or openly sympathetic to the cause. The scholars who chose the Soviet Union, even those who migrated directly from Germany and Central Europe after the Nazis came to power, had far more diverse motives.

Ideology and the Migration of Displaced Scholars to the Soviet Union

The Soviet government categorized university teachers as foreign experts or specialists. Their migration was not contingent on Communist Party membership but rather on their qualifications and specialized skills. There was no single reason why refugee academics chose the Soviet Union. An examination of the individual scholars' flights reveals that ideology was only one factor, and not the most important one. Only seven of refugee scholars who migrated to the Soviet Union have been identified as members of the German Communist Party (KPD) – Julius Schaxel, Wolfgang Steinitz, Gerhard Harig, Fritz and Charlotte Houtermans, Herbert Murawkin, and Fritz Lange. Konrad Weisselberg joined the Austrian Communist Party in 1933, just a few months before departing for Ukraine. Many of these Communists were deeply committed to the party, the Soviet Union, and the spread of the proletarian revolution, so much so that a majority were engaged in espionage for the KPD or the Soviet government.

Charlotte Houtermans, in a rare, candid admission, would later reveal the full extent of her and her husband's involvement in illegal activities not just for the KPD, but also for the Communist International. The Comintern was the branch of the Soviet Communist Party that worked "by all means available, including armed force, for the overthrow of the international bourgeoisie."[2] Fritz Houtermans joined the KPD in 1926 but only became active three years later. By then he had persuaded his wife Charlotte to join the party. Fritz "became involved with KPD's Industrial Reporting Agency (*Betriebsberichterstattung, BB-Apparat*)." Much of the information they acquired was sent to Moscow via encrypted messages. Charlotte assisted her husband's efforts by coding and decoding messages and typing reports.[3] As we shall see, despite this deep commitment to the cause, the couple did not choose to go to the Soviet Union in 1933.

There were likely other party members among those who settled in the Soviet Union, but it is difficult to distinguish between the majority who were sympathetic to left-wing causes and socialism and those who were deeply committed Communists. It is certain that most of these émigrés were not KPD members. While few of the displaced scholars chose the Soviet Union strictly for reasons of ideology, many had certainly flirted with "social" Communism, a popular fad among intellectuals in the interwar years. There was, however, enough knowledge of the Soviet Union to discourage migration even among those who were sympathetic to Communist ideology.

Hermann Borchardt, the philosopher and high school professor, was an active member of the left-wing intelligentsia in Weimar Berlin. He may have been a member of the Socialist Democratic Party (SDP), and his membership in the KPD seems to have been limited to six months in 1923.[4] Borchardt later claimed that his own political philosophy was based on that of Ferdinand Lassalle, the German political philosopher and activist, who argued for a state socialist system. As early as 1919, Borchardt wrote a pamphlet advocating the union of socialists and Communists to overturn the German political system against right-wing opposition. In 1924 he began contributing to *Der Knuppel*, the satirical magazine of the KPD. At the magazine he became close friends with the famous Weimar cartoonist and painter George Grosz. He also became friends with the playwright and poet Bertolt Brecht. Borchardt sketched out the overall concept and plot for Brecht's play *Saint Joan of the Stockyards.*[5]

Another example is Edgar Lederer, who was forced to leave Germany in 1933, heading west to France, his wife's birthplace. Despite being a "rabid communist" and "spiritually very attracted by the USSR," he was not formally a party member.[6] The Hungarian physicist Laszlo Tisza offered one explanation for why so many of his generation were attracted to socialist ideology: "Being influenced by friends and by the deepening economic depression were rather typical motivations for turning politically left in those times. I also had a more personal motivation that pushed me in the same direction. As a physicist I tended to look at relativity and quantum mechanics as products of scientific revolution, and to consider 'revolution' as a legitimate engine of change."[7]

It is difficult to determine the ideological leanings of many of these scholars. Most of the documents about them are dated after their expulsion from the Soviet Union. Having been victimized by the Soviet state, as they scrambled for positions in countries where declaring their Communist sympathies would place them at a distinct disadvantage, none admitted to being a party member. Decades after his sojourn in the Soviet Union, throughout the Cold War, Tisza continued to hide his youthful flirtation with Communism. In a 1987 oral history interview almost fifteen years after his retirement from the Massachusetts Institute of Technology, Tisza claimed that his motives for going to Ukraine were strictly economic and intellectual:

So at that time there was the deepest depression, and I was not established. You see, all the others, Teller and Peierls and Bethe were all established and receiving fellowships, but I was sort of not really in the crowd, and I had nothing of the sort. I stopped in Budapest for a while. I worked in my father's shop, and I got a job at the insurance company in actuarial mathematics, and I was sort of drifting around in a sense.[8]

Almost twenty years later, approaching his hundredth birthday, Tisza confessed to another factor that had left him both unemployed and willing to accept Soviet physicist Lev Landau's offer to join him in Kharkiv.

The tendency to draw a parallel between the two kinds of revolution cuts both ways. At that period in Budapest it rendered me open to political entrapment … Two months after I defended my thesis I was arrested, and although I was not a Party member I was sentenced to 14 months in jail. In Hungary the Communist Party was illegal at that time, and any Communist activity was heavily apprehended. After that it was out of the question for me to obtain an academic position in Hungary.[9]

Soviet academic leaders perceived the dismissals from German universities as an opportunity to recruit world-class scholars, regardless of their ideological leanings. Ideology, therefore, influenced only a few of the refugees' decisions to go to the Soviet Union; far more important were the often close institutional, personal, and professional relationships between Soviet and Central European scientists. German–Soviet scientific and intellectual cooperation predated the First World War. After 1919, Germany and the Soviet Union were pariah states; thus they found it mutually beneficial to work together on several levels.[10] The Treaty of Rapallo, signed in 1922, contained provisions for cultural collaboration, including scientific exchanges. In the 1920s, Soviet science had an "extremely tight connection" with German scholars.

Soviet scientists were allowed to publish in Western journals, and they usually chose German-language publications. By the middle of the decade, Soviet scientists were authoring 16 per cent of the articles in the leading German journal of physics.[11] Between 1925 and 1932 the Rockefeller Foundation provided around fifty post-doctoral fellowships to Soviet scholars, allowing them to spend a year at a Western university.[12] These Soviet fellowship holders spent time at the best scientific centres in Europe. A photograph taken in 1928 in Gottingen shows Fritz Houtermans sitting at a table surrounded by four visiting Soviet physicists, including George Gamow, his scientific collaborator. Until 1934, physicist Pyotr Kapitsa was allowed to travel back and forth between the Soviet Union and Cambridge, where he was director of the Mond Laboratory.

In August 1930 the All-Union Congress of Physicists was held in Odessa. Some eight hundred delegates attended, including many from Germany, some of whom would later seek refuge in the Soviet Union. There were regular visits by Western European scientists to the Soviet Union. During one of the visits of Lev Landau, the brilliant Soviet theoretical physicist, to Niels Bohr's Laboratory in Copenhagen, he had met the equally brilliant Hungarian Edward Teller. The two men forged a friendship based on their mutual respect for each other's scientific abilities. In 1934, Teller decided to help his school friend, Laszlo Tisza, who could not find work after his stint in a Hungarian prison for Communist activities, "although he had no sympathy" for his political views. Teller recommended Tisza to Landau as a possible assistant at UFTI in Kharkiv. Landau invited Tisza to join the renowned Niels Bohr and some British and French scientists at "a small but prestigious theoretical meeting" held at UFTI. Tisza was encouraged by what he saw in Kharkiv:

A noteworthy aspect of the meeting was that the general political atmosphere could not have been better. A good harvest seemed to have ended the agricultural crisis of several years, and the friends of the regime were optimistic as to the future. These expectations were to some extent fulfilled.

> Not only was the shortage of bread ended, but even gourmet food had become easily available in the stores of the grocery chain Gastronome. However ... the expected mellowing of the political suppression did not materialize; the situation went from bad to worse.[13]

Tisza was sympathetic to Communism, but he agreed to go to UFTI only after he had an opportunity to learn first-hand about life in the Soviet Union. Unfortunately for Tisza, and the other refugees who agreed to migrate there, they were making their choice at a time when the Soviet Union was enjoying a brief interlude of relative prosperity and political tranquility, between the famines and social upheavals of the 1920s and the start of the purges in the second half of the 1930s.

Soviet scientists and academics recruited most of the displaced scholars who came to the Soviet Union, but the handful of Central European experts already in the country played a role as well. Alexander Weissberg no doubt supported the appointment of his friends Fritz Houtermans and Fritz Lange to join him at UFTI, and he arranged a job offer for Charlotte Schlesinger. Hermann Muntz "helped, sometimes directly and sometimes indirectly, mathematicians from Central Europe obtain positions in the Soviet Union." He arranged for the appointment to the University of Moscow of mathematician Stephan Cohn-Vossen, who would die of pneumonia shortly after arriving in Moscow. With others, Muntz helped Stephan Bergman become a professor of mathematics at the Universities of Tomsk and, later, Tbilisi.[14]

Direct Migration to the Soviet Union by Communist Refugee Scholars

Four of the seven members of the KPD – Schaxel, Steinitz, Harig, and Murawkin – made the Soviet Union their first place of refuge. Julius Schaxel, a professor of biology at the University of Jena, was the best-known and most senior scholar to migrate directly there. Schaxel was an outspoken "red professor" during the Weimar Republic. In 1924 he became Minister of Adult Education in the short-lived Socialist/Communist coalition government in the German state of Thuringia. When that government was deposed by the Reich government, Schaxel fled temporarily to the Soviet Union. When he returned to Jena in 1925 he cultivated contacts between that regime and Germany.[15]

In May 1938, while languishing in Lubyanka Prison in Moscow, Schaxel was given twelve sheets of paper and a pencil, with which he wrote his autobiography. While he did not discuss his time in the Soviet Union, he did provide details on his forced exodus from Germany. In January 1930, as a politically active Communist, Schaxel became the target of Wilhelm Frick, the education minister in Thuringia. Frick was the first member of the Nazi Party to be made a state minister in Germany. By Schaxel's account, Frick targeted him because he had been campaigning against Nazi race theory (ethnogeny) since 1920. Frick wrote in Schaxel's personnel file: "What is this fellow doing here? Why do we need this Russian Jew?"

Frick did not remain in power long enough to dismiss Schaxel, but the next two years, during the lead-up to Hitler's rise to chancellor, were anything but pleasant. Schaxel recounted: "Very often I received threatening letters and other good wishes. Stones were thrown at my car. My front door was painted with swastikas in the morning. There were minor disturbances during my lectures even though national socialists with badges were sitting in the room. In March 1933 I was forbidden to lecture."[16] Schaxel was barred from teaching even before the first, formal dismissal of German professors, which commenced the following month. He escaped to Switzerland, where, unlike most academic refugees, he almost immediately received an invitation to undertake research at the Academy of Science in Leningrad. Schaxel enthusiastically migrated to the Soviet Union in October 1933 and was likely the first refugee scholar to find sanctuary there as a foreign expert.[17]

Wolfgang Steinitz was born in 1905 in Breslau into a middle-class family whose grandparents had converted from Judaism. He was a rebellious youth who eschewed his father's desire that he pursue a law degree, choosing instead to research folklore and linguistics. From 1923 to 1928, he studied Finno-Ugric languages and ethnology at the Universities of Berlin and Breslau. In 1924 he became a research assistant at the Berlin Museum of Ethnology. Two years later he took a similar position at the Hungarian Institute of the University of Berlin. While at that institute, he worked on his doctorate under the supervision of Professor Ernst Lewy. While only a teenager, Steinitz joined the Freideutsche Jugend, the Free German Youth movement. In 1919 he joined the Sozialdemokratische Partei Deutschlands (Socialist Democratic Party; SDP). His belief that revolutionary reform of society was necessary was reinforced by early research trips to Finland and Hungary, where he witnessed government suppression of workers' movements. In 1926, a visit to the Soviet Union to study the Udmurts, a Finno-Ugric people, exposed Steinitz to Communism. Soon after his return to Germany, he met Minna (Inge) Kasten, a committed member of the KPD. Steinitz soon joined the KPD as well and became involved in the party in Silesia, then Berlin. He was militantly active in the party, so much so that in 1929 he was arrested and briefly held for participating in illegal political activities. The following year he went to Finland, where, according to some sources, he spied for the Comintern. Inge followed him to Helsinki, where they were married in April 1930. Two months later, Steinitz's clumsy efforts at espionage were reported to the Finnish government, and the couple fled to Tartu, Estonia, then to Riga, Lithuania. During their exile in the Baltic states, Inge persuaded her husband to abandon espionage and politics and focus on his academic studies. They returned to Germany only in 1932, just in time to watch the Nazis come to power.

Steinitz's political activities did not affect his productivity as an academician. He published eighty-two articles, reports, reviews, and translations before receiving his doctorate in 1934. In April 1933, just before completing his doctorate, he was dismissed from his position at the institute, as much for his political activities, it seems, as for his Jewish heritage. Before his firing, however, he had successfully defended his thesis on the Finnish-Karelian folk poetry of Arhippa Perttunen. Lewy, despite having been dismissed from his professorship, was able, with the cooperation of the remaining faculty, to arrange for the rest of the formalities to be completed for his degree. By this time, Steinitz and his wife had migrated to Estonia or Finland, where he looked for an academic position. Steinitz's dissertation was published in Estonia, and this finalized his doctorate, since German regulations required publication as a last step prior to the formal awarding of a degree. In October 1934, Steinitz, unable to find a position elsewhere, accepted a position as a visiting German expert at the Institute of the Peoples of the North (INS) or Nordic Institute in Leningrad.[18] His biographer believes he never intended to go anywhere but Leningrad and that his Baltic hiatus was a deception to allow his doctorate to be awarded *in absentia*. The university administrators were informed by a faculty member that Steinitz was conducting research, not fleeing Gestapo arrest. The doctorate was required for him to work in the Soviet Union, and as soon as it was officially conferred, he moved to Leningrad.[19]

While Schaxel and Steinitz, like most other refugee scholars, arrived in the Soviet Union as foreign experts, Gerhard Harig and Herbert Murawkin did so as Communist refugees. Harig was born in 1902, in the small town of Niederwürschnitz. His father died when he was a young child, and his mother Else moved the family to Leipzig. Harig studied experimental physics at the Universities of Leipzig and Vienna; the former awarded him a PhD in 1929. While still at university, he developed an interest in the history and philosophy of science. In 1927 he was hired as an assistant at the Institute of Theoretical Physics at the Technical University of Aachen.

Harig's interest in the relationship between science and society, and the social turmoil in Germany in the 1920s, led to him forging close friendships with students in the left-wing youth moment. He did not formally join the KPD while in Aachen, but by 1931 he was a member of the Society for the Friends of New Russia, which was closely associated with the Communist Party. In 1933, Harig was the secretary of the Friends of New Russia as well as a prominent figure in Aachen University's anti-fascist student organization. Immediately after the Reichstag fire on 1 March, Harig was arrested; he was held "in protective custody" until 8 April. For being a public figure in local left-wing and anti-fascist movements, he was dismissed from his position at the university. While still in jail, Harig arranged to marry Katharina (Kathe) Heizmann, with whom he had been living for some time. His marriage was one factor that led to Harig's release. Upon his discharge, Harig moved back to Leipzig so that he could live with his mother. In Leipzig he was admitted as a member into the KPD, which by that time had been declared an illegal organization. Harig and Kathe joined a group of underground anti-Nazi intellectuals who distributed illegal publications and promoted socialism and Communism.

Harig realized that these underground activities were exposing him and his wife to serious danger and were unlikely to succeed in overthrowing the Nazi regime. Since 1932, he had been corresponding with Abram Ioffe, head of the Leningrad Physical Technical Institute (LFTI). In 1933, Ioffe offered Harig a research position in nuclear physics, and the German scientist decided to immigrate to the Soviet Union. Harig needed a new passport, since his had been taken from him when he was arrested in Aachen. Harig applied at the passport office in Leipzig, where officials were unaware of what had transpired in Aachen, and issued him the necessary document. He quickly secured an Intourist visa to travel to Leningrad. He had arrived in the country as a Communist refugee, but almost immediately on his arrival Ioffe kept his promise and hired him to work at LFTI. Harig's time at the Leningrad institute was brief, certainly less than a year. As we shall see, he quickly transitioned away from physics and carved out a new niche for himself as a historian of science.

Herbert Murawkin (sometimes spelled Muravkin), a young physicist who had just started his academic career when Hitler came to power, fled to the Soviet Union in April 1933. Murawkin was born to Russian-Jewish immigrants in 1905 in Berlin. In 1914, the Murawkins moved to Saint Petersburg, but at the end of the Russian Civil War, in 1922, they returned to Germany. In 1925, Murawkin entered the Technical University of Berlin, later transferring to the University of Berlin, where he completed his doctorate in 1931.[20] When not working as a physicist,

Murawkin was involved in the international movement promoting Esperanto, a language invented in the late nineteenth century. Many in the Esperanto movement viewed the language as an apolitical auxiliary tongue that might unify all people of the world; socialists and Communists saw it a useful tool for promoting international solidarity among the working classes. While still a student in the late 1920s, Murawkin had been active in promoting Esperanto as a means to organize the masses for the proletarian revolution. In 1930 he was one of eight German Communist Esperantists expelled from Worldwide Non-National Association (SAT), the umbrella organization that linked all supporters of the language, whatever their political ideology. Two years later, in August 1932, Murawkin was a founding member of the German branch of the Proletarian Esperantist International (IPE), a Communist group that had broken with the SAT. The IPE declared itself to be "an organization of the masses" and the collective voice of 14,000 proletarian Esperantists.[21]

After the summer of 1931, Murwakin also engaged in industrial and scientific espionage for the Soviets. Abraham Ioffe, the scientific head of LFTI, was introduced to Murawkin by Fritz Lange, his doctoral supervisor and a fellow Communist. Ioffe offered Murwakin a cash subsidy of 300 marks a month to send him reports on the design and construction of high-energy particle accelerators at Lange's laboratory. These accelerators were required for cutting-edge research into atomic structure. Murawkin began to provide regular reports on this research to Berlin-based Soviet intelligence officers. Murawkin's information was important for Soviet scientists at UFTI in Kharkiv, who were building their own very large particle accelerator. Lange suspected Murawkin of leaking information to the Soviets, but he was enticed not to pursue his concerns about his protégé by invitations to see Soviet atomic research facilities at LFTI and UFTI.[22]

Murawkin's espionage activities went undetected by the German government, but his work in the Communist Esperanto movement was too public to ignore. Because of the language's international character, all Esperanto organizations in Germany were targeted by the Nazis. The IPE's close association with the KPD led to the expulsion of Murwakin and other leaders of the organization soon after the Nazis came to power in the spring of 1933.[23] In April, Murawkin fled to Moscow as a Communist refugee. He remained unemployed until November, when the Technical Bureau of the Economic Department of the OGPU (secret police) arranged for him to work at the Research Institute of Communications and Electro-Mechanics of the Red Army. There he would "work on ultra-high voltages and the atomic nucleus."[24]

The Austrian chemist Konrad Weisselberg also went directly to the Soviet Union, but from Austria, not Germany. Weisselberg had completed his doctorate in chemistry at the University of Vienna in 1930 and was still looking for his first university position. His family has a story that Weisselberg was offered a post at a major Western university, either Manchester or Harvard, but had not decided to accept the offer before he became caught up in Austria's turbulent politics. From the limited sources available, it is impossible to be certain why Weisselberg migrated to the Soviet Union. The memories of Weisselberg's family, together with his NKVD interrogation files, provide some information. Alexander Weissberg's account only says that "[Weisselberg] was not a Communist, but he had come to the Soviet Union to work because he had believed that a juster system of society was being built up." Weissberg is, however, incorrect, since Austrian Communist Party records show that Weisselberg joined the party in late 1933.

Transcripts of NKVD interrogations must be used with caution, since all confessions were extracted under some form of torture. Many transcripts consist of clearly fabricated confessions, with fantastic and elaborate stories of intricate conspiracies against the Soviet Union and Stalin. A few secret police files, however, may contain relatively reliable accounts provided by the prisoners. These testimonies have only occasional embellishments, as demanded by the NKVD to prove their case against the accused. As we shall see, the information contained in the interrogations of medical scientist Kurt Zinnemann and his wife Ena are well corroborated by other sources. Weisselberg's interrogation seems to have been part of this latter group and provides much useful information.

In 1922, at the age of seventeen, Weisselberg joined the Union of Underground Communists in Vienna. Three years later he left the Communist Party for the Socialist Student Union; he remained active in the Social Democratic Party until he rejoined the Communists. In the late 1920s, Weisselberg was twice arrested after street battles with fascists. In 1930 he completed his doctorate in chemistry at the University of Vienna. Unable to find an academic position in Austria because of rampant antisemitism, Weisselberg found employment as a chemical engineer at the Kreditel Mainer oil refinery in Drösing, just outside Vienna.

In May 1933, Weisselberg travelled to the Soviet Union as a tourist. He spent a little over two weeks in Moscow and in Kharkiv, where he visited his school friend Alexander Weissberg. Weisselberg was impressed by what he saw in the Soviet Union and asked Weissberg to find him a scientific job in Kharkiv. Upon his return to Austria, Weisselberg became increasing involved in the growing political unrest in the country. In March 1933, Chancellor Engelbert Dollfuss overthrew the country's democracy and replaced it with a Catholic fascist dictatorship. In February 1934, violent unrest broke out, pitting socialists against fascists, the police, and, eventually, the Austrian army. The rebellion was suppressed after four days, after which most opposition groups were outlawed by the Dollfuss government.

Weisselberg rejoined the Communist Party at the end of 1933, no doubt influenced by his sister Erna, who was a member of the Austrian Communist Party's Comintern. Weisselberg was appointed a cell leader in Drösing, in the district of Gänserndorf, where he was assigned to organize Communist cells among the workers at the oil refinery. Although Weisselberg took no part in the rebellion against Dollfuss's government, his party activities brought him to the attention of the police. He was searched for illegal weapons and accused of distributing Communist literature. He was fined for possessing the weapons, but not arrested. In the meantime, Weissberg had not forgotten his friend; in July 1934, he arranged for Weisselberg to receive a job offer from the Ukrainian Coal Chemical Research Institute in Kharkiv. Fearing arrest at any moment, Weisselberg immediately accepted the offer. He left Austria so quickly that he failed to secure permission from the Austrian Communist Party to resign from his party position. Before his departure, he attempted to persuade his fiancée to come with him or to follow him to Kharkiv; she hesitated, wanting more information about life in the Soviet Union, and Weisselberg left without her.[25]

Direct Migration to the Soviet Union by Non-Communist Scholars

Weisselberg, Murawkin, Harig, Schaxel, and Steinitz took a more or less direct route to the Soviet Union; other Communist scientists did not. In fact, as many non-Communists as party members sought refuge in the Soviet Union. They all had their own personal reasons for seeking refuge that were not connected to ideology.

The mathematician Fritz Noether, who along with Schaxel was one of just two full professors to migrate to the Soviet Union, fought heroically to maintain his position at the University of Breslau. Noether argued that despite being Jewish, he was exempted from the racial provisions of the Civil Service Laws of 1933 because of his military service in the First World War. Noether had served Germany honourably in the conflict; he had been wounded, and awarded the Iron Cross. But this sacrifice did not save Noether: Nazi-dominated student organizations at the university demanded his removal from the teaching staff because of his political activities, which included his support of the League of Human Rights.

Image 5. Fritz Noether with his wife Regina (right) and sister Emmy in
Germany prior to their flight. Emmy and Regina would both die in the 1935;
Emmy in the United States after surgery to remove cancers, and Regina in
Germany from unknown causes after returning to Germany after suffering
a bout of depression in Tomsk. Fritz was arrested by the NKVD in 1937 and
executed in 1941.

Despite Noether's vehement denial of any political activity, he was forced
to resign in 1933 in a desperate effort to save his pension.

Almost immediately after his dismissal, the Soviets offered Noether
a professorship at the Institute of Mathematics and Mechanics at the
University of Tomsk. It is not known why he accepted that offer, but
he did not pursue any other possible positions. There is no record
that he applied to any of the refugee academic assistance organizations.

In the summer of 1934 he bade farewell to his sister Emmy, who after being dismissed from Gottingen had secured a position in the United States. The two did not know that this was the last time they would see each other. In 1935, Emmy died of post-operative complications following cancer surgery.[26]

Stefan Bergman, another mathematician, also went directly to Tomsk after being dismissed from his position in Berlin for reasons of race. Bergman was born in 1895, into a Jewish family in Częstochowa, then part of the Russian Empire, now in Poland. In 1920, he received an engineering degree from the University of Vienna. The following year, he began his doctoral degree at the Institute of Applied Mathematics at the University of Berlin, under the supervision of Richard von Mises. In 1922 he completed his thesis on "the development of all analytic functions in a given domain in terms of a fixed set of orthogonal functions." The results could be applied, "on the one hand, to fluid dynamics, conformal mapping and potential theory and led, on the other hand, to the 'Bergman kernel function' which is one of his major achievements in pure mathematics." Bergman's research career continued to flourish in the 1920s. In 1930, he habilitated and received a "very rare" joint appointment as *Privatdozent* at the Institute for Mathematics and the Institute for Applied Mathematics in Berlin. After his dismissal he accepted a position at the University of Tomsk. Why he chose to go to the Soviet Union remains a mystery.[27]

Hans Hellmann also moved straight to the Soviet Union. In his case the motivation appears to have been strictly personal: his wife Viktoria was from Ukraine, and her family still lived in the Soviet Union. At the urging of Viktoria, he had previously applied for positions in the Soviet Union, but the authorities there had refused to issue him an entry visa. In 1934, having lost his academic position in Germany, he found himself in desperate straits. He met with A.N. Frumkin, deputy director of the Karpov Institute, to solicit for a research position at the Moscow-based research Centre of Physical Chemistry. Frumpkin was impressed by Hellmann, and he was able to convince the NKVD that the German was a politically suitable candidate. Hellmann arrived in the Soviet capital later that year.[28]

Of all of the academic refugees who went to the Soviet Union, Herbert Fröhlich, seems to have led the most charmed life, but he was also, perhaps, politically the most naive. In 1930 he had completed his doctorate in theoretical physics at the University of Munich under the supervision of Arnold Sommerfeld. Like so many young physicists of his generation, Fröhlich focused his work on the applications of quantum mechanics. He served as Sommerfeld's assistant for two years, and in January 1933,

Summerfeld arranged for him to become a *Privatdozent* at the University of Freiberg, where he was to teach modern theoretical physics. Fröhlich was thus just beginning his academic career when Hitler came to power. He thought nothing of the Nazis' electoral victory. He had left for a skiing trip before assuming his duties in Freiberg and thus was away from Munich when the Hitler came to power. He returned home only to discover that his father and brother had been briefly arrested and his father badly beaten, perhaps because the family were not just Jewish, but Zionists as well. Fröhlich's sister had migrated to Palestine even before 1933. The rest of the family, except for Herbert, fled Germany for France, intending to follow her to Palestine, which they did one year later.

Herbert stayed behind in Munich, where he attempted to negotiate with Nazi officials for the return of some of the family's property. These negotiations were dangerous, and the risks were compounded by Fröhlich's correspondence with the Soviet physicist Yakov Frenkel for a position at LFTI in Leningrad. Fröhlich survived another year in Munich, and later recorded that he "enjoyed the danger."[29] In early 1934, he finally left Germany, stopping briefly in Strasbourg, but he was reluctant to accept Frenkel's offer. Like so many others, he was not politically active, although he had been involved with a socialist-leaning Zionist hiking club in his teens. After a brief trip to London and Paris brought no immediate prospects of employment, he felt "forced" to depart for Leningrad.[30]

If Fröhlich went to the Soviet Union out of political naivety, fellow physicist Werner Romberg later mused that his doctoral supervisor arranged for him to go there as a way to cure him of "my leftist illusions." Romberg, active in the anti-Nazi Socialist Workers Party, went directly from the completion of his PhD at the University of Munich to a position at the Physical Technical Institute in Dnipopetrowsk.[31]

The medical researcher Kurt Zinnemann's flight to the Soviet Union best demonstrates the complexity and diversity of the motives that drove scholars to the East. Zinnemann had run out of options when he fled Germany with his fiancée in 1935. He did so only after enduring two years as an internal refugee, a fate shared by Jews unable or unwilling to leave Hitler's Germany. Zinnemann, a young medical researcher just starting his career in Frankfurt in the early 1930s, delayed his departure because he was in love.[32] In 1931, he had met Irene (Ena) Loesch, his young laboratory technician. The couple were soon in love, but Zinnemann was Jewish and Ena a Lutheran. To make matters worse, Ena's father and brother were either members or strong supporters of the Nazi Party. The brother was also a close associate of Hermann Göring. Ena's father threatened to renounce her if she married a Jew. Despite this, Kurt and

Ena's relationship blossomed. They met in secret in the town wood, near a well-known café.

When Hitler came to power in 1933, Kurt and Ena continued their relationship in secret. Kurt was laid off from his research position in April; according to his dismissal letter, the hospital claimed it was for financial reasons. There can be little doubt, however, that he was terminated simply for being Jewish.[33] He was forced to seek employment in Frankfurt's segregated Jewish hospital, were he worked as a surgeon. As restrictions placed on German Jews escalated and hostilities mounted, life became increasingly difficult. One evening, as Kurt left the Jewish Hospital, a woman approached him on the street "rattling a tin collecting for the Nazis." Kurt told the woman she should be ashamed of herself "collecting for such a cause." At that very moment, four SS men appeared from around the corner to arrest him. Miraculously, before the SS could act, "a crowd of ordinary Frankfurt people" formed around Kurt, moving forward spontaneously, and pushing him onto a tram. Long after the Second World War, Kurt's daughter Pam asked: "Father, do you hate Germans?" Kurt replied, "I owe my life to them."[34]

Kurt now realized that flight from Germany was the couple's only option. Professor Hugo Braun, who had been overseeing Kurt's research before both men were fired, suggested that Kurt write to some medical researchers in the Soviet Union, one of whom was Ernst Simonson in Kharkiv. Dr Franz Gugenheim, Kurt's best friend and a Communist, helped Kurt mail the letters safely, via a contact in Saarbrücken, which was under the control of a League of Nations mandate. Only Simonson replied, offering to find Kurt a position and suggesting that he visit the Soviet consulate in Berlin to secure a work visa. Without official documents to confirm that he had a specific job offer in the Soviet Union, the consulate refused to issue a visa. Kurt then travelled to Paris to visit his brother, who helped him apply for a visa through the Soviet embassy in Paris. Kurt's application was again rejected. Desperate, Kurt secured an Intourist visa to visit Moscow and Kharkiv in the autumn of 1934. In Kharkiv, Simonson warmly greeted Kurt and introduced him to Professor G.B. Melnikov, the director of the Bacteriological Institute in Kharkiv. Melnikov hired Kurt to conduct medical research into disease and arranged for him to obtain all the necessary paperwork for a work visa.[35]

Back in Germany, Kurt and Ena carefully planned their escape. On 1 December, Kurt and Ena boarded a train heading eastward, accompanied a short way by Kurt's parents, Leah and Lazar. Kurt's mother was distraught: her son was leaving her and, worse, was marrying a *shiksa* – but she understood their need to escape. Ena's mother was aware of the couple's plans but remained at home, broken-hearted that she would

Image 6. Ena and Kurt Zinneman in their winter coats shortly after their arrival in Kharkiv in December 1935 (reproduced by permission of Pamela Zinnemann-Hope).

probably never see her daughter again. She knew that if Ena and Kurt were caught, they would both be "broken." She kept the news from her husband, but somehow Ena's father learned of his daughter's flight. Family stories state that he called Hermann Göring asking him to close the borders to the couple. If Göring did anything, it was too late: Kurt and Ena were able to cross into Poland. A few days later, they continued on to Kharkiv, where they soon began to settle into their new life free of the Nazi racial laws that had separated them. They married and purchased heavy winter coats to ward of the cold Ukrainian winter.[36]

Conclusion

Determining the motivations for academic migration to the Soviet Union is not as straightforward as it may be for the vast majority of German Communists who fled their home country after the Nazis came to power. Ideology was an important influence for only a handful of the displaced academics who went directly to the Soviet Union. Of those scholars who went there straight from Germany, only four were members of the KPD. Much more significant was the personal recruitment by Soviet and Soviet-based scholars of their Central European colleagues. Those sympathetic to Communist beliefs may have been more inclined to go to the Soviet Union, but even this appears, at best, uncertain. Ultimately, none of these factors led most of these scholars to flee Hitler's Germany for Stalin's Soviet Union. The vast majority did so out of necessity. We now turn to their stories.

4 Scientists in Flight to the Soviet Union

In the early years of the academic refugee crisis, the Soviet Union was viewed as one option among many. SPSL and AEC application forms specifically asked displaced scholars whether they were willing to take positions in exotic locales like the Soviet Union, the Far East, or South America. Such places were far from home, and their alien cultures posed their own difficult challenges to scholars and their families, such as primitive or non-existent research facilities, political instability, and the very real prospect of death from endemic diseases. The Soviet Union was different from other exotic places only because it offered scholars opportunities to renew their research careers. The vast majority of refugee academics wanted to remain in Western Europe, Great Britain, or the United States. Most non-Communist refugees viewed the Soviet Union in same light as places like Ecuador, China, or East Africa, as an option of desperate last resort. This chapter examines why so many were forced, even so, to seek refuge in Stalin's Soviet Union.

The Academic Refugee Crisis, 1933–1937

Personal and professional friendships between Central European scientists and their counterparts in the Soviet Union provided opportunities for the former to visit the latter, and this helped persuade some to migrate to the Soviet Union; however, many other displaced scientists were not persuaded that Stalin was any better than Hitler. They saw the danger signs and rejected job offers from Soviet institutions. In 1931 the Austrian physicist Victor Weisskopf, then a graduate student, had rented a room in Eva Striker's Berlin apartment. He soon became close friends of Striker and Alexander Weissberg. Not surprisingly in that environment, he began to consider whether socialism or Communism might provide an answer to society's disparities. A year later, while waiting for a Rockefeller post-doctoral fellowship to begin, he spent eight months in the Soviet Union, which included visiting Weissberg and Striker in Kharkiv. Weisskopf travelled widely in the country and saw first-hand the final months of the horrific famine that killed millions of Soviet citizens. He would return to the Soviet Union for shorter visits in 1933 and 1936.

By 1936, Weisskopf had completed his post-doctoral fellowship and a three-year stint in Zurich as an assistant to the great theoretical physicist Wolfgang Pauli. He was seeking an academic position when he made his third visit to the Soviet Union. When he visited his friend Weissberg in Kharkiv, he encountered a diverse group of expatriate scholars: "[There were] Viennese who went there, Communists who moved over. It was a very bad time. A lot of people who were either communists, half-communists or non-communists who just went there because it was the only place where you could stay alive."[1] Weisskopf was offered a full professorship at the University of Kiev, with very good pay. Around the same time, the University of Rochester in upstate New York offered him an instructorship, though it would pay far less (only $200 per month). Despite the disparity in pay and position, Weisskopf chose Rochester. He would recall that it was an easy decision: "It was very clear to me that I would not go to Kiev under any condition[s], and I think that if I would have accepted that I am not sure that I would be alive today ... I certainly saw the backside of Stalinism; it was visible things were going the wrong direction."[2] Weisskopf was fortunate to have had the choice; many did not.

Before going to the Soviet Union, at least fourteen of the scholars had attempted to find positions elsewhere; only in desperation had they turned to Soviet universities. This desperation was understandable. The AAC's annual report for 1934–35 outlined the sad reality of the odds against displaced scholars finding a permanent safe haven. By May 1935, only 287 out of an approximately 650 refugee academics had found long-term employment anywhere in the world. Three hundred thirty-six, including all of those who would eventually make their way eastward, were in often precarious temporary jobs. At least another 650 remained in Germany, where they faced unemployment and growing persecution.[3]

Most of the refugees had few or no options once they fled Germany. The lack of opportunities available to them was in part a result of their comparative youth – most were in the earliest stages of their careers and had not completed their habilitation, which was a necessary qualification for a permanent research position at a university. The AEC and, to a slightly lesser degree, the SPSL in Britain discriminated against these younger Central European scholars, out of fear that recruiting them would anger American and British graduate students and unemployed academics. An antisemitic backlash against refugee scholars haunted both academic rescue organizations throughout the 1930s.[4]

Having been cut off from the two countries that offered the best prospects for continuing a university research career, the refugee scientists now listened to the siren call of the Soviet Union, which was offering world-class research facilities as well as an opportunity to participate in the great Communist social experiment. Although there were abundant signs of impending disaster, the allure of Stalin's Soviet Union was simply too strong for these individuals to resist. Very few were like Weisskopf, who had a choice. For those who aspired to remain in academe, the temptation to take any opportunity that offered the even remotest hope of a permanent research and teaching position was too great not to pursue. Add to this the pressure they felt to provide for their families – often not simply the immediate family, but other relatives facing increasingly horrific conditions in Germany. Many tried first to go west because they wished to resettle in a liberal democracy. Migrating to the Soviet Union posed other challenges besides: one could not simply travel there to seek employment. First there had to be a job offer, and then the Soviet state had to approve it. One could neither enter nor – as these scholars would discover later – leave the Soviet Union without a visa. And for those living in Germany after the Nazis came to power, direct contact with Soviet officials and scholars was dangerous.

In mid-1934, Gerhard Herzberg, a future Nobel laureate for chemistry, was courted by Soviet scientists to take a position at the Leningrad Institute of Chemical Physics, even though he professed no interest in Communism. Herzberg, a gentile, was still employed in Germany, but his position there was precarious because his wife was Jewish. Herzberg used the AAC office in London to forward and receive letters while he was being courted by the University of Leningrad. In the end, he rejected the lucrative Soviet offer, having decided that the risks of going to Leningrad outweighed the dangers of staying in Germany. Fortunately, Herzberg, like Weisskopf, was able eventually to escape to a new life in North America. A former Canadian post-doctoral research student who had worked with Herzberg in Germany found Herzberg a position in the chemistry department at the University of Saskatchewan. The winter climate in Saskatoon was little better than Siberia's, and it was only slightly less out of the way, but for Herzberg and many others, any position in North America was far better than the most prestigious position in the Soviet Union.

Great Britain as an Initial Place of Refuge

In the early years of the academic refugee crisis, Great Britain hosted more displaced scholars than any other country. Even though the AAC was formed within weeks of the first dismissals of professors, teachers, and researchers from German post-secondary institutions, and the civil service, the council's limited resources allowed it to offer only a few two-year fellowships. Some British universities and colleges were able to raise their own funds to hire a handful of scholars, but despite massive efforts by the AAC to acquire grants from American and British foundations and corporations, little assistance was forthcoming. Among those who first sought refuge in Britain were the physicists Hans Baerwald, Fritz Lange, and Fritz Houtermans. Charlotte Houtermans accompanied her husband, but she became a full-time parent at home. All three of these physicists worked in temporary industrial research positions, though only Baerwald had experience at this type of facility prior to their forced migration.

Charlotte Houtermans provides an insightful account of her husband's rationale for accepting a position in the Soviet Union. In the summer of 1933 the Houtermans, fearing arrest after the Gestapo visited their apartment, fled Berlin for England. Fritz Houtermans's reputation as a brilliant young experimental physicist led to Isaac Shoenberg, the head of research at Electric and Musical Industries (EMI), offering him an industrial research position. Scientists at EMI's brand-new, state-of-the-art industrial research laboratory in Britain were conducting pioneering investigations into television technology. The Houtermans had rented a house in Hayes, a suburb of London, but despite opportunities to attend scientific gatherings in London, and notwithstanding frequent visits by physicists from around the world, Fritz Houtermans felt isolated from the scientific mainstream. Fritz chafed at the nine-to-five, five-day-a-week work schedule; he hated getting up early in the morning, and he was used to working until late at night to solve pressing research problems. Supported by a grant from the Friends (Quakers) Committee for Refugees and Aliens, Fritz spent his free time finding covert ways to smuggle information into Germany, including sending micro-negatives of stories from the *Times* hidden under postage stamps. He also helped his fellow refugees find positions elsewhere. This work was soon superseded by a number of international academic relief agencies, including the SPSL.

Soon after Fritz was informed that his grant from the Quakers was to be reduced in order to support additional refugees, the Soviet physicist Alexander "Sasha" Leipunski (sometimes spelled Lejpunsky) arrived from Kharkiv with a two-year stipend to work in Cambridge at the

Image 7. Fritz and Charlotte Houtermans with their first child, Giovanna (from the collection of Professor Mikhail Shifman).

Cavendish Laboratory. Charlotte recalled that "[Leipunski] spent all his weekends in Hayes. Sasha was loved by everybody. He was 'simpatico,' gentle, entertaining, pleasant and very intelligent. It was he who told Fritz about the scientific work in the Physics Institute in Kharkiv. His interests were the same as Fritz's. Sasha painted the scientific possibilities in such rosy colors that Fritz began to anticipate a possible renaissance there for his scientific ambitions."

Leipunski offered Houtermans a good salary, half of which would be paid in foreign currency. In addition, he would be provided with a modern apartment or house, as well as annual vacations abroad. Despite numerous warnings from friends about the dangers of going to the Soviet Union, Houtermans found Leipunski's offer too seductive to resist. The theoretical physicist Wolfgang Pauli, a close friend of the Houtermans, tried especially hard to deter them. Charlotte explained:

> When Pauli visited us and warned over and over again not to go, he [Fritz] did not listen any more. Our political instincts were dormant. What was alive was the hate of the Nazi regime, the danger and the terror of the Third Reich, the possible dangers and risks awaiting us in Russia were minimized

by comparison. It was the time of the Kirov murder, and if nothing else this alone should have warned us and prevented us from leaving England.[5]

Charlotte and Fritz did not heed the warnings. Both of them felt that their presence in England was ephemeral; they did not purchase furniture or carpets, and they only bought curtains because the landlord and some neighbours complained. They laughed that they were "camping indoors." Charlotte, often unable to leave their home because of the baby, struggled against the social isolation of Hayes, which felt all the more acute when she looked back on their boisterous former life in Berlin. Fritz spent as much time as he could with other scientists in London or Cambridge. The Houtermans seemed determined to hate everything British; Fritz even loathed their food and their sense of humour. This sense of transience and of isolation, combined with Fritz's vanity, his desire to return to an academic research institute, and his impatience for better opportunities, resulted in the couple accepting Leipunski's offer, with little hesitation. The family arrived in Kharkiv in February 1935.[6]

In 1933, Fritz Lange, a pioneer in high-energy atomic physics, left his position at the Physical Institute of the University of Berlin and fled Germany, fearing arrest by the Gestapo. Lange was a KPD activist and, as we have seen, had made a number of visits to the Soviet Union. Despite his ideology, he headed first to Britain, where he worked at EMI laboratories with Houtermans, his old school friend and fellow Communist. After his temporary position at EMI ended, Lange spent some time in the United States. In 1934, he inexplicably returned to Germany, where he worked for AEG and the chemical company IG Farben. Only in 1935 did he accept a research position at UFTI in Ukraine. There is no known explanation for his sudden departure from Germany for a second time. He may have been one step ahead of the Gestapo, or he may have been lured away by the chance to undertake basic research, live in the socialist workers' paradise, or join his old friends Houtermans and Weissberg.[7] The only hint of an explanation for Lange's bizarre travels before finally heading for the Soviet Union comes from Charlotte Houtermans, who mentions in her diary that Lange was under a delusion, shared by many, that Hitler would soon be murdered, and that displaced scholars would all be able to return to Germany. So Lange, like many displaced scholars, initially refused to settle down and instead waited to be recalled to a German university.[8]

Lotte (Charlotte) Schlesinger, the musicologist and composer, began her exile from Germany with a brief stay in Czechoslovakia, visiting her father. She then sought refuge in Austria, where she worked for a Montessori and Social Democratic Labor School in Vienna. The Austrian government, however, did not want refugees from Germany living there, and in the spring of 1934 she followed her good friends the Houtermans to Britain. While she looked for work, she lived with Fritz and Charlotte in Hayes. The Academic Assistance Council tried to help her, but she did not even get an interview for a post. In the spring of 1935, Schlesinger's brother found her a six-month position as a piano accompanist at the Vienna Conservatory. When that temporary position ended, she scraped by giving private piano lessons, but the sense of impending doom was growing in Vienna. Only a chance encounter with Alexander Weissberg, another old friend from Berlin, offered a route out of Austria. Weissberg was on one of his regular trips to Central Europe buying industrial machinery and plans for the factory in Kharkiv. Having learned of Schlesinger's plight, within two weeks Weissberg arranged for her to be hired as a piano accompanist at the Kharkiv Conservatory. The job came with a room in a house and a pass to a special store where she could buy food and other goods generally unavailable in the Soviet Union. Schlesinger immediately accepted the job, though she later confessed that the main attraction was the chance to be reunited with her beloved Fritz.[9]

After his dismissal from the German patent office in 1933, Helmuth Simons had one of the more interesting sojourns in Britain before heading to the Soviet Union. Simons fell into a grey area – he saw himself as both an academic and an industrial scientist. This distinction was important, since the different categories were assisted by different refugee assistance organizations. Professionals like lawyers, doctors, engineers, and dentists had to overcome many more obstacles to secure a position in their areas of expertise. National professional regulatory bodies objected to giving licences to potential foreign competitors or even allowing them to sit for qualifying exams. For industrial scientists, and for civil servants with doctorates and research experience, the boundaries between different refugee assistance organizations were often unclear.

In November 1933, Simons fled Germany for London. He applied to both the Jewish Professional Committee (JPC) and the AAC for financial and placement assistance. In January 1934 he agreed to become an unpaid voluntary assistant to Professor James Munro at the Royal College of Science at Imperial College London. He was supported by small grants from the JPC. Simons did a search of literature dealing with the use of vitamins to control noxious pests. Munro was impressed with the German's work, and he spent the next eight months attempting to persuade Imperial Chemical Industries (ICI) to hire Simons to work with the company's pest control committee. By June, Simons had been asked to send copies of his publications to senior executives of ICI, and he was under the mistaken impression that soon he would be offered a position. Munro, however, was far less optimistic about Simons's prospects, due to cutbacks in research funding and the low numbers of appointments available due to the economic depression. Munro felt that Simons had better prospects of employment in technical journalism rather than in industrial research.[10]

While Simons waited for a job offer from ICI, he became embroiled in efforts by some refugees to expose the still-secret German rearmament program. This work brought him to the attention of the British domestic intelligence service (MI5). In June 1934, journalist and editor Wickham Steed, in the lead story in the periodical *Nineteenth Century and After*, claimed that even before Hitler came to power, the German government had been experimenting with how best to use aircraft to deploy chemical and bacteriological weapons on cities. He reported that some of these experiments had been secretly carried out in the London Underground and the Paris Metro. The Germans, Steed claimed, had used harmless microorganisms and growth cultures to examine methods of introducing weapons of mass destruction into these vital transportation systems.[11]

Steed's claims generated a minor public sensation and drew the attention of numerous national and international newspapers. His assertions were also noticed by the British government, which sent agents to investigate the story's veracity. The security service quickly ascertained that those secret German documents had been provided by Otto Lehmann-Russbueldt, a pacifist and political activist. Lehmann-Russbueldt had been one of the first people arrested after the Nazis came to power. A bureaucratic snafu resulted in Lehmann-Russbueldt's release just ten days later. Lehmann-Russbueldt used that opportunity to escape Germany by posing as a psychiatric patient out for a walk near the Dutch border. He arrived in Britain six months later and moved into the same residence in London as Simons.

It is unclear whether the two men had known each other in Germany, but they soon began to collaborate to expose German militarism to the British government and people. It was Simons who acquired the German documents from a former business partner during a clandestine meeting in The Hague; Simons was an expert on microorganisms and the spread of disease and was, therefore, well-connected to some of those who might be working on or have knowledge of the secret military experiments. MI5 put Simons, Lehmann-Russbueldt, and Steed under investigation. Their mail was intercepted and copied. The security service soon concluded that the story of German experiments in Britain and France was, at best, dubious. Even so, MI5 continued to monitor the two Germans.

Simons acted as Lehmann-Russbueldt's assistant, because the latter could neither speak nor read English. They worked together on Lehmann-Russbueldt's new book, *Germany's Air Force*, "an exposé of secret German aerial rearmament in contravention of the Versailles Treaty."[12] In May 1935, Simons and Lehmann-Russbueldt alleged that a Gestapo agent, posing as a member of Scotland Yard's Special Branch, had attempted to trick his way into their flat to steal documents related to the book. Simons, who was home at the time, had thwarted the agent, refusing to believe he was a British policeman. Two real Special Branch officers, the counterintelligence arm of the British police, interviewed Simons about the alleged attempted theft. The officers concluded that the story was a fabrication. Both men, the police reported, "have a 'persecution complex,' see spies everywhere, and are constantly warning refugees to be on their guard." They concluded, however, that neither man had done anything inimical to British interests.

It is impossible to ascertain the truth of the matter, but the policemen's conclusion that the Germans were suffering from a "persecution complex" suggested irrational paranoia. Simons and Lehmann-Russbueldt, however, had good reason to fear for their safety, and this incident illustrates the rational fear of all those who fled the Nazis. Gestapo agents had killed anti-Nazi activists in Austria and Czechoslovakia and had lured another from France to Switzerland, where he was kidnapped and dragged back to Germany. Closer to home for the two men, just one month before the alleged attempted theft from their flat, two anti-Nazi refugees, Dora Fabian and Mathilde Wurm, had been found dead under suspicious circumstances in Fabian's London apartment. The coroner's inquest into the deaths concluded that the women had committed suicide; many in the German expatriate community were unconvinced of this, believing that Gestapo agents had murdered them. Fabian had provided documents to Lehmann-Russbueldt for *Germany's Air Force*, and her suspicious death must have terrified Lehmann-Russbueldt and Simons. Their fears for their own safety were further aroused by demands by some conservative MPs and newspapers that the government expel refugees who were using Britain as a platform to mobilize resistance to Hitler.[13]

Simons continued to support Lehmann-Russbueldt's work on his next book, a second volume on the international armaments industry, which would update and expand on an earlier monograph published in Germany in the 1920s. As part of this work, they interviewed an ICI chemist, whom they believed was a fellow pacifist. The chemist became alarmed when the Germans' questions would have required him to give them classified information. The queries were phrased in such a clumsy way as to convince the ICI chemist that Simons and Lehmann-Russbueldt were "either fools or knaves." Still, the chemist reported the interview to Scotland Yard. Special Branch determined, however, that the Germans, rather than security threats, were merely naive.[14]

While assisting Lehmann-Russbueldt, Simons continued to seek a long-term scientific appointment. He kept working in Munro's lab, but in September 1934, after eight months of lobbying, neither a post at ICI nor another possible opening at the Flour Millers' Association materialized. Simons was "stranded" and in a desperate straits. It cannot be ascertained whether his becoming a person of interest to MI5 played a role in his inability to find work.

Finally, in late October, Simons was offered an appointment as assistant bacteriologist in Dr Chaim Weizmann's laboratory in London. This was a nine-month appointment, funded by a £150 grant from the JPC. Weizmann, a chemist of Russian-Jewish heritage, had made a fortune during the First World War, when his company developed techniques for mass-producing chemicals vital to the Allied war effort. He is best remembered as a prominent Zionist who would go on to become the first president of Israel. Simons was certain his research would convince Weizmann to hire him full-time at one of his four international research facilities, including his new one in Palestine. Simons was so confident about his prospects that he wrote to the Academic Assistance Council stating: "You can cancel my name from your list and destroy copies of all my testimonials."[15]

Details about Simon's final months in Britain are scarce. In February 1935, Weizmann wrote to the JPC expressing his admiration for Simons's work ethic and his "great experimental skill." One year later, however, Simons was no longer working for Weizmann. We do not know why. The JPL continued to search for an opening for Simons in a British or Commonwealth research facility, but without success, and by the spring of 1936 the scientist was expressing to relatives a growing depression and a sense of despondency. He finally turned to what must have seemed to him his last hope, a position in the Soviet Union.[16] He may have received assistance in contacting the Soviet Union from bacteriologists David Keilin, a Russian-born biology professor and director of the Molteno Institute at the University of Cambridge. Interestingly, one of the people Keilin suggested Simons write to was Julius Schaxel, who by then had been in the Soviet Union for more than two years. In mid-February 1936, Simons sent a carefully worded letter offering his services to E.N. Pawlowsky, a professor of zoology and comparative anatomy at the Military Medical Academy in Leningrad.[17] Simons confessed in letters to a Berta Simons – which were intercepted by MI5 – that he had lied and altered his curriculum vitae to make himself more scientifically attractive and politically acceptable to the Soviet scientist. He told Berta that his statement claiming he knew "the section technique of Anopheles-flies for malaria plasmodia is not quite true," but he had written this because it "will not fail to make an impression on the people there." If necessary, he could learn the technique before departure. He hid the fact that he had worked for Weizmann for nine months, because Zionism had been officially condemned by the Comintern. He only stated that, since leaving Munro's laboratory in the autumn of 1934, he had been in close touch with Keilin, from whom he had learned the sectional technique, which he applied to his own research into "parasitology and protozoa." To sweeten the pot, Simons offered to bring to Pawdowsky's institute his "valuable optical instruments, my collection of microscopic preparations and my countless books, as well as special editions."[18]

It would take several months for the Soviets to agree to hire Simons. To ward off his growing anxiety and uncertainty, he participated in one more major project. He was the scientific adviser for another expatriate German anti-Nazi pacifist, the journalist Heinz Liepman, for his book *Death from the Skies: A Study of Gas and Microbial Warfare*.[19] *Death in the Skies* was published in 1937, by which time Simons was already in the Soviet Union. He had finally departed for Moscow in the summer of 1936, his trip to the Soviet border paid for by one last grant from the JPL. Upon crossing the Polish border, Simons disappeared from all of his friends and associates in Britain. They would hear from him again two years later, when he reappeared in Paris.

France as an Initial Place of Refuge

France, with its liberal refugee policy, took in many of those who fled Germany, but it was considerably more difficult in France than in Britain for displaced scholars to find any type of financial assistance or employment. An academic assistance organization, the French Committee for the Reception and Organization of the Work of Foreign Scientists (Comité français pour l'accueil et l'organisation du travail des savants étrangers), was founded a few weeks before the AAC in Britain. Due to its structure, it had far less money to assist refugee scholars. Moreover, all French universities were state institutions, and civil service rules precluded the permanent hiring of foreigners. By the summer of 1935 the committee had collapsed due to fiscal shortfalls and ideological divisions.[20] A new, more effective academic assistance organization was established in July 1936, which was too late for the first wave of academic refugees from Germany.[21]

In April 1933, Hermann Borchardt, the philosopher, high school professor, and left-wing intellectual, became an early victim of laws banning Jews from the civil service. After the director of his school was visited by the Gestapo (Geheime Staatspolizei), Borchardt was dismissed for racial reasons. He was advised by an unnamed person that he should not return to his apartment, so he withdrew 200 marks from his bank and signed a power of attorney giving his wife Thea access to his bank account. Then, to secure a German exit visa, he arranged for a doctor's certificate stating that he had to immediately travel to the spas at Bad Lindewiese, Czechoslovakia. But instead of Czechoslovakia, on 6 April he arrived in France, via Basel.[22] He languished in Paris, unable to find work. In July, he received an offer from the Teachers College at the University of Minsk, as "a foreign specialist" in German language. The offer of a job at Minsk was not a surprise: in the 1920s he had twice rejected similar offers. Borchardt was unsure how to proceed. he had been reluctant to go to the Soviet Union in 1933. In an account he wrote for a right-of-centre Catholic journal in 1945, by which time he was a devout anti-Communist, he explained his feelings: "I knew very well that the last remnants of freedom of thought and expression had already been completely extinguished throughout the Soviet Union under Lenin, and that even in minor issues only one opinion was allowed at any time – the opinion of the Communist Party."[23]

In an anti-Communist tract written two years later, however, Borchardt admitted he was less critical of the Soviet Union before his arrival at Minsk: "Before going to the 'new Russia,' I had a vague sympathy for the revolution, like so many open-minded foreigners subjected to the continuous Soviet propaganda."[24] In his private account of his time in the Soviet Union, which was not published until 2021, he indicates that in 1933 he was even more sympathetic to Soviet Communism. Only later would he view Communism, Nazism, and fascism as "the three best known brands of the modern totalitarian state."[25] Communist or not, desperation would lead Borchardt to move his family to the Soviet Union. His friends recommended that he approach Professor Paul Langevin, one of France's greatest physicists and a leading left-wing intellectual, to discuss the possibility.

Langevin played an important role for many of the academic refugees who went to the Soviet Union, for among Western scholars, he was one of the best-connected to the Soviet scientific and political leadership. In 1924 he was elected a Corresponding Member of the Academy of Sciences of the USSR, and in 1929, he was made a Honorary Member. In the early 1930s he was not a member of the French Communist Party, but he was certainly a fellow-traveller and an apologist for Soviet violations

Image 8. Hermann Borchardt's German passport issued for "health reasons" on 4 April 1933. Two days later he arrived in France (reproduced with permission of Christophe Hesse).

of human rights. After Hitler came to power in Germany, any scholar in France wishing to establish contacts in the Soviet Union or to appeal to Stalin's regime looked to Langevin for assistance. The only other scientist approached as frequently was Albert Einstein, but only because of his international status as a genius, not because he had any real sway in Moscow. Langevin did have considerable influence, at least until the purges began.[26] Borchardt's visit to Langevin was brief and to the point. Borchardt explained that he hoped to find a post similar to the one being offered in Minsk, in either France or its empire. Langevin replied that, with so many French-born teachers unemployed, there was no chance for Borchardt to find any such position in France. "If I were in your place," Langevin advised, "I would accept the Russian invitation." With no other option to support himself and his family, Borchardt accepted the offer from Minsk.[27]

As Borchardt prepared for the move to the Soviet Union, he received an unanticipated job offer to teach at an academy in Dijon, France. Herman and Thea were unsure how to proceed, but the consensus among the couple's circle of French and German friends in Paris was that they were fortunate to have a chance to live in the workers' paradise. One friend, an influential Parisian millionaire, wrote from her castle in Nice: "I congratulate you, to be able to live and work in the home of the world proletariat, the only country in which one feels comfortable." A doctor who had spent months visiting the Soviet Union as a tourist told the couple: "You will be immediately bewitched and carried away by this swing of revolutionary idealism." A third friend told them: "Oh, you lucky ones, you can take part in building the new world." Reading the letters of support, Thea became genuinely enthusiastic about Minsk, though not because she had any interest in the Soviet utopia. Instead, she explained to Hermann, it was her interpretation of the letter from the millionaire: "The thing about the millionaire is simply that she can't feel at home in Nice, because it is teeming with millionaires, especially American ones, where she certainly can't keep up. But in Moscow she is a queen. And I – even if you don't allow me to do it either, and I know that you won't allow me to do it! – I will be a queen with two flags from the 'Bon Marché' in Minsk." Pleased with Thea's endorsement, in February 1934, Hermann headed to the Soviet Union, leaving Thea and their two children in France. He would send for them when he was securely settled in Minsk.[28]

Despite his affinity for Communism, Edward Lederer, the Austrian biochemist, naturally chose France as his refuge of choice when he fled from his post-doctoral fellowship in Heidelberg. While in Germany, he had married Héléne Fréchet, the daughter of Maurice Fréchet, a well-known mathematician and professor at the Sorbonne. Though not yet a French citizen, Lederer easily found a temporary research position at the Pasteur Institute in Paris. In 1935 he was offered a long-term fellowship, but because he was still a foreigner, the annual stipend offered by the institute was a mere 2,000 francs, just one third the amount that would have been provided to a French citizen. This meagre stipend was too little for Lederer to support his wife and two children. Like Borchardt, he turned to Langevin for advice. Langevin introduced him to the Soviet cultural attaché, who in turn arranged for a three-year contract to be offered to Lederer at the Vitamin Institute of Leningrad, a perfect fit with the biochemist's research interests. Lederer and his family arrived in Leningrad in October 1935.[29]

Siegfried Gilde, the young medical researcher, was given a leave of absence from his Berlin hospital clinical research position in March 1933, then formally dismissed for being Jewish in July. He too fled to Paris, where Professor Karl Weissenberg helped him land a one-year research scholarship from a private, Lichtenstein-based foundation. Weissenberg had been a physics professor at the University of Berlin before he too sought refuge in the French capital. He was a prominent experimental physicist who had made important discoveries in the field of X-rays and medical imaging. It may have been as a result of his medical research that he came to know Gilde in Berlin. Weissenberg provided him with the opportunity to make the unusual shift from medicine to applied chemistry and physics. In October 1933, Gilde became a doctoral student at the General Chemistry Laboratory (Laboratoire de chimie générale), at the University of Paris, under the supervision of Professor Georges Urbain. Weissenberg was a guest of Urbain's laboratory and assisted with the oversight of Gilde's research.

Gilde believed that his new expertise could be applied to "physiological" research. In late 1934 he completed his "doctorat ès sciences de l'university" in the field of "electro-chemistry." Weissenberg then arranged for Gilde to work in his brother's laboratory in Vienna, examining techniques in electro- or shock-therapy. Even before leaving Paris, however, Gilde's financial situation had become precarious. In desperation, he applied to the Academic Assistance Council in London for a scholarship of 1,000 francs per month to pursue his studies. The foundation, which had funded his first year in France, did not have any more money to renew his fellowship. He told the AAC: "As a personal fortune doesn't exist, I beg for the granting of a scholarship for another year so as to be able to continue my studies."[30] The ACC did not consider Gilde "a strong case"; he was too young and had no publications, and moreover, he was asking for post-doctoral support, something outside the AAC's mandate. They rejected his application that autumn.[31]

In late February 1935, a second appeal was made to the AAC on Gilde's behalf, this time by M. Roland de Margerie, First Secretary to the French ambassador to Great Britain. It is uncertain how de Margerie became involved in Gilde's case, but it was likely through Georges Urbain, who was well-connected to the French political establishment. The AAC held firm to its earlier decision: C.M. Skepper, the AAC's assistant secretary, explained to de Margerie that at that moment, the council had no further funds for new grants. Skepper held out some hope that a new fundraising drive starting in March might make it possible to reconsider

Gilde's application. What Skepper did not say, however, was that even if the council had money, it would be unlikely to select Gilde for one of its precious few fellowships. Also unasked was why the French could not themselves find a way to assist Gilde.[32] Skepper's claim that the council lacked funds was true; in early 1935, the organization was teetering on bankruptcy. The March fundraising efforts would rescue the council, and, along with additional grants from British Jewish charities, it would soon be able to resume its work. The new funding, however, came too late for both Skepper and Gilde. Skepper himself would soon be laid off as the council struggled to drag itself out of the red. Gilde, now destitute, had no choice but to accept the offer of a research position at a Moscow hospital. He and his wife Ruth moved to the Soviet Union in the late spring.[33]

Initial Flight of Refugee Scientists Elsewhere in Europe

Other displaced scholars were sprinkled across Europe, wherever they could find a place of initial refuge. Physicist Fritz Duschinsky, after leaving Germany in early April 1933, received a temporary unofficial fellowship in Brussels from his doctoral supervisor, Professor Peter Pringsheim. Pringsheim was himself a scientific refugee, having fled from his position at the University of Berlin. Pringsheim was a convert to Lutheranism, but his family was Jewish. The Pringheim family was wealthy, and although the family fortune had been greatly diminished as a result of the First World War and postwar hyperinflation, he had enough money to fund himself while working as a guest in the laboratory of Professor Auguste Piccard at the Belgian Université Libre in Brussels.[34] Prinsheim had just enough resources left to offer Duschinsky a fellowship, paid out of his own pocket, but these funds would run out in May 1935. Prinsheim found the young Czech "a very intelligent young physicist" who specialized in television, cinematography, and the construction of electrical sending and receiving stations.[35] Duschinsky was an early applicant to the AAC and the AEC. He requested help in July 1933, stating that he was willing to undertake any research work that offered him a chance of employment. The AAC provided Duschinsky with numerous leads for positions in the United States, New Zealand, and Ecuador; only when none of these opportunities materialized did he migrate to the Soviet Union. In April 1936, after a three-year search for work, Duschinsky accepted an invitation from the State Optical Institute in Leningrad, to work as a researcher with the rank of a university professor.[36]

In the spring of 1933, Stefan Cohen-Vossen, the expert in geometry, sought refuge in Switzerland after being placed on leave by the

University of Cologne. While seeking positions in the United States and Great Britain, and applying to the AAC, he found temporary employment as a high school math teacher in Zurich. In early October he received official notice that he had been permanently dismissed from Cologne. One year later, his wife Margot Maria Elfriede (Friedel) gave birth to their son Richard. With a new baby, and without any job prospects in Western Europe or the United States, Cohn-Vossen was in an increasingly difficult position. However, he was aided in his employment search by Heinz Hopf, a mathematics professor at the Swiss Federal Institute of Technology. Hopf had excellent connections in the United States; in 1927 he had spent a year at Princeton University, funded by a Rockefeller fellowship. Hopf had turned down a faculty position at Princeton and in 1931 had moved from Germany to take up his position in Zurich. By the end of 1934 he realized his efforts to find Cohn-Vossen an academic position in the United States had failed. So he turned to Pavel Alexandrov, a mathematics professor at Moscow University. At Princeton, Hopf had developed a close friendship with the Soviet mathematician, who was also there on a Rockefeller fellowship. In the 1920s, Alexandrov had studied under Cohn-Vossen's co-author David Hilbert at the University of Gottingen, and his research interests were similar to Cohn-Vossen's. Alexandrov was delighted to offer the young German mathematician a professorship at Leningrad University, where he would work at the Steklov Mathematical Institute of the USSR Academy of Sciences. Like so many others, Cohn-Vossen had no other option but to go to the Soviet Union. No doubt he was relieved that he would be able to support his young family and continue his research and teaching career.[37]

In 1933, Heinrich Luftschitz, the chemical engineer and expert on concrete, was working at his recently acquired position teaching engineering at a Dresden technical college. When Luftschitz was dismissed for racial reasons, he returned to Trieste, where he had attended high school between 1899 and 1903. Trieste had been part of the Austro-Hungarian Empire when Luftchitz was born but had been annexed by Italy in the early 1920s. Luftschitz applied to the Academic Assistance Council for help in early 1934. He expressed, perhaps better than any other displaced scholar, his great dismay of being deprived not just of his job, but his avocation: "I was always inspired with the wish to teach academic youth and for this reason I acquired the right to give lectures at the university in order to devote myself to the academic career. Unfortunately this has unexpectedly and artificially become impossible as a result of the conditions in Germany according to which it is no longer possible for a Jewish aspirants to obtain professorship."[38]

Luftshitz's application to the AAC is among the most detailed submitted to the organization. It includes testimonials from former employers and scholars around the world as well as reviews of some of his many publications. All of these documents attest to his being "a name very much esteemed amongst the experts."[39] Luftshitz's preference was for a teaching and research position at a technical college, but he understood that he would have to be flexible, and he was willing to consider industrial work. He had practical expertise that should have made him employable. The AAC helped him apply for multiple positions as far afield as Ceylon, Bolivia, Ireland, and Portugal, but he received no job offers. There is no clear explanation for this failure, although Luftschitz's age may have been a factor. He was a highly qualified academic and research engineer, but engineers, like many other professional groups, were unwilling to accept people they perceived as foreign interlopers. Luftschitz also appealed to the council for assistance in getting permission to migrate to Palestine. In this regard, as with other refugees who made similar requests, the council firmly declined to intercede with the British Colonial Office. So Luftschitz lacked a visa for that region. The council would help when a refugee scholar was offered an academic job in Palestine, but it would not risk disrupting its relations with the British government by advocating for visas not linked to specific positions.[40] Like so many others who made the fateful decision to migrate to the Soviet Union, Luftschitz did so only after a long and futile job search. Nothing is known about his recruitment to the Soviet Union, but among his testimonials are extracts from several letters from Professor Adam Warschavsky, a Moscow-based engineer. In a note written after the Second World War, Luftschitz mentions that a Swiss-based committee for displaced scholars helped him find a position at a Building Research Institute at Sverdlovsk-Urals. The committee that helped him was probably the Notgemeinschaft Deutscher Wissenschaftler im Ausland, a German academic self-help organization founded in Geneva in 1933.[41]

The United States and the Refugee Crisis

In the early days of the academic refugee crisis, the Emergency Committee in Aid of Displaced German Scholars (AEC) was formed in New York to facilitate the permanent placement of displaced scholars in the United States. Using funds provided by two New York–based Jewish foundations, with matching grants from the Rockefeller Foundation, it provided money for one-year fellowships to American universities and colleges. The AEC's hope was that these temporary grants would lead to offers of permanent positions for German academics. By March 1934 the AEC had found positions for only thirty-six scholars; it planned to terminate its work when just fifty had been placed. This small number was supplemented by some independent efforts to aid German scholars, most notably at the New School of Social Research in New York. Although the AEC would eventually expand its efforts and place more scholars than any other agency, before 1936 most displaced academics found the United States to be an unobtainable utopia. The Great Depression had led to the layoffs of more than 1,000 American faculty; moreover, antisemitism was well-entrenched in American post-secondary institutions. Three of the Ensnared first tried to find refuge in the United States; despite good connections with American scholars, they failed in their quest for permanent refuge there.[42]

In late 1933, Marcel Schein, a Jewish Czech astrophysicist, faced an uncertain future. Schein was born in 1902 in the Austro-Hungarian town of Tristena, the son of a provincial banker. With antisemitism everywhere in the empire, Schein, like many of his contemporaries, became an agnostic and did his best to deny his own Jewishness. He showed an aptitude for science, and he did his undergraduate education at the University of Heidelberg, where he focused on physics. In 1925 he moved to the University of Zurich to pursue a doctorate. In 1927 he married Hilde Schoenbeck, a Lutheran German engineering student. One year later, their son Edgar was born. That same year, Schein received his doctorate summa cum laude and was immediately offered a position in the physics department at Zurich.[43]

In 1929, Schein won a prestigious Rockefeller fellowship, which enabled him to work for the next two years at the famous X-Ray Laboratory of Arthur Holly Compton at the University of Chicago. Compton had won the Nobel Prize in Physics in 1927 for his discovery of the Compton effect, which proved the particle nature of electromagnetic radiation. Much of Compton's work focused on understanding the nature of cosmic rays, those high-energy atomic nuclei or other particles travelling through space at close to the speed of light. Schein began studying cosmic rays with Compton, and this work would become the focus of his later research. Although Compton had hoped to keep Schein at Chicago, Depression-related budgetary cutbacks made this impossible, and Schein was forced to leave once his fellowship ended.[44]

When he returned to Europe, Schein become a *Privatdozent* at the University of Zurich, and it seemed likely that he would have a long and prosperous career in Switzerland. However, he became the victim of a little-known diplomatic dispute between Czechoslovakia and Switzerland. In 1934, the Czech Ministry of Social Welfare decide to expel some Swiss workers so that Czech citizens could be employed instead. The Swiss

Image 9. The Schein family – Hildegard, Ed, and Marcel – in Switzerland prior to their expulsion (reproduced with permission of Ed Schein).

ambassador to Prague informed his government that the only way to make an impression on the Czechs was to expel some of their citizens living in Switzerland. On 28 March 1934 the Swiss Federal Police informed Schein that he was going to be expelled. When Schein appealed the decision, the police offered to rescind the immediate order for expulsion if he agreed to leave the country by 31 July 1935. Schein appealed to university authorities for assistance, but despite strenuous efforts by them, nothing could persuade the government to allow Schein to stay.[45] It seems that by late 1933, Schein had some inkling that his position at Zurich was threatened. He began to explore other opportunities, but he was cut off from employment at German universities because he was a Jew, and the economic crisis in the United States seemed to have closed doors there as well. Schein received an offer to teach at the University of Nanking in China. China was already in turmoil – just two years before, the Japanese had seized Manchuria, and the civil war between the Nationalists and the Communists continued unabated. Schein could have returned to Czechoslovakia, but there were no opportunities for him there, a fact he learned from his old friend, Guido Beck.

Beck, the up-and-coming Czechoslovak theoretical physicist, had been assistant to Werner Heisenberg at Leipzig from 1928 to 1932, until he left for a professorship at the German University in Prague. Soon after Beck's arrival, the Czech government rescinded the promised long-term funding for his position. By the end of 1933, Beck could not return to Germany; the only offer for employment came from the University of Kansas, funded by an AEC fellowship. Kansas, however, informed Beck that, due to the fiscal restraints imposed by the Great Depression, his position could only be guaranteed for one year.

Before leaving for Kansas, Beck travelled to the Niels Bohr Institute in Copenhagen, where he met Schein. The two discussed their plight with James Franck, the Nobel laureate physicist, himself in self-imposed exile from Germany. Franck suggested they ask Yakov Frenkel for a job in the Soviet Union. Frenkel was on the faculty of the Leningrad Physico-Technical Institute; in 1933 he had been a visiting scholar at Copenhagen. Beck had visited the Soviet Union in 1933 and was less than keen, but according to Beck, "Frenkel was quite enthusiastic about the job. I didn't want to go along, but Marcel Schein who was in Zurich was in trouble and wanted to get out. So I combined with Marcel Schein that we would go to Odessa. The initiative was taken by Frenkel. He wanted something to be done there. So Schein went to Odessa and I went for a year to Kansas."[46]

While Beck remained cool to the idea of migrating to the Soviet Union, Schein found the Soviet offer too enticing to resist. Foreign scientists were treated like "dignitaries," and Schein was no exception. He was offered a large, well-equipped laboratory and the opportunity to set up a high-altitude observatory on Mount Elbrus, the highest mountain in Europe. Cosmic rays could be more easily detected at high altitude, where silver-based photographic plates would record the passage of the rays. Schein was also offered a good salary, a large modern apartment, and access to the special stores. Schein and his family set off for Odessa in November 1934.[47] Beck hoped never to have to go to Odessa; the University of Kansas had arranged for him to receive permanent immigrant status in the United States. However, Depression-era budget cuts prevented any extension of his fellowship. In June 1935, Beck booked passage for a remarkable round-the-world journey via Hawaii, Japan, Vladivostok, and the Trans-Siberian Railway.

Michael Sadowsky, long stateless after his family was stranded in Finland at the end of the Russian Civil War, likely could not imagine that his journeys as a refugee were only just beginning when he became a German citizen in 1928. Citizenship came with his appointment as *Privatdozent* in mathematics at the Technical University in Berlin. He had married Hilde Bein, a German Jew. Sadowsky's previous experience as a refugee and his wife's religion seem to have made him acutely sensitive to the growth of German nationalism and antisemitism long before Hitler came to power. In 1929, convinced that he needed to seek his fortunes elsewhere, he took an extended leave of absence from Berlin to become an assistant professor in the mathematics department at the University of Minnesota. There he immersed himself in American academic life, culture, and habits, something he "liked very much." In 1933, however, he became a victim of the mass lay-offs that had beset so many American universities. Under the principle of "last hired, first fired," Sadowsky was rendered not only unemployed but also, since he was not yet a permanent resident, an illegal alien. After he exhausted all possibility of further employment in the United States, he returned to Europe.

A German of Russian heritage, married to a German Jew, he could not return to Germany. As long as he remained married, the Technical University would "not admit him to any kind of activity." He chose to reside in Belgium, because he and Hilde spoke French and the country was the least expensive place to live on the entire continent. The Belgian people treated the Sadowskys with great kindness, but while Michael was hired to teach a few courses at the University of Brussels, he was not allowed to seek permanent employment.[48] His resources having dwindled, in April 1934 Sadowsky applied to the AAC for help. He informed them that, while he preferred an academic position, he would happily accept any work, even if it provided only room and board. A few leads were offered to him by the council; most promising was a position at a Black college in Atlanta. By late August, however, no offer had arrived from Atlanta, and his position was now desperate. He informed the AAC that on 4 September, "the Secretary of Education in Germany, Herr Rust" had withdrawn "permission to participate in public and private instruction in Germany." Moreover, his visa to re-enter the United States would soon expire. By late summer, Sadowsky had word that he might be offered a position at an unnamed major university, but it would not be in the United States. His first choice remained returning to any post there.[49]

Sadowsky remained in limbo for a few more months. His US visa expired on 22 November, and he had heard nothing from Atlanta or from the mysterious major university. In what must have seemed like his darkest hour, the unnamed major university suddenly offered him a research and teaching position. With no prospects for returning to the United States, he immediately accepted the job offer. On 17 December, Sadowsky wrote to the AAC from Leningrad. The letter is unique, as it is the only one written so soon after arrival in the Soviet Union. It conveys the profound relief and joy that must have been shared by so many of those who had been forced to flee German universities and who believed the Soviet Union might allow them to rebuild their lives. The letter is worth quoting in full:

> I am happy to inform you I have just been appointed by the University of Leningrad. I feel very much honoured by becoming a member of the Faculty of this leading Institution of Science, Research and Learning and I certainly am going to stay with this university. Therefore, thanking you sincerely for all the efforts you did for me, I beg you to cancel my name in your list of unemployed people.[50]

Conclusion

Like so many others, Sadowsky accepted a position in the Soviet Union as a desperate last resort. The movement of displaced Central European scholars to the Soviet Union was driven less by ideology than by dire necessity. Some of the hazards of migrating to the Soviet Union were by then widely known; even so, for many refugee scholars that course presented the best and in some cases the only hope. For those who had been forced to flee Hitler, the risky decision to move to Stalin's Soviet Union seemed a better option than returning to Germany. The migration of refugee academics to the Soviet Union was not exceptional; indeed, it was closely analogous to the experiences of other European Jewish scholars who, in the 1930s, found themselves forced to migrate to countries considered to be "the back of beyond."[51] No one, however, could have predicted the terror that Stalin would soon unleash on everyone residing in the Soviet Union. The migrant scholars would soon discover that they had jumped from the frying pan into the fire.

5 Living in Stalin's Soviet Union

In late March 1934, the Hellmann family – Hans, Viktoria, and Hans junior – arrived at Moscow's Belorussian-Baltic Station. They were met there by Viktoria's aunt, who, when she learned of the family's plan to settle permanently in the Soviet Union, declared: "How did you decide to come here? You are crazy!" The Hellmanns were not swayed by the aunt's words. Viktoria had long dreamed of reuniting with her family and had retained her Russian/Soviet citizenship. Hans was looking forward to working at a major research institute in a city that was one of the two centres of scientific research in this great nation of 150 million people. The Hellmanns, like most of the other academic migrants, believed that the Soviet Union offered them real hope, after Nazism and fascism had driven them from Central Europe. Viktoria and Hans felt certain they had brought their young son, just four and a half, to a land of promise and opportunity. Tragically, as we shall see, they could not have been more mistaken.[1]

A Survey of Academic Refugees in the Soviet Union

German and Central European scholars were located throughout the Soviet Union, from Moscow to the Siberian city of Tomsk, some 3,600 kilometres to the east of the capital. These scholars were concentrated in Leningrad and Kharkiv. One third of them were based in Ukraine, which included Kharkiv. Most were employed in universities and research institutes, although Siegfried Gilde, a medical scientist, was employed as a researcher in Moscow-area hospitals as well as at the Nutrition Institute.

We know remarkably little about the lives of refugee academics in the Soviet Union. With some important exceptions, those who survived the purges and the Holocaust were reluctant to talk about their experiences. As well, most of the refugee scholars stayed in the Soviet Union only briefly – an average of no more than two and a half years. Most of the voluntary migrants, like Alexander Weissberg, Martin and Barbara Ruhemann, and Herman Muntz, were in the country for much longer than the academic refugees. Emanuel Wasser, however, was forced to leave Leningrad in 1935, less than three years after his hiring. Wasser was one of two Central European scholars dismissed during the first round of expulsions of foreigners from Leningrad after the 1 December 1934 assassination of Sergei Kirov, the Communist Party leader of the city. This first round of expulsions took place before many of the refugee academics had even arrived in the country.

Because they were spread out across the vast nation, and divided by disciplines and institutions, there was no refugee scholar "community" *per se*. This was true even in Leningrad. Three Central European mathematicians, Herman Muntz, Stephan Cohn-Vossen, and Michael Sadowsky, were employed at the University of Leningrad in early 1935, but the latter two would soon leave for positions elsewhere; it is likely that Muntz transferred them out of Leningrad before they too could be expelled. Cohn-Vossen moved to Moscow, while Sadowsky assumed the post of professor of engineering mechanics and head of department at the Engineering College of the Industrial Institute in Novocherkassk. The eight scholars who remained or who came after the Kirov expulsions were at seven different universities or research centres. Besides Muntz, the other Leningrad-based scholars were Fritz Duschinsky, at the State Optical Institute; Ernst Emsheimer, at the Hermitage Museum's music collection; Gerhard Harig, first at the Leningrad Physical Technical Institute (LFTI) and after 1934 at the Institute for the History of Science and Technology; Herbert Murawkin, at the Leningrad Electro-Physical Institute; Edgar Lederer, at the Vitamin Research Institute; Julius Schaxel, at the Experimental Zoology and Morphology Laboratory; and Wolfgang Steinitz, at the Institute of the Peoples of the North.

Those in more remote places, the scholars themselves and also their families, must have felt profoundly isolated. Hermann Borchardt was the only German expert at the University of Minsk and acutely felt his complete separation from other German intellectuals. Fritz Noether was on the staff of the mathematics department at the University of Tomsk in Siberia. The move from Germany proved to be too much for his wife Regina. She suffered from severe depression and in 1935 was forced to return to Germany, where she died soon after of unknown causes.[2] Two other refugee scientists called Tomsk their home. Stefan Bergman, one of the mathematicians whom Muntz helped find employment in the Soviet Union, also worked briefly at the university before transferring to the more congenial climate of Tbilisi, in the Soviet Republic of Georgia.

Hans Baerwald worked as a physicist and electrical engineer at the State University and the Siberian Physical-Technical Institute.

Only in Kharkiv did a community of expatriate scholars exist, and only as part of a larger community of Soviet scientists working to establish a leading international centre for physics. This focus on physics was why so many of the refugees were based in Ukraine. Four research institutes in the field had been established in the Soviet Union in the late 1920s, the most important and largest of these being the Physico-Technical Institute of the Ukraine (UFTI; sometimes styled UPTI). UFTI was located in specially built facilities in Kharkiv. Its young, dynamic staff focused on the new and exciting fields of low-temperature and atomic physics, which made UFTI more attractive to the younger refugee scientists. Seven physicists, Charlotte and Fritz Houtermans, Fritz Lange, Martin and Barbara Ruhemann, Laszlo Tisza, and Alexander Weissberg, were on the staff of UFTI. Konrad Weisselberg was employed at the Coal and Chemistry Institute in Kharkiv. After Weisselberg's dismissal in 1936, Weissberg arranged for his friend to work part-time at UFTI.[3] One other experimental physicist, Hans-Jurgen Cohn-Peters, was also apparently based in the city, although none of the accounts of UFTI mention him. Cohn-Peter was formerly a doctoral student of Fritz Lange. Kurt Zinnemann and Ernst Simonson are also known to have been part of the larger circle of German academics and professionals who regularly socialized in Kharkiv.[4]

The sense of isolation felt by the academics and their families was compounded by the profound culture shock they experienced as a result of their forced transfer to this strange new world. Prior to 1914, Germany and the old Austro-Hungarian Empire had been among the world's great centres of scientific, technological, artistic, and intellectual culture and learning. While Central European societies had been almost destroyed by the cataclysm of the First World War and the subsequent turmoil, the inherent strengths of these societies allowed them to rebuild much of their former glory by the end of the 1920s. This was particularly true in Germany, which survived largely intact as a political entity. During the Weimar Republic, Germany remained a world leader in physics, chemistry, and most of the social sciences. Great artists, including the writers Heinrich and Thomas Mann, the playwright Bertolt Brecht, the filmmaker Fritz Lang (no relation to the physicist), and the composers Alban Berg and Arnold Schoenberg, were leaders of a cultural renaissance. The vibrancy of the café and cabaret life of Berlin was legendary throughout much of Europe and North America. Many of the academic refugees had enjoyed this cultural milieu before 1933.

Compared to the Central European powers, Russia had been backward and undeveloped prior to 1914. Its vast size and population were its strengths, but far too many of its citizens lived lives little different from a century before. The Russian centres of science and learning, such as Moscow and Saint Petersburg, were certainly as good as any in Europe, drawing heavily, as we have seen, from German universities for education and training. The First World War had been even more devastating for the Russian people than for most Central Europeans. The subsequent Civil War and the Bolsheviks' efforts to reshape Russian society took an equally disastrous toll on the citizens of the new Soviet Union. Stalin's first Five-Year Plan, launched in 1928, was a centrally managed program of forced rapid industrialization and farm collectivization. It was this plan that had opened the doors to foreign experts, whose knowledge and skills were desperately needed. Although great progress was made in the production of basic industrial commodities like steel, that progress led to one of the greatest mass murders in history. In a policy designed to break the will of the peasants to resist the collectivization of agriculture, Stalin and the Soviet government deliberately created widespread famine in the rural areas of Ukraine, the lower Volga, and the northern Caucasus. The death toll of the great famine of 1932–33 is estimated to have been between 6 and 7 million people. That famine was accompanied by the mass arrests and executions of the Ukrainian intellectual elite and even many Ukrainian Communist Party members. Stalin had ordered this purge to crush any remnants of Ukrainian nationalism. The NKVD, the state police, ruled the region with extreme ruthlessness. Even foreign scientists risked imprisonment if they challenged the authority of the party or the police. The worst excesses of the First Five-Year Plan and the Great Famine were over by the time the refugee scholars arrived, but by then, the voluntary migrants, particularly those in Kharkiv, had witnessed the horrors first-hand.

Voluntary Migrants and the Great Famine

In January 1932, Alexander Weissberg was joined by Eva Striker, and the couple were soon married. The reasons why Eva married Alex remain unclear. According to Arthur Koestler, Laura Striker, Eva's mother, would only agree to visit the couple if they married for "propriety's sake." The couple had been engaged prior to Weissberg's migration to the Soviet Union, and certainly Alex was deeply in love with her. Some of Striker's biographers suggest that the marriage was one of convenience, a ploy to enable her to remain in the Soviet Union. Whatever the case, the marriage lasted only two years: Striker left Kharkiv for Leningrad and then Moscow, to become senior designer in some of the country's largest ceramics factories.

Weissberg and Arthur Koestler, who visited his friends at Kharkiv several times in 1932 and 1933, provide bleak accounts of life in Ukraine at the end of the Great Famine. The work of setting up the low-temperature physics laboratory and nitrogen production facility was hampered by shortages of electricity – often only two hours a day was supplied to Kharkiv. The hunger in the countryside spilled over into the city; there was almost no food in the markets or the stores. Both saw the far more horrific conditions in the countryside. They witnessed the mass murder of the peasantry, yet both men were devout Communists who denied it was happening, claiming that it was only the kulaks, the rich landowners, not the peasants, who were suffering. They engaged in heated debates about the situation with Victor Weisskopf, a non-Communist, who also visited Kharkiv around that time. Much later, Weissberg and Koestler would renounce their former views, but in 1932 they believed that defending the socialist revolution from counter-revolutionaries was much more important than the truth.[5]

In the spring of 1932, Martin and Barbara Ruhemann arrived at the Ukrainian capital after a forty-hour train ride from Berlin. They were greeted at the station by Alex Weissberg, who brought a horse-drawn cart to transport the couple to their new apartment. "In 1932 the country was experiencing the worst year since the end of the civil war"; all the Ruhemanns "could see was chaos, poverty and hunger." Martin Ruhemann recounted that when they arrived, the situation was rather miserable, even in Kharkiv, where "there was bread but little else." Soviet workers at UFTI ate "'sandwiches' for lunch – dry bread with a little sugar strewn on it." The city's residents survived on a ration of 2 kilograms of dark brown bread per day; other than bread, the shops were mainly empty of food. Foreign specialists had access to a special store, but even there, supplies were "still pretty meagre." The foreigners could get butter and eggs and "occasionally green vegetables and rather tough meat." If the food situation was grim in the city, the peasants in the countryside were facing catastrophe. Seed "stocks had been depleted, and it was rumoured that the peasants were too weak to reap even if they had sown." Urban "collectives," including UFTI, tried to help rural collectives by sponsoring them, offering whatever expertise or aid they had to offer. The institute sent out Yasha Kahn, a young Russian-Jewish staff physicist, to investigate the institute's sponsored rural collective. At a meeting of some three hundred UFTI employees held after his return, Kahn reported on the situation at the rural collective. He explained that the sowing of the wheat had been carried out "in spite of the bitter resentment of the peasants." There were still two months to go before harvest, however, and the peasants had nothing to eat. Kahn urged UFTI to send the peasants bread, lest they die before the harvest.

Kahn's proposal was discussed at length. To Ruhemann's surprise, most of the Soviet staff spoke against offering food to the peasants. The prevailing view was that the peasants would rather "cut off their nose to spite their face," or that they would rather starve than submit to the state collection of grain. "If we send them bread," said a number of staff members, "the peasants will eat the bread and not collect the harvest." At the time, these arguments, Ruhemann recollected, "seemed absurd to me, but I had to confess that I had no knowledge of the Ukrainian peasants." The meeting concluded with a voting down of Kahn's proposal, in what Ruhemann observed, perhaps naively, was "a clearly democratic decision, taken voluntarily, without pressure from outside, by the majority of the 300 odd members of the community."[6]

Faced with harsh working and living conditions coupled with the horrors in the countryside, the Ruhemanns regularly threatened to leave the Soviet Union during their first six months at UFTI. Weissberg and other believers worked to persuade the Ruhemanns of the righteousness of the Communist cause, but it was the vital nature of their work at UFTI that won the couple over:

> [Martin Ruhemann] had never thought much about the significance of social revolutions, but once in the Soviet Union he grasped what was really at stake and was swept up in a new and larger movement than he had ever known before. His horizons enlarged and he began to understand that we were all architects working on the plans for a new world, and that the history of our century would depend on our success or failure.[7]

Barbara went even further than her husband, becoming a devout disciple of Stalin and his brand of Communist ideology. She gave herself completely over to the view that whatever evil the state might inflict on individuals, the goal was the long-term common good; that, or the criminal acts were being perpetrated by counter-revolutionaries without Stalin's knowledge. Given the recent rise of a new wave of political extremism in the world today, it may be easier to grasp that intelligent, well-educated people can be seduced by malevolent ideologies. Weissberg wrote of Barbara:

> From the time she embraced the cause of socialism as a young woman Barbara had believed fervently in Stalin. She was quite convinced that all evil in the Soviet Union – and she realized that there was quite a lot – was due

to the wicked apparatus and not to Stalin's deliberate policy. She used the word "apparatus" as though it had some life of its own and were a thing apart. It was difficult to understand how she managed to believe that in a country so politically centralized as the Soviet Union it was possible for an apparatus to carry out a policy independent of and opposed to the will of an all-powerful dictator.[8]

If Barbara Ruhemann became an unrepentant ideologue, there was more behind her beliefs than Weissberg's view that she suffered some sort of personality flaw associated with her Prussian roots. Among contemporary members of the Western intelligentsia, she was not alone in subscribing to and retaining such beliefs long after the purges and even after the Nazi–Soviet Non-Aggression Pact of 1939. As with many Communists and fellow-travellers, there is little doubt that her experiences growing up in the tumult of war, revolution, and disease, with all the economic and social dislocation that ensued, made her susceptible to these views. This is something she shared with her husband.

Many of the people with whom Martin and Barbara associated on a daily basis, particularly Weissberg himself, were master proselytizers for the new religion. Thus it is easy to understand how the Ruhemanns were seduced. Weissberg's later anger toward Barbara stemmed from her refusal to repent even after he had been imprisoned and tortured by the NKVD for four years, after which she and her husband were forced to leave the Soviet Union. Still, the Ruhemanns are unique among foreign scholars in that they were the only ones to become Communists while in the Soviet Union, and they would retain their belief in Stalinism long after their forced exit from the country.

The famine lingered through the winter of 1933–34. On a freezing night in February 1934, Hermann Borchardt arrived in Minsk; he was one of the first academic refugees to migrate to the Soviet Union. On the train station platform, porters took his bag and ushered him into the station. The porters had to force open the doors, and having done so, they were unable to proceed inside "because ragged men, women and children lay close-packed on a grimy stone floor up to the threshold." Borchardt later learned that the people on the floor were peasants who had fled the famine areas. Borchardt was shocked, having "never before seen such a tableau of human misery."[9] By the spring of 1934, when a much larger wave of refugee academics began arriving in the Soviet Union, the worst of the Great Famine was over. According to Martin Ruhemann, there was food in the shops and "there was less nervous tension; people were relaxed. Life became normal."

The Lives of the Foreign Specialists

In the ever-shifting reality of Soviet life, any normalcy people found was at best fleeting. Even with access to special stores, food remained in short supply for the foreign scholars. When he arrived in Kharkiv in January 1935, Laszlo Tisza discovered that since his first visit to the city in 1934, the food supply had been drastically curtailed. The wives of the staff spent much of their day trying to purchase black market supplies, and the bachelors, like Tisza, took turns missing work to do the same. "In practice only one satisfactory meal per day was possible for almost all grown-up people."[10]

Hermann Borchardt provided the most detailed account of the life of a refugee scholar in the Soviet Union. His recollections, we must remember, are contained in anti-Communist tracts and are often self-contradictory. Even so, he can be a useful source, especially when his account backs up those of others. Borchardt considered his access to the special department store for foreign specialists "the equivalent of a 200 per cent increase in salary." In 1936, when the education ministry made his continuing employment conditional on his renouncing his German citizenship and becoming a Soviet citizen, he refused, in large measure because he feared this would end his special status as a foreign specialist. By becoming a Soviet citizen, he would risk "automatically" losing all his "rights and privileges" and descending "the social ladder to the low rank of Russian teacher."[11]

Alexander Weissberg corroborated Borchardt's observation on the importance of the special stores to foreign specialists. Those stores allowed them access to a variety of goods that were unobtainable for Soviet citizens, and they were far less expensive. Weissberg estimated that the purchasing power of foreign specialists was ten times greater than that of the equivalent Soviet worker.[12]

There was also a critical shortage of decent housing throughout the Soviet Union. Most Soviet families lived in single rooms, often in shared apartments. Foreign specialists and, to a lesser extent, Soviet scientists were given privileged access to accommodations.[13] In Leningrad, Edgar Lederer found that conditions were "very good": the family was provided with "a lovely apartment overlooking the famous canals."[14]

Wolfgang and Inge Steinitz lived in a hotel in Leningrad for two months until their apartment was ready. The apartment was in a side building at the Institute of the Peoples of the North; it had two rooms with parquet flooring, a kitchen, and a bathroom. Inge spent the next several months furnishing the apartment, having cupboards and chests made, and scouring stores for suitable lamps, vases, and other finishing touches. Inge complained to her mother that Wolfgang was only interested in book-shelves to house the library he had brought from Germany, along with rare volumes he was acquiring in local antiquarian bookstores. Despite Inge's efforts, and the comparative luxury of the apartment, when Inge's mother visited in 1936, she was appalled by the primitive living conditions her grandchildren were forced to endure.[15] Notwithstanding her complaints, the Steinitzes, when healthy, enjoyed a comfortable lifestyle that most Soviet people could not even have imagined. The family even managed to rent a dacha or cottage in the countryside in the summers.[16]

At UFTI, specially built apartments were incorporated into the insti-tute's campus. By Soviet standards, life was relatively comfortable for the privileged at Kharkiv. The Houtermans were given a rather large apart-ment with rooms on two different floors, where, for a time, they lived with Charlotte Schlesinger, their old friend from Berlin. They were able to hire a maid, and when their second child was born, a nurse as well. The Ruhemanns had a similar apartment. Both couples lived alongside the senior Soviet scientific and administrative staff.

Not all of the scholars were as fortunate. Siegfried Gilde and his wife enjoyed their stay in Moscow, except for their cramped apartment, which became even more overcrowded after the birth of their daughter Marina.[17] The Hellmanns, also in Moscow, lived in a small two-room apartment.[18] Hermann Borchardt had negotiated his contract with the University of Minsk so that he and his family would receive a three-room apartment. When he arrived in Minsk, he found that there were no apartments avail-able anywhere in the city, and the family had to live in a hotel. Only after months of badgering by Borchardt did the university find them suitable accommodations. To Borchardt's dismay, he learned that the apartment had come available only after the police summarily evicted the previous tenants. He later claimed that when he tried to protest, officials told him to relent since there was little pity for such "bourgeois prejudices."[19]

By Western criteria, even the best housing in the Soviet Union was substandard. The Houtermans's, Ruhemanns', and Borchardts' apart-ments were all incomplete when they moved in. Bouchardt's apartment had no window panels, and there were no railings installed on any of the staircases. Borchardt paid five hundred rubles to have a shower installed in a storeroom adjoining the kitchen. It was considered a "European wonder" by those Russians who had a chance to see it first-hand, even if it only provided cold water.

A bonus many of the foreign scholars enjoyed was that they were able to bring with them a large portion of their household goods and schol-arly collections. The Borchardts arrived with large wardrobes, beds, chairs, rugs, more than a thousand books, four tables, and their china and glassware. The Houtermans too arrived with much of their worldly possessions intact, all packaged in a single large box, 3 metres long, by 1.5 metres high and 1.5 metres wide. Inside was everything they owned, from Fritz's precious books to the family's mattresses. Included among Fritz's books was a small collection of rare Bibles, which caught the attention of a customs agent. Fritz and Charlotte had to persuade the agent that the books would not be used to proselytize during their stay in the Soviet Union.[20]

Marcel Schein's family lived in a spacious new apartment in a complex owned by the University of Odessa. The apartment easily accommodated the family's furniture and Oriental rugs acquired in Zurich. Edgar Schein, just six years old when the family moved to the Soviet Union, was able to bring all of his favourite toys, including his Marklin electric trains and his working model steam engine. The toys fascinated two local boys, who soon became Edgar's friends. Edgar was tutored at home by German-speaking teachers, but he quickly learned enough Russian to communicate with his new friends. The boys played on the yet unlandscaped grounds of the apartment complex; Edgar remembers burning insects with a magnifying glass and capturing tarantulas to admire "their big hairy bodies."[21]

Most of the foreign specialists enjoyed three other privileges that gave them much more freedom than their Soviet counterparts. By contract, the foreign specialists often received part of their remuneration in foreign currency, and they were allowed to take a vacation outside of the Soviet Union at least once a year. Herman Muntz "traveled abroad widely in the company of his wife, visiting Finland, Germany, Switzerland and Poland. Generally, visits were related to vacations or had to do with mathematics research or meetings, but sometimes they were motivated by his and his wife's health."[22] Edgar Lederer's contract allowed for visits to France for holidays. While there, he was provided with funds to buy any required laboratory equipment, such as "reagents, glassware and appliances."[23] In July 1936, Fritz Noether was one of eleven Soviet mathmaticans nominated to attend the International Congress of Mathematics in Oslo. Noether was the sole delegate actually allowed to attend the congress in person.[24]

Another privilege enjoyed by the foreign scholars was that family members could visit them. Eva Striker's mother made several visits to see her, both in Kharkiv and, after her separation from her husband, in Moscow. In fact, she was visiting when her daughter was arrested by the NKVD. Three other scholars, Hans Hellmann, Wolfgang Steinitz, and Guido Beck, were visited by their mothers. Beck's mother apparently lived with him in Odessa and may even have accompanied him during

Image 10. Konrad Weisselberg and (Anna) Galia Mykalo. This badly
damaged photograph was likely taken at their wedding on 21 April 1936, and
miraculously survived the harrowing ordeals of Konrad's arrest and murder,
Galia's life as an internal exile, and the German occupation of Ukraine. Galia
is wearing a traditional Ukrainian costume as a wedding dress. (From the
collection of Art Kharlamov, Konrad's grandson; restoration of the photograph
was done by the author.)

his round-the-world trip from Kansas to take up his new post. All of this
international travel helped make life in the Soviet Union far more bear-
able than it might otherwise have been.

Finally, many of the refugees took advantage of opportunities to travel
within the Soviet Union. For those in Ukraine, this meant trips to the
Crimea, the Caucasus, and the Black Sea resorts in Georgia. Edgar Schein
recalls week-long voyages from Odessa on "well-appointed" passenger
liners on the Black Sea, visiting the ports of Sochi, Sukhumi, and Batumi.

The liners were often well-stocked with luxuries, including caviar. Edgar took a particular liking to caviar, impressing the waiters with how much he could eat. Edgar quickly acquired the reputation for being a "young gourmand."[25]

The scholars and their families interacted closely with their Soviet colleagues, as well as with other members of society. Some intended to make the Soviet Union their permanent home. In 1933, during his first visit to Kharkiv, Konrad Weisselberg met a beautiful nineteen-year-old Ukrainian woman, (Anna) Galia Mykalo. Galia had a coveted job working in the canteen at UFTI. When Galia accidentally spilled a bowl of borscht on Konrad, ruining his suit, she burst into tears, fearing she might be fired for her clumsiness. Konrad comforted her, reassuring her that he would not make a complaint and that she would not lose her job.

In 1934, Konrad moved to Kharkiv, leaving his fiancée behind in Vienna. He and Galia began a relationship and were soon in love. They married on 21 April 1936, and the couple moved into Konrad's apartment. Unfortunately, in August, Konrad was suddenly dismissed from his job at the Coal Institute, and as a result, the couple were evicted from their apartment. The loss of his job and the subsequent loss of accommodation would normally have compelled a foreign expert to leave the Soviet Union, since there was almost no other way to find housing. Konrad was desperate not to be separated from Galia. Galia had no desire to immigrate, and even if she was willing to do so, they both knew there was little chance of her obtaining an exit visa from the NKVD. Fortunately, Alex Weissberg invited them to stay at his apartment in UFTI, even offering Galia a job as his housekeeper. Konrad and Galia's son Alexander, named after Weissberg, was born in December 1936, just a few months before Konrad's arrest. In late 1936, Konrad had decided to apply for Soviet citizenship, believing that by renouncing all ties to Austria, he might be allowed to reside permanently in the Soviet Union. He became a citizen in early 1937.

Weisselberg was not the only refugee scientist to opt for Soviet citizenship. At the urging of his wife Viktoria, Hans Hellmann also became a citizen. As we shall see later, Weisselberg's and Hellmann's efforts to integrate fully into Soviet society would help seal their fate. Other scientists tried to establish permanent roots in the Soviet Union. Walter Zehden also married a Russian woman, and they too had a child. A number of the scholars or their wives had children in the Soviet Union, among them Inge Steinitz, Barbara Ruhemann, Charlotte Houtermans, Hélène Lederer, Sophie Simonson, Hilde Sadowsky, and Ruth Gilde. Katherina Harig gave birth to two children while in Leningrad, one of whom died in infancy.

Disease in the Soviet Union

The foreign specialists lived lives of comparative affluence but could not escape all the hardships of Soviet life. Public health there was far behind the West.[26] Shortly after arriving in Moscow in 1936, Stephan Cohn-Vossen, just thirty-four years old, died of pneumonia. Kurt and Ena Zinnemann arrived in Kharkiv while the city was in the grip of a severe diphtheria epidemic caused by poor sanitary conditions in the city. The Zinnemanns decided to investigate which strains of diphtheria bacilli were present so as to determine which ones were causing the most severe health problems. Kurt began exchanging information on the epidemic with Professor J.W. McLeod at the University of Leeds, one of the world's foremost experts on the disease. The Zinnemanns' research, and the correspondence with McLeod, would have important ramifications for the couple in the years ahead.[27]

As part of his duties with the institute, Kurt Zinnemann was sent on a medical mission to the Donetz Basin, in eastern Ukraine, to administer vaccinations in the rapidly expanding industrial region. His first stop was the city of Stalino (now Donetsk), the region's main city. Zinnemann was astonished at the vast numbers of new factories casting a "pall of heavy smoke" over the city. There were many well-stocked shops, and cultural activities were considerable. He was impressed with the large number of bookshops and the wide choice they offered. All of this in a district "too well known before 1918 for its backwardness and illiteracy." Zinnemann then travelled by bus along rough roads to the nearby steel town of Makeevka (Makiivka), where he would be working. He found a new town of 30,000 inhabitants, at the centre of which was a giant steel mill. Makeevka had been almost completely rebuilt since the revolution, and workers had been provided with almost everything: a modern, well-staffed hospital, schools, daycare centres, clubs, and "good and cheap canteens." Most workers, however, were living in crude mud huts they had built themselves, without windows or chimneys. Enteric disease was rampant in the town, likely caused by the absence of sewage drains, and sanitation was "in the good old country style," presumably with festering open sewers. The problem was compounded by the lack of a clean, reliable water supply. Zinnemann administered inoculations for typhoid, paratyphoid, and dysentery. He came away with a great admiration for what Soviet workers had accomplished, but also with a sense of just how far they had yet to go to make their communities safe from endemic disease.[28]

Diphtheria was not the only endemic disease in Kharkiv. Soon after their arrival in the city, both Fritz Houtermans and Lotte Schlesinger

contracted dysentery. Lotte almost died of it; only the personal intervention of some friends at the institute saved her. The friends called
in the medical supervisor for the district, who immediately sent an
ambulance with two doctors. Charlotte Houterman would recall that
"we were all scared to death, never having seen doctors with floating
black beards." "Bimbus," as Schlesinger was affectionately nicknamed
by her friend, was whisked away to a special dysentery hospital, where
she was kept in total isolation for several months until she was no longer
infectious.

Meanwhile a special truck arrived, carrying a team of fumigators, who
disinfected everything and everyone in the apartment. Fritz was declared
a carrier of the disease, and for several months he and Charlotte were
forced to live in isolation in the lower floor of the apartment, separated
from their daughter, who lived upstairs, looked after by a maid. Fritz
was limited to a bland diet of rice, boiled chicken, toast, and tea. In
the midst of their isolation, workers arrived to complete the staircase
connecting the two floors. Despite their best efforts, all of the residents
of the apartment were soon covered in a fine dust of cement and mortar. Charlotte endured all this while pregnant with her second child.
When Fritz had recuperated, "the siege ended and we emerged literally
as from a dungeon and had a reunion with our small child." They discovered, much to their dismay, that during their period of isolation, the
upstairs maid had given her boyfriend three of Fritz's suit jackets and
pants. Worse still, the suits and jackets did not match, so she effectively
ruined six suits. The maid was promptly fired.[29] Schlesinger was not as
fortunate. While she was eventually released from hospital after several
months, she suffered permanent health problems as a result of her prolonged illness. Shortly after her release from hospital, she moved from
the Houtermans's apartment to assume a position teaching music at the
Kiev Conservatory.

Inge Steinitz fought a never-ending war to eliminate bedbugs from the
family's bedding. She complained in letters to her mother-in-law and,
when he was away in Siberia, to Wolfgang, bemoaning the filth and the
lack of basic hygiene in the streets of Leningrad. The Steinitz family suffered from "waves" of illness in 1936. First, young Klaus came down with
scarlet fever with severe complications. In June, Inge caught the same
disease and was hospitalized. Soon after Inge's release from hospital,
both the Steinitz children, including their infant daughter, developed
whooping cough. Inge, exhausted by her own illness and the strain of
caring for the children, contracted a severe flu, which resulted in a "heart
neurosis" (likely an irregular heartbeat). It was several months before
everyone in the family fully recovered their health.[30]

Life in the Soviet Police State

The efforts by the scholars to integrate into Soviet society were hampered by the reality of living in a state whose security apparatus increasingly reflected Stalin's growing paranoia. Mutual surveillance of everyone in society was the norm. Despite being dedicated Marxists, Ernst Emsheimer, the Leningrad-based musicologist, and his wife Mia, opted to leave the Soviet Union for Sweden before the purges began, in part because "they were fed up having to leave their flat, where someone clearly overheard their conversations, every time they needed to discuss anything vaguely political."[31]

Those who worried about being spied on were not being unreasonably suspicious. Anyone they met, even another foreign scholar, might be a potential spy. At least three of the refugees were gathering intelligence for the Soviet government or for the Comintern. Wolfgang and Inge Steinitz became intelligence agents for the Soviet government in Finland and Estonia in the late 1920s. Despite almost being caught by the Finns, and Wolfgang vowing to focus on his scholarship rather than on secret party work, the couple continued working for Soviet intelligence. When they arrived in Leningrad, the Steinitzes were instructed to hide their party affiliation and to pretend that Wolfgang was simply an apolitical foreign scientist. At the Institute of the Peoples of the North, only the director knew that Steinitz was a Communist. The couple was given a code name and a contact who would transmit any intelligence they gathered to the Comintern in Moscow. No one else in Leningrad, even in the party or the police, knew of their assignment. The Steinitzes' main task for the Comintern was to report on what they overheard at receptions they attended at the German consulate in the city. Soviet citizens and foreign members of any Communist Party were forbidden to go to these functions, which were normally attended by diplomats and German specialists working under contract in the Soviet Union. The Steinitzes also spied on the staff and students at the Institute of the Peoples of the North. The extent of their espionage is unclear, and it is unknown whether anyone was arrested as a result of their efforts.[32]

In 1934, Gerhard Harig, the physicist turned historian of science, was recruited by the NKVD as an intelligence source. Little is known about his snooping in Leningrad, but Inge and Wolfgang Steinitz believed that Harig had something to do with the October 1937 decision not to renew Wolfgang's contract at the Peoples of the North Institute. Why they came to believe this, however, is not explained in any of the accounts written about them.[33] Even less is known about Harig's work as a spy in Leningrad. According to Harig's own account, his arrest by the NKVD in 1937 as a suspected German espionage agent was camouflage to allow him to return to Germany to spy for the NKVD. If this is true, the deception was a very poor one indeed. Harig was arrested by the Gestapo almost immediately after being "deported" from the Soviet Union. He was never formally charged or tried, but was sent to the Buchenwald concentration camp as a political prisoner. Miraculously, he survived seven years in the camp before being liberated by the US army in April 1945.[34]

It is quite possible that other foreign scholars were working for Soviet intelligence or security agencies. Martin and Barbara Ruhemann had arrived in the Soviet Union travelling on German passports. After Hitler came to power, Martin decided to renew his long-expired British passport and to seek British nationality for Barbara. When Barbara received her British passport, Martin offered their still valid German passports to the NKVD. The NKVD was more than happy to receive the documents and thanked Martin profusely for the gift. Barbara was later accused of being an informer for the Kharkiv NKVD.[35] After they migrated to Britain in 1937, British intelligence services suspected the couple of being Soviet spies – especially Barbara, who would remain under close MI5 surveillance for almost thirty years. Swedish security police kept close tabs on the Steinitzes and the Emsheimers after they migrated to that country in 1937. A 1949 Swedish government report suspected Ernst Emsheimer of being a Soviet agent before and during the Second World War, although it does not specifically say that his espionage activities commenced in the Soviet Union. Emsheimer remained under suspicion until 1966, when a lack of evidence finally led the security police to drop their investigation.[36]

Despite the fear of being spied on, close friendships were forged between the foreigners and the Soviet citizens with whom they interacted. As they became trusted confidants, some of the scholars learned of the secret views and lives of many of those who, in the light of day, appeared to be loyal supporters of the party and the state. The most vivid account of the secret lives of those in Stalin's totalitarian state is provided by Hermann Borchardt. His account is tainted by his anti-Communist rhetoric but is supported by the recollections of other refugee scholars, notably Guido Beck. According to Borchardt, in public he never made "any criticism or unkind remarks" of the Soviet system. The one exception to his silence was in "my tightly closed apartment during those late and joyful evenings when trustworthy friends of mine, after a few vodka had induced a Dostoevskyian mood, and sought to outdo each other in revelations of the back biting, bloodthirsty, and satanic Bolsheviks." Borchardt's friends were a combination of old Russian Liberals and Social Democrats, and they all delighted in "making fun of the Communists" – "those ambitious upstarts, criminals, hypocrites, and fools who, by enslaving ninety-nine per cent of the Russian people, hoped to build a 'better world.'" The following day, Borchardt's "trustworthy friends" would be "wearing the masks of perfect Communists and Stalin-apostles." Like Borchardt, his friends never publicly voiced a word of dissent – only deep in the night, in a closed room, would they "exchange our sad and comical experiences."[37]

For Borchardt, one of the most troubling aspects of life in the Soviet Union was the effect it was having on his family. At some point in the 1920s, like many other German Jews seeking to assimilate into German society, Borchardt had foresworn Judaism completely, converting to Lutheranism. In the Soviet Union, any form of religion except dialectic materialism was frowned upon. His and other apartments were regularly searched by authorities; even so, Borchardt risked putting up a small Christmas tree in one of the bedrooms. He insisted that his children learn about Christmas and other Christian traditions. Nothing, not even the closest family supervision, however, could stop his children being influenced by Communist ideology.[38]

Borchardt's main professional criticism of the Soviet system was the way the Communist theology of dialectic materialism had infused every aspect of the education curriculum, even language instruction. Borchardt considered it a futile exercise to try to teach German while simultaneously trying to instruct students on the Marxist world view. This was an aspect of Soviet life encountered by a number of the foreign scholars: they met many students who believed that time spent studying dialectic materialism was detrimental to their education. As we shall see, Borchardt's efforts to reform the teaching of German at Minsk would be the eventual cause of his expulsion from the Soviet Union.

Life and Intrigue at the Ukrainian Physical Technical Institute

While Borchardt and his friends were forced to criticize the Soviet system in a dark apartment, in the wee hours of the night, quite the opposite was true at UFTI. One of the unique aspects of life at Kharkiv was that it was the only place where a large group of voluntary migrants, refugee scholars, and some of the most Westernized Soviet scientists in the country worked and lived together. From its inception in 1928, UFTI was the most closely connected to the West of all Soviet scientific centres. The brief history of UFTI as a world-class centre for physics, and its precipitous fall from grace, is the one part of the academic refugee experience in the Soviet Union that has been extensively explored by historians prior to this study. The story of the research centre was first told by Alexander Weissberg shortly after the Second World War, although his account is concerned mainly with his own horrific experiences after his arrest at the hands of the NKVD, rather than his scientific career at Kharkiv. Since 1992, with the collapse of the Soviet Union and the creation of an independent Ukraine, at least six articles and books have been published about UFTI and the physicists who worked there.[39]

The experiences of those German and Central European scientists at UFTI was in many ways markedly different from those of other foreign scholars in the Soviet Union. While the accounts of UFTI-based physicists will be integral to this chapter, it will, as much as possible, examine the totality of the experiences of the much larger group. No attempt will be made to write yet another narrative of the rise and fall of UFTI, which is now well-travelled ground. Among the leaders of UFTI was a group of young physicists, all of whom had spent considerable time in the West, funded mostly by Rockefeller scholarships. This group included Lev Shubnikov, leader of the low-temperature group, Alexander Leipunski and Kirill Sinelnikov, leaders of the nuclear research group, and Lev Landau, head of the theoretical physics group after 1930.

In the late 1920s, while working at the Cavendish Laboratory at Cambridge University, Sinelnikov had married an Englishwoman, Edna Cooper. Leipenski had recruited the Houtermans to Kharkiv while on his own Rockefeller-funded research trip to Cambridge. Landau had invited Tisza to work with him while he visited Niels Bohr's laboratory at Copenhagen. According to one recent study, "by the early 1930s UFTI had the highest percentage of foreigners employed among Soviet Academic institutions."[40]

Weissberg paints an idyllic picture of life at UFTI prior to December 1934:

> The administrative apparatus and the Party apparatus in the Institute were both quite small. The center of interest was the actual research work. No attention at all was paid to political questions and not a great deal even to industrial problems. The atmosphere was free and unconstrained. The scientists talked to each other quite openly and the Director was just one of us. There were no bureaucratic obstacles in our relationship to him, and quite often it came about that the majority of the physicists at the Institute overruled him and he accepted the majority decision.[41]

L.J. Reinders, in his recent biography of Lev Shubnikov, presents a more nuanced and accurate picture of UFTI. He views the institute as the scene of a struggle: Westernized scientists and their foreign colleagues on one side, and on the other Soviet workers at UFTI, who opposed the power, prestige, and freedom of the scientists. One constant source of friction between the two groups was that the scientists, including all the foreigners, received three to four times the pay of engineers, laboratory assistants, and other technical workers. They also enjoyed access to the special stores.

This conflict became part of another, greater struggle – for control of UFTI, fought between the leadership of People's Commissariat of Heavy Industry (Narkomtjazhprom) in Moscow and the local Ukrainian party and NKVD in Kharkiv. The commissariat had been given control over UFTI and other Soviet scientific research institutes in 1932. While it emphasized the requirement that all Soviet science support industrial development, its leadership also helped shield the scientists, allowing them to conduct basic research. Sergo Ordzhonikidze, head of the commissariat, had been a close friend of Stalin before and during the Russian Revolution. In 1926 he was appointed to the Politburo (the Central Committee of the Soviet government). As long as Ordzhonikidze remained in charge of scientific research, scientists enjoyed some measure of protection from the secret police.

There was constant skirmishing over control of UFTI. Many of the scientists were openly critical of some aspects of Communist approaches to work and education. Chief among the critics was Landau, the brilliant theoretical physicist, who according to Weissberg "was head and shoulders above" the rest of the physics establishment. He was also the "enfant terrible" of Soviet science, an iconoclast with a wicked sense of humour, equally disdainful of dialectic materialism and old-fashioned experimental physics. His group at UFTI was the only independent theoretical physics group in the Soviet Union. The role of the theoreticians in furthering industrialization was never clearly articulated. Landau's personality, and the dangerous tightrope he walked in thumbing his nose at authority, is most apparent in one incident in the early spring of 1934. A government commission had been appointed to ensure that each scientist was assigned the appropriate rank and pay levels. A subcommittee was sent to UFTI and spent a week investigating the scientific staff.

One day, an official decree appeared on the bulletin board outside the director's office announcing the results of the investigation. "The list was simply grotesque. Leading physicists were set down as 'Underscientists,' and modest assistants found themselves 'Doctors of Science.'" When a group of demoted scientists stormed into the director's office to protest, they learned that the director had no knowledge of the decree. Upon further investigation, it turned out that the decree had been brought in by Landau the night before and that he had convinced the secretary that the director had ordered the document to be officially stamped and posted. Only later did people remember that it was April Fools' Day and that they had been well and truly fooled.[42] Landau's prank had clear political overtones, and the fact that he got away unpunished in the spring of 1934 did not mean it was forgotten. Instead, it was just one of many examples of the scientists' belief that they were an elite group who were above party discipline.

In 1934 the Communist Party organization at UFTI complained that only 7 per cent of the scientists were members and that there were almost no prospects for increasing this ratio "since among physicists in the USSR there were almost no party members." The party was deeply suspicious of the scientists because their "work in laboratories is not carried out during firmly established hours." The party established methods of control that were typical of Soviet workplaces; these included a wall newspaper, where notices of transgressions and new administrative or state policies could be announced. Another method of control was a series of self-criticism meetings, which all the staff were expected to attend in order to admit their sins.

At the end of 1934 the NKVD delivered a report to a general meeting of UFTI personnel. The report declared UFTI "'ideologically unhealthy,' since it lacked ideological consistency; there was no collective responsibility; and the party organization failed to exert any party influence on the work" being undertaken.[43] In a remarkable display of independence indicative of the attitudes of the senior scientists at UFTI, Lev Shubnikov spoke out at the meeting, openly criticizing the NKVD report. While he did not deny the shortcomings in the party organization at UFTI, he chastised the report for leaving out "the most important fact of all, the impressive accomplishments, which in just four years, had seen UFTI become the best equipped institute, with the highest and quickest scientific work, in the entire country." In a year or two, Shubnikov explained, "we shall be among the top in Europe."

In fact, the NKVD report did recognize that the scientists had made remarkable progress, singling out Landau, Shubnikov, Sinelnikov, and Martin Ruhemann, as well as other scientists as "the best shock workers of 'socialist emulation.'"[44] In the short term, Shubnikov escaped any consequences for his defiance, and the scientists maintained their independence. The scientists were, however, already being challenged by the surprise appointment of a new director, Semen Abramovich Davidovich (sometimes spelled Davidovitch), who replaced Leipunski on 1 December 1934, when the latter was conducting research at Cambridge. Davidovich had some scientific training but was primarily an administrator, with close links to the Communist Party. There is no clear explanation as to why Davidovich was appointed, although the order appears to have come from Moscow in an effort to make UFTI's research program more accountable to the direct, practical needs of the state. Around the time of Davidovich's appointment, several top-secret scientific teams were set up at the institute to undertake military research for the Soviet military. The scientists at UFTI were not consulted about setting up these new research groups, nor were they allowed to participate in or have any knowledge of the military work.

Just as Davidovich became the director, the local NKVD pressured UFTI to expand its security force dramatically and to strictly limit visits by foreigners. Weissberg considered all of these restrictions part of serious efforts by the authorities to "Sovietize" UFTI. The foreign scientists could not avoid being caught up in the struggle for control of UFTI. Weissberg, despite being a more devout Communist than most of the Russian scientists at the institute, took the lead in organizing opposition to the new director. He believed Davidovich had disregarded the basic principles of Soviet industrial management. Barbara Ruhemann, now a devout party member, stuck to Weissburg "through thick and thin, but at the same time she was perfectly convinced that we were the real Stalinists, and not Davidovitch and his friends."[45]

Weissberg, who had been placed in charge of building a new experimental low-temperature industrial gas extraction station, fought a prolonged bureaucratic battle with Davidovich for control of the project. Weissberg and the other scientists lobbied Moscow to remove Davidovich. Davidovich in turn informed the NKVD in Kharkiv that Landau and Weissberg were the leaders of a conspiracy to sabotage defence research at UFTI. Neither the Communist Party cell at UFTI nor the local NKVD was willing to take action against either scientist; instead, they waited to see who would prevail in the power struggle. On 25 November 1935, the scientists' campaign proved successful: the People's Commissariat of Heavy Industry fired Davidovich and ordered Leipunski to resume the directorship. Some scientists naively believed that they had preserved their scientific freedom; others, though, wondered how much longer UFTI could continue to serve as a unique haven in the storm then sweeping the country.

This very brief summary of the complex and potentially dangerous political situation swirling around UFTI reveals the world in which the foreign scientists lived, whether as volunteer migrants or as refugees. Weissberg considered that before Davidovich's appointment, "UFTI was an oasis of freedom in the desert of Stalin's despotism." The scientists lived "in a separate community on the territory of the Institute," and they had very little to do with outsiders. It was "an idyllic and exceptional state of affairs."[46]

Lazlo Tisza, who as a young post-doctoral student stayed out of the internal and external political events affecting UFTI, had fond memories of his life in Kharkiv. When he arrived in January 1935, in the dead of winter, there was snow on the ground. He "was soon escorted to a Dynamo sporting-goods store," where he bought a pair of cross-country skis. On Sundays a group of UFTI staff would go out to the country to ski. In the late spring and summer, a pickup truck fitted with wooden benches provided them with transportation for swimming parties on the Donets River. There was also a lively social life that, despite social divisions, brought all the institute's employees together. There were occasional dances and singing parties. The evenings "were enlivened by amateur productions of Anton Chekhov's one-act plays." These plays were mainly comic masterpieces, little known in the West but loved by Russians.[47]

The Houtermans arrived at UFTI in February 1935, at the start of the struggle against Davidovich. Charlotte and Fritz could not immediately see the dangers, which were not obvious, but they were disconcerted by the armed guards with rifles and fixed bayonets who now guarded the gates to the institute. They were overjoyed to be reunited with old friends, including Weissberg and most of the Russian physicists.

In a small way, the scientists re-created the life they had enjoyed so much in Berlin before 1933. Within the closed compound where the senior scientists and their families lived, there was a lively social scene. The Houtermans opened their apartment to frequent gatherings of UFTI scientists and their spouses, with parties reminiscent of those they had held in Berlin. Weissberg's and Landau's homes were also scenes of frequent gatherings for special occasions like birthdays. The Shubnikovs and the Sinelnikovs also kept open houses. These gatherings were made more interesting by visiting foreign physicists, who arrived at Kharkiv in a steady stream until the end of 1936. At these gatherings, just as in Berlin, discussions on physics were mingled with conversations about politics and current events.[48] The political discussions were initially quite open.

For the first eighteen months or so, Fritz Houtermans, having recovered from illness, felt glad he had decided to move the family to the Soviet Union. By mid-1936, it seemed that the efforts to reign in the scientists' freedom at UFTI were ending. Davidovich's dismissal in late November 1935 was just the first. One day after Leipunsky's reappointment, however, Moisej Korets (or Koretz), one of Landau's assistants, was arrested by the NKVD, ostensibly for "disrupting defence related work." Korets was low-hanging fruit, not important enough to cause a major incident, but his arrest was intended to hurt Landau and his friends.

Korets had already been expelled from the Communist Party for hiding his social origins. Both charges were false.[49]

Korets was convicted by a court and sentenced to eighteen months in prison. At a meeting of all staff at UFTI, Korets was publicly denounced by many of his colleagues as "an enemy of the people" and a "foreign spy." Weissberg claims he was reluctant to attend but that he had been persuaded by a fellow Communist that his absence would be viewed as an act of defiance. Weissberg later wrote: "I allowed myself to be persuaded. I did not speak against Koretz, but when the resolution condemning him was put I raised my hand with the rest. I still experience a flush of shame and discomfort every time I think of it. When I look back on my whole Party career, including the time I spent in the Soviet Union, I can find nothing I am more ashamed of."[50]

Landau and Weissberg wrote to authorities in Moscow and Kharkiv requesting Korets's release on these false charges. In June 1936, Korets's conviction was rescinded and he was allowed to return to UFTI. Some of the scientists believed they had won their struggle against Davidovich, the NKVD, and the Communist Party.

Conclusion

For the foreign and Soviet scientists at UFTI the return of Korets was just one of several encouraging signs that Stalin might be willing to liberalize society. The new constitution of 1936 "promised a dazzling array of civil rights to Soviet citizens, including freedom of assembly and freedom of speech."[51] Perhaps, some dared hope, the Soviet Union could become the land of opportunity for the foreign scholars. They believed that at the very least they would be allowed to pursue their scholarly research careers in peace. Whatever the hardships and risks of life under Stalin's rule, most of the refugees were happy conducting research and teaching. By 1936 they had already achieved great things in their precarious haven from Nazism. It is the history of their scholarly accomplishments that we must now explore.

6 Refugee Scholarship in the Soviet Union

In the extensive literature on academic refugees, the major recurring theme is their remarkably positive influence on the countries in which they settled. Some of these scholars won Nobel Prizes or other prestigious awards, and many were instrumental in major scientific discoveries such as nuclear fission and the building of the first atomic bomb. Still others established entirely new fields of teaching and scholarship at the institutions that employed them. Forced mass migration brought about a vast transfer of knowledge and skills. That transfer was felt mainly in the United States and Great Britain but was significant wherever the refugees found safe haven. The professional work of academic migrants in the Soviet Union might have been just as transformative as in the Western democracies had they not been killed, arrested, or forced by Stalin's madness to abandon their newly adopted country.

The professional lives of the foreign scholars in the Soviet Union varied a great deal, depending to a large degree on the type and location of the institutions that employed them. Whether they were refugees or voluntary migrants, they all hoped to continue their careers, including research, publishing, and teaching. In the Soviet Union they encountered a society that had been so disrupted by its recent tumultuous history that while students there were eager to learn, they often lacked the necessary background, knowledge, and skills. The near total collapse of the educational system during the 1920s had resulted in acute shortages of supplies and equipment as well as a lack of the skilled workers needed to maintain the few complex scientific devices that were available.

The Scholarship of the Voluntary Migrants

The voluntary migrants spent more time in the Soviet Union than the refugee scholars and thus contributed more to the country's scientific development. Tragically, most of what they achieved was either swept away by the purges or destroyed after the Germans invaded during the Second World War. Of all the migrants, be they voluntary or refugee, it was probably Herman Muntz who had the most rewarding experience in the Soviet Union. Having finally been provided with a long-deserved permanent academic position, as professor of mathematics and head of the Department of Differential Equations at Leningrad State University, Muntz thrived. In 1931 he was also placed in charge of mathematical analysis at the Scientific and Research Institute in Mathematics and Mechanics. "Muntz's position in Russia seems to have been quite firm, since [one year later] he was given the singular distinction of being sent to the International Congress of Mathematicians in Zurich as one of four of the Soviet Union's official delegates." In 1935, as further reward for his scholarly work, Muntz, who had been denied his habilitation in Germany, was awarded his second or higher doctorate by Leningrad State University, "without requiring the submission of a written thesis." His biographers commented that "all the above testifies to the fact that without doubt Muntz held a senior position in Leningrad State University and had the respect of his colleagues. He had fulfilled his ambition." Muntz carried a heavy teaching and administrative load. F.I. Ivanov, a graduate student, recalls Muntz "as being very actively occupied with science, both accessible and sociable." Despite his heavy workload, Muntz remained an active scholar, publishing the important textbook *Integral Equations* and half a dozen scientific papers.[1]

Ernst Simonson also flourished during his time in the Soviet Union. In 1930, when he arrived in Kharkiv, he was appointed vice-director and chief of the Physiological Laboratory of the Institute of Labour of the Ukrainian Commissariat of Labour. He oversaw numerous studies for improving hygiene and worker efficiency in factories and coal mines throughout Ukraine. He also undertook "theoretical" research in the physiology of muscles, examining such aspects as "adoption, fatigue, coordination of voluntary movements, and the effects of high temperature." Five years later he was promoted to chief of the Physiological Laboratory at the Ukrainian Institute of Physical Culture and Education. His research there shifted to examining the physiology of running. One year later, while retaining his director's position, he was made professor of physiology at the First Medical School at the University of Kharkiv. Finally in 1937, just a few months before fleeing the Soviet Union, he was elected to the Professors' Council, which gave him the rank equivalent to professor with habilitation at a German university. He was placed in charge of all metabolic laboratory and theoretical research at the university. He was also responsible for giving lectures on all aspects of physiology. Simonson had a remarkable publication record while in the Soviet Union. He wrote or co-authored some fifty scholarly articles between 1931 and 1938, covering a dizzying range of applied and theoretical research in physiology and related fields. Many of these papers were published in Russian, indicating Simonson's desire to fully immerse himself in Soviet scientific culture. All of that, however, would not be enough to prevent the persecution that resulted in Simonson and his family fleeing the country. In fact, they left in such haste that a number of his articles written in Kharkiv were not published in the Soviet Union until months after his departure.[2]

Like a number of the foreign scholars, the musicologist Ernst Emsheimer changed the focus of his research when he moved to Soviet Union. During his time in Leningrad, his interests shifted from classical and contemporary popular music to folk music. While he worked at the Hermitage, with its large collection of recordings, his task was the transcription of hundreds of folk songs, previously known only from oral traditions. The songs originated in various regions and cultures in the Soviet Union, especially Georgia and neighbouring South Ossetia. In July and August 1936, Emsheimer joined two Soviet music ethnographers conducting research in the South Caucasus. They recorded hundreds more of that region's traditional songs. It was during this expedition that he mastered the techniques for preserving this music for posterity and began addressing the need to collect the instruments required to perform it. "His experience in Leningrad taught him the necessary skills in dealing with the legacy of the oral traditional musical cultures."[3] Emsheimer's experience in the Soviet Union would guide the rest of his professional career. His shift to the study of folk songs and traditional instruments was perhaps a logical development from his musicology training in Germany and Austria; other refugee scholars went through much more fundamental transformations.

Walter Zehden had arrived in the Soviet Union in April 1932, just two months after completing his doctorate in physics at the Kaiser-Wilhelm Institute for Physical Chemistry and Electro-Chemistry in Berlin. He provided no explanation for his migration to the Soviet Union, although no doubt his Jewish heritage played a role. There is no evidence that he was a member of the Communist Party. He was hired to work with Professor Grigory Landsberg in the Optical Laboratory of the Physical Institute of Moscow State University. Landsberg was a pioneer in the field of atomic and molecular spectral analysis, especially in the spectroscopy of organic molecules.[4]

Zehden found Landsberg's laboratory an exhilarating place to work. He collaborated with Landsberg and another Soviet scientist to develop a spectral method "for the quick identification of different kinds of steel" and to design the necessary apparatus for it. The concept was patented, and the men were awarded a prize of 20,000 rubles from the Commissariat of Heavy Industry. Zehden had arrived in Moscow with what he believed was "a special ability for laboratory technique of any kind," and "in particular, for the construction of highly sensitive physical apparatus with very delicate parts, often scarcely visible to the naked eye." Since there was "a noticeable lack of sufficiently qualified special workshops for the construction and repair of apparatus" in the Soviet Union, he needed to become a master builder of new devices or attachments to existing ones. He also taught himself to repair and adjust the precision electrical instruments of all the major Western firms that were supplying scientific equipment to the Soviet Union. He even learned how to repair watch mechanisms, and to blow glass in order to make the small vessels needed for his optical experiments. While looking for a job in Britain after his expulsion from the Soviet Union, Zehden boasted that "I have unusual comparative knowledge of the qualities of instruments made by the world's leading firms."

Zehden's skills were highly sought after, and he provided technical advice to numerous scientific institutes in Moscow, Leningrad, Minsk, Odessa, and other cities throughout the Soviet Union. He also developed a course in laboratory techniques. In addition to all these accomplishments, he found time to author or co-author three scientific papers.[5] His considerable contributions to Soviet science and industry were noted by the pioneering British science writer and journalist J.G. Crowther in his 1936 book *Soviet Science.* Crowther was the scientific correspondent for the *Manchester Guardian,* and he was especially intrigued by the relationship between Communism and science. Crowther travelled throughout the country several times in the late 1920s and mid-1930s and wrote two books about scientific developments in the Communist paradise. Zehden was the only German scientist in the Soviet Union singled out by Crowther for a detailed discussion because of his singular, direct contributions to Soviet industry.[6]

Saul Levy, Zehden's compatriot at Landsberg's laboratory, also prospered during his time in the Soviet Union. No longer a mere assistant, Levy was able to pursue his own research. He co-authored four articles related to spectroscopy in 1933 and 1934. Landsberg considered Levy a first-rate scientist, and he promoted him to be much more than an assistant. In 1935, the institute awarded Levy a doctorate, equivalent to the German habilitation, and made him a full member. Levy was able to supervise his own graduate students and to employ his own assistants. Among Levy's undergraduate research students was Vitaly Ginzburg, who in 2003 shared the Nobel Prize in Physics for his work on superconductivity. Ginzburg stated that Levy's research "produced shoots" that would lead to the discovery of lasers. Ginzburg recalled that Levy was a "pleasant and educated man" who had a profound influence on the budding scientist.[7] Levy taught a course on spectroscopy and conducted special, advanced optical training. While continuing his work at the Optical Laboratory, he was appointed to head the laboratory at the Moscow Mineralogical-Geological Institute. There he developed techniques for applying quantitative spectral analysis to find concentrations of key metals, such gold, platinum, palladium, and lead, in ore samples.[8]

The Research Achievements of the Voluntary Migrants at UFTI

The voluntary migrants at UFTI also made important contributions to Soviet science and industry. Soon after arriving at UFTI in 1931, Alexander Weissberg noticed that most Soviet physicists were publishing their findings in foreign journals, mainly in Germany, because Russian was not an international language of science. Weissberg set out to establish a journal that would attract the best Soviet physicists to submit articles accessible to the international physics community; they would be published in English, French, and German. This plan was quickly approved at UFTI, but it required the permission of the central authorities in Moscow. There, Weissburg met with Nikolai Bukharin, a member of the Central Party Committee and protector of Leipunski. Bukharin agreed to support Weissberg's proposal, and the Central Committee approved the scheme soon after.[9] In 1933, *Physikalische Zeitschrift der Sowjetunion* (Physics Journal of the Soviet Union) published its first issue. Abram Ioffe, elder statesman of Soviet physics, was chair of the editorial board. Weissburg, Leipunski, and other physicists at UFTI oversaw the journal's day-to-day operations. Barbara Ruhemann and, later, Charlotte Houtermans worked as editors and translators.

This journal could not have been founded at a more opportune time. Hitler's rise to power had made it politically dangerous for Soviet scientists to publish in German journals, and German journals were reluctant to print articles by Soviet citizens; in fact, the number of articles by Soviet physicists in one leading German journal, *Zeitschrift fur Physik*, declined from around seventy per year in 1930 to only a dozen in 1935. In its six years of existence, *Physikalische Zeitschrift* published around 800 articles in addition to short notes, letters, and announcements. Though many leading Soviet physicists would continue to publish brief accounts of their research in foreign publications, such as *Nature*, to ensure that their work was noticed in the West, Weissberg's journal became, almost overnight, the dominant physics journal in the country.[10] Many of its articles were by Soviet physicists, but a number of foreign scientists in the Soviet Union, including Hans Hellmann, Gerhard Harig, Fritz Noether, Saul Levy, Lazlo Tisza, Barbara and Martin Ruhemann, Marcel Schein, and Fritz Houtermans, also published their research there.

Weissberg's other major project in the Soviet Union involved overseeing the construction of a small-scale industrial production plant to explore methods for improving low-temperature techniques for extracting gases for industrial use. This project brought Weissberg into conflict with Davidovich. Weissberg's main goal was to manufacture ammonia for agricultural fertilizer. Ammonia is "produced by reacting hydrogen with nitrogen and, while the nitrogen was obtained by separation of air, the principal source of hydrogen at the time was the gas from a coke oven." Around 1930 the Soviet government had purchased several hydrogen extraction plants from Germany; these plants had been surreptitiously copied by the Soviets. The Soviets could build new plants, but they lacked detailed information on the complex properties of the coke oven gas necessary to improve their designs and techniques.

In the early 1930s, Soviet authorities urged Shubnikov's low-temperature physics group at UFTI to focus their research on practical industrial problems, such as the analysis of coke oven gas. Shubnikov, however, was far more interested in pursuing cutting-edge low-temperature physics research, including superconductivity; like others at UFTI, he resisted undertaking work on applied scientific problems. Into this void stepped Weissberg, as much an engineer as a physicist, and the recently arrived Martin Ruhemann. The latter's work in Germany studying chemical engineering prior to his migration drew him to pursue applied research in this field. By mutual agreement, it was decided to split the low-temperature group, with Shubnikov continuing his pure research and Ruhemann organizing an applied research team. Ruhemann's team, which included three Soviet women scientists and Barbara, worked to understand gas separation techniques and other cryogenic industrial processes, including helium extraction.

The laboratory facilities at UFTI were excellent. Even so, by 1935 Weissberg and Ruhemann realized that they needed a much larger facility to test actual industrial production techniques. Weissberg, now very well-connected politically at the Commissariat of Heavy Industry, and a master of working the Soviet system, convinced Moscow to approve an industrial research centre separate from UFTI. Thus was born the deep-cooling research station, OSGO (Opytnaja stantsija glubokogo okhlazhdenija). Weissberg would oversee the construction; Ruhemann would assume the directorship of the new facility once it became operational.

OSGO opened in 1936 in a small village just outside Kharkiv, with "well equipped research laboratories and living accommodations for staff, and adequate factory space for the erection of pilot plant[s] for gas liquefaction and gas separation." The Ruhemanns moved to the OSGO site in early 1937. A pilot plant for air separation was installed soon after. A model plant for coke-oven gas separation was to have been built next, but the Ruhemanns were forced to leave the Soviet Union before it could be installed.[11]

Barbara and Martin Ruhemann made one more important scientific contribution while in the Soviet Union: they co-authored a pioneering textbook, *Low Temperature Physics*, a brilliant summary of the history of low-temperature research and the then current state of scientific knowledge in the field. The book became an instant classic when it was published by Cambridge University Press in 1937 in England and by Macmillan in the United States. In 2014, Cambridge republished the book, a fitting tribute to its importance in this subfield of physics.

Refugee Physicists at UFTI

Refugee physicists began to arrive at UFTI in 1934. Since all but Fritz Lange had either fled the Soviet Union or been arrested there by the end of 1937, they would not have enough time to accomplish a great deal scientifically. Laszlo Tisza, who had fled Hungary after his release from prison, likely benefited the most from his time in Kharkiv. Tisza's most important scientific research, which he undertook in France between 1938 and 1940, would build on what he had learned at UFTI. As a postdoctoral student, he had a somewhat different experience than most of the other refugee scholars. He had arrived in the Soviet Union with a doctorate from Budapest, but Lev Landau had insisted that the degree was not sufficient to meet his rigorous standards. When Tisza arrived at UFTI, he soon realized that Landau was correct. Landau was one of the great original thinkers in the history of physics. By the time Landau arrived at UFTI in 1930, he was "already famous as a theoretical physicist and represented the main scientific attraction for the young German speaking physicists looking for a place to live and work outside Hitler's political sphere."[12]

The new physics, discovered in the first third of the twentieth century – special relativity and quantum mechanics – were generally viewed by most physicists as quite separate from the classical traditional physics that had dominated science before 1900. Landau, however, believed that the old with the new concepts of physics could be unified: "To use special relativity (SR) to integrate classical electrodynamics (CED) with the canonical mechanics of particles (CMP); to use quantum mechanics to integrate macro- with microphysics and chemical thermodynamics with atomic mechanics."[13] He carefully laid out a plan to prove as much. In the long term, he would write a multivolume text on theoretical physics to explain his concepts. In the short run, he planned to train a cadre of young physicists who would assist him in producing his great work. Landau laid out an eight-part syllabus for his students, with accompanying oral exams. Tisza, with his Hungarian PhD, was exempt from these exams but decided to take the program anyway to improve his basic understanding of the latest developments in theoretical physics.

Tisza initially failed the exam on thermodynamics, because his understanding of the subject was based on seminars given by Max Born at Gottingen. He recalled that Born's view of the subject was classical, "quite separate from both statistical physics and quantum mechanics. For Landau thermodynamics was statistical, quantal, and in a state of evolution." Tisza later recalled that when Landau quizzed Tisza on the subject, "I did not know what he was talking about, and failed the test." Only after a member of Landau's group took pity on the young Hungarian and gave him an "informal summary" of Landau's approach was he able to pass the exam. On his second attempt, Tisza became the fifth person, and the only foreigner, to pass Landau's program. In 1936, after solving a special problem posed by Landau, Tisza was awarded a "candidate" degree, equivalent to a PhD in the West.[14]

Landau made sure that the theoretical physicists in his group kept on top of the latest developments in the field. UFTI had a first-rate research library, with subscriptions to all of the most important academic journals. Each week, Landau reviewed the new journals and selected the most important three or four articles for discussion in a group seminar. Landau assigned each article to a different physicist or student to present at the seminar. In those seminars, Landau "expressed his appreciation or dismissed with sovereign assurance [each article], and his judgement was accepted without any question."[15]

Now a full member of Landau's team, Tisza was charged with translating some of Landau's recent scientific articles from Russian into German for publication in *Physikalische Zeitschrift der Sowjetunion*. The articles dealt with the phase transition between solid, liquid, and gaseous states of matter. As well, Tisza began to publish articles in the same journal based on research supervised by Landau. Looking back on his scientific experiences at UFTI more than seventy years later, Tisza felt that Landau had provided him with "a solid apprenticeship, and I was prepared for both research and teaching."[16]

Fritz Houtermans was initially overjoyed with his position in Kharkiv. Freed from the drudgery of an English industrial research laboratory, he found himself once again able to pursue his own independent research. By the time his family arrived in Kharkiv in February 1935, it had been almost two years since he had been able to do his own research, and he decided to take a close look at the rapidly evolving world of physics to decide on the direction of his work. In 1934, while he was labouring at EMI at Hayes, Enrico Fermi's research group in Rome "had discovered the abnormally large reaction cross section of slow neutron absorption by the nuclei of certain elements. Every physics center in the world immediately started studying this type of reaction." Work on this topic was already under way in the Leningrad Physical Technical Institute, and Houtermans decided to focus on this new and exciting aspect of atomic physics.[17] He joined Leipunski's Nuclear Research Group and was given his own laboratory as well as an assistant, Valentin Fomin. Fomin's father was a Soviet diplomat who in 1924 had been assigned to the Soviet mission in Berlin. Fomin completed high school and university during his eight years in the German capital. He attended the Technische Hochschule Charlottenburg at the same time that both Houtermans and Weissberg were there. In an NKVD interrogation, Fomin would recount that the former had been one of his instructors. Fomin returned home in 1932 and was hired that same year to work as a physicist-engineer in the high-voltage group at UFTI. It is possible he was hired because of Weissberg's intervention or on Houtermans's' recommendation to his old Berlin friend.[18]

Houtermans soon established himself as a leading scientific figure at the institute. He earned the friendship and respect of Landau, who recommended that his team consult Houtermans on any question related to atomic physics.[19] In 1937, while awaiting interrogation by the NKVD, Weissberg visited Houtermans's laboratory one day and was impressed by his friend's calm demeanour and his ability to stay focused on his scientific work:

> Houtermans chattered away about modern physical problems. He always talked well and amusingly as though the discovery of new physical laws were a sort of parlor game specially designed for him and his friends. To listen to him you might have thought that the world's physicists formed a little family of bright people occupying themselves with the problems of the universe as a sort of hobby. There was no trace of sentiment in his approach and at the idea that he was serving the cause of human progress he would have smiled tolerantly.[20]

Beginning in March 1936, during his brief productive time at UFTI, Houtermans authored or co-authored seven scientific papers on slow neutron absorption. Most of the articles were published in *Physikalische Zeitschrift der Sowjetunion*, although one was published in a relatively obscure Ukrainian physics journal. The Russian historian of science Victor Frenkel summarized the significance of this work: "These findings stimulated the successful development of nuclear theory (in the USSR, by Ya. I. Frenkel, and abroad, by N. Bohr and D. Wheeler), so it's entirely fair to say that in the USSR, F. Houtermans participated in research that was commensurate with the leading laboratories in the world."[21] If Houtermans was at the cutting edge of atomic physics in 1936, he was destined to miss the incredible progress of the next three years, which culminated in the discovery of the basic concept of nuclear fission. By the time Houtermans was allowed again to participate in scientific research, the world was at war, and physicists around the world were contemplating the atomic bomb.

Secret Military Research by Refugee Scholars

Houtermans's work at UFTI, like all atomic research at the time, was public, and he published his results in academic journals. While before 1940 atomic research seemed to have little practical military value, two or perhaps three refugee scientists were involved in secret defence research in the Soviet Union. Houtermans's friend Fritz Lange, though employed at UFTI, was not part of the main community of German physicists in Kharkiv. Instead he worked in one of the secret laboratories set up during Davidovich's tenure as director. Lange had been recruited to continue his investigations into high-energy physics, specifically particle accelerators, a subject that it was believed would have direct military applications. Very little else is known about his time there before 1939. Lange may have conducted research into atomic structure, but it is unlikely that his investigations examined the possibility of constructing an atomic bomb. By 1940, however, with the great strides made in nuclear physics over the previous four years, Lange had become a pioneer of Soviet research into nuclear weapons. He, along with several Soviet scientists, wrote the first letter to Soviet officials pointing out the possible military applications of atomic fission. His research shifted to the military applications of atomic physics, and he continued to study the fissionable properties of uranium until 1942, when he was forced to flee Kharkiv just ahead of the approaching German army.[22]

Lange's former student Herbert Murawkin, the physicist, Esperantist, and spy, outlined his proposed research program to the NVKD shortly after his arrival in Moscow, in the spring of 1933: "I would like to work at an institute that is most adapted and provided for the entire complex for the generation and utilization of high voltage currents. There should be built a powerful pulse generator with branches for use in military affairs, medicine, agriculture and chemistry." Murawkin's proposal to develop a death ray must have met with some scepticism, which may account for the eight-month delay before he was hired to commence his research at the Institute of Communications and Electro-Mechanics of the Red Army. The army's research centre, however, lacked the necessary high-energy generators, and Murawkin was soon transferred to work in his own secret laboratory at the Leningrad Electrical Physical Institute (LEFTI). There is little information on Murawkin's research, but for good reason the senior management at LEFTI doubted the project would succeed. It was theoretically possible to develop a death ray, and various efforts to do so had been made by researchers worldwide in the 1920s and 1930s. They had all failed, and some of them had been outright fraudulent. The most famous (or infamous) death ray invention had been marketed by the elderly inventor Nikola Tesla, but his never-seen machine was dismissed by every government he approached as likely a hoax. After two years of no discernible progress, and with increasing misgivings about his scientific abilities, LEFTI dismissed Murawkin. Appeals by the NKVD's Economic Department to LEFTI to reinstate Murawkin fell on deaf ears. Instead, in the summer of 1936, Muravkin became a research assistant at the All-Union Electro-Technical Institute in Moscow, where he worked until his arrest in 1937. It is unclear whether he continued his death ray program in Moscow.[23]

Mystery surrounds Helmuth Simons's time in the Soviet Union, in part because he broke off all communications with the agencies that had supported him in Britain and had facilitated his migration. After leaving the Soviet Union, he never mentioned any aspect of his time in the Communist state. All that is known is that the biochemist was hired by Professor Eugenius N. Pawlowsky, professor of zoology and comparative anatomy at the Military Medical Academy in Leningrad. The academy would have some association with post–Second World War biochemical warfare research. While there was a Soviet biochemical warfare research program in the 1930s, there is no evidence that any such research was conducted at the Military Medical Academy prior to 1945. Still, given Simons's interest in biochemical warfare, and his absolute silence about his work in the Soviet Union, it is possible he was involved in this highly classified military research. His interest in this type of science was demonstrated by his involvement in the Wickham Steed scandal in Britain and, as we shall see, by his wartime efforts to convince the Allies that the Germans had a biochemical warfare program. Pawlowsky's scientific focus, however, was parasitic infections and communicable diseases, which closely matched Simons's own research. It seems more likely that Simons continued his research into diseases, work that he would continue in France after he fled the Soviet Union.[24]

Hans Hellman: The Forgotten Pioneer of Quantum Chemistry

Of all of the refugee scientists in the Soviet Union, Hans Hellmann had the most productive and one of the longest research careers there. The Karpov Institute, which Hellmann joined in May 1934, was the country's leading institute of physical chemistry, employing 150 scientists and 250 support staff. Hellmann was appointed head of a theory group because he was an "expert in theoretical physics and theoretical chemistry." Hellmann enjoyed living in Moscow, the centre of science in the country; he also enjoyed the prestige accorded to scientists, the pay, and the privileges granted to foreign scholars. Writing to family in Germany in August 1934, he compared his situation in the Soviet Union to what he had left behind in Hannover: "There is quite a difference between sitting isolated in some provincial place, and working at an Institute together with scientific elite from a people of 150 million. This is where I had the scientific contacts I need."[25]

During his first year in Moscow, Hellmann was able to attend two international conferences held in the Soviet Union, in Kharkiv and Leningrad, and to meet leading scientists from the West, including Niels Bohr and Paul Dirac. The Kharkiv conference –the same one that young Laszlo Tisza attended – helped convince him to move to UFTI. Hellmann was amazingly productive as a scholar, producing on average two scientific papers per month; he continued to produce groundbreaking work that explored how quantum mechanics could be used to understand chemical bonding. He also supervised a large number of doctoral students and post-doctoral fellows.[26]

If this was not enough, Hellmann continued work on a project he had started in Germany with Wilhelm Jost: the first textbook on quantum chemistry. Hellmann was unable to find a German publisher willing to take on a project by an exiled scientist. He completed the book in Moscow, and three young Soviet scientists helped translate it into Russian. The translation was difficult because they had to find "suitable terms for scientific terminology and expressions that were also new in the German language."[27] *Quantum Chemistry* was published in Moscow in 1937 and quickly sold out. Hellmann continued work on an updated, abridged German text, which was finally published in Vienna in 1937. Despite its

Image 11. Hans Hellmann with Hans Jr. (reproduced with permission of Hans Hellmann Jr. and Petra Netter).

importance, few Germans were willing to purchase it, and it sold poorly. An English-language version appeared in 1944 in the United States. It was a remarkable groundbreaking work that, owing to circumstances, was all but forgotten. Even today, scientists who read the book continue to find "astonishing discoveries" in its pages.[28] A recent biography explains the continuing significance of Hellmann's book:

> In some respects the book bears inspection even today: its instructional pro-
> cedure, its scientific content and the clarity and precision of its assertions.
> There is, of course, a great deal which today would be presented differently
> and expressed in a less antiquated way. There are parts which can be criti-
> cized. All in all, however, we have in Hellmann's "Quantum Chemistry" a
> significant and admirable offering by someone who by today's standards

would be regarded as a really young scientist. Here he had tried to portray a wide scientific area, of only 7 years standing at the time, advanced also through his own efforts and contributions, in a work of such quality that we can still benefit from reading it more than half a century later.[29]

Early in 1935, Hellmann was rewarded for his scientific endeavours when the Karpov Institute awarded him a doctorate equivalent to the German habilitation he had been denied in 1934. He was awarded a number of special prizes for his research and teaching, and in November 1936 his salary was increased from 700 to 1,200 rubles per month. A month later, Hellmann was invited to address the Soviet Academy of Sciences. On New Year's Day 1937, Hellmann became a full member of the Institute, a rank equivalent to professor at a Western university. Finally, in the autumn of 1937, the Karpov Institute declared Hellmann a leading or senior scientist.[30] Hellmann enjoyed all the hallmarks of a young scholar destined to enjoy a long and successful scientific career; however, his promotion to senior scientist would prove to be the last scientific honour he would ever receive.

Refugee Scholarship Elsewhere in the Soviet Union

The refugee scholars made numerous important contributions, and for many of them their experience in the Soviet Union opened an entirely new chapter in their professional careers. Julius Schaxel, the notorious "red professor" of the Weimar Republic, was perhaps the first refugee academic to arrive in the Soviet Union. He reached Leningrad in October 1933 after a brief sojourn in Switzerland. He initially worked at the Experimental Zoological and Evolutionary Morphology Laboratory at the Leningrad Academy; two years later, he was asked to establish and head a new laboratory in developmental mechanics at the Academy of Sciences in Moscow. Between 1923 and 1933 he had done little scholarly research and publishing, focusing instead on a brief and failed political career and on public education. His arrival in the Soviet Union reinvigorated his scholarly career. He conducted experimental and theoretical research, publishing a number of papers on determinism, evolutionary development, and genetics. He also wrote a piece on biology and Leninism for the Academy of Sciences of the USSR, as well as a paper exploring the main idealistic theories of contemporary biology.[31]

Like Schaxel's, Wolfgang Steinitz's research career prospered during his time in the Soviet Union. The Institute of the People of the North brought him into contact with peoples from forty indigenous cultures of the vast hinterland of the Soviet Union. Founded in 1930, the institute had a dual purpose. The first was to train students from these indigenous peoples to become teachers, administrators, and technical experts in these remote regions. Their main task would be to mobilize indigenous people to achieve the goals of the Five-year Plan, including collectivization and industrialization – in other words, impose state control on them. This threatened the very existence of indigenous cultures and societies. At the same time, the staff of the institute were mandated to create a written language for these mainly illiterate societies; this task included developing grammars and other textbooks for use in the new, mandatory, state residential schools. Hand in hand with this work was the study of these endangered ancient societies and the preservation of their cultural practices.[32] It was in this latter work that Steinitz found himself immersed upon his arrival in the Soviet Union.

Steinitz spent hours interviewing students from the Khanty (Ostyak) and Mansi (Vogul) people, listening to their folk tales, songs, proverbs, and rituals while trying to unravel the mysteries of their proto-Finno-Ugristik languages. The interviews were carefully recorded and studied. This work formed the basis of the dictionaries and grammars that Steinitz intended to write for these two language groups; the interviews also formed the necessary preparation for a research trip to Siberia to visit the lands of these indigenous peoples. In July 1935, Steinitz began a field expedition to the Khanty, who inhabited a vast region in west-central Siberia, some 2,000 kilometres from Moscow. The region was isolated, travel was rugged, and schedules were unpredictable. Steinitz was frequently ill, yet he had to carry by himself all of his luggage and heavy recording equipment. He went weeks sleeping on floors or tents and without a bath. Still, he was able to observe and record the still living elements of the Ostyak culture, while trying to master the intricacies of their language. He witnessed their festivals and their hunting rituals, including their unique bear festival, and their funeral rites, and he collected artefacts such as musical instruments and birch bark cradles. He filled thirty notebooks with information on fairy tales, songs, mysteries, and family surnames; he also filled thirty-one phonographic reels with recordings of Ostyak songs. Above all else, he investigated the Ostyak philology in order to develop an alphabet that could be used to turn their pre-literate language into one that could be written without any loss of the rich linguistic heritage of their ancient tongue.

Everywhere in the region, Steinitz observed the intrusion of Russian culture and the Soviet state into the traditional culture of the Ostyak. Alcohol was a scourge to indigenous people, one he had already observed at his institute in Leningrad. While in Siberia, he often waited many hours for tribal leaders who failed to arrive for interviews, usually because they were too drunk or were recovering from a night of excessive drinking. In one district, he observed that vodka had been made available for the first time in seventeen years, perhaps to hide the signs of an impending food shortage. Steinitz recorded the damage caused by forced resettlement and by Soviet officials' efforts to suppress the traditional shamans. Despite the region's isolation, Soviet officials were everywhere. Many bureaucrats questioned Steinitz's research, viewing it as potentially disruptive to efforts to enforce Soviet conformity on the indigenous populations. At no time, however, did he ever indicate that he grasped that the imposition of Communist ideology was incompatible with the preservation of ancient traditions.

Steinitz was forced to end his research trip in October, some two months earlier than he had planned. The reasons for this decision will be explored in the next chapter. Despite its premature ending, the expedition to Siberia was, for Steinitz, a formative event that would strongly impact the rest of his academic career. Although his interest in the linguistics of Finnish language groups was well-established before his forced migration to the Soviet Union, it was there that he developed both his skills and his passion for traditional folk music, which would become the other focus of his later scholarship.[33]

Gerhard Harig's hiring by the Leningrad Physical Technical Institute (LFTI) in 1933 should have been an important step forward in the career of the thirty-one-year-old physicist. The Leningrad Institute remained the Soviet Union's premier institute of physics, whatever the pretentions of the young upstart institute in Kharkiv. For six years before his forced migration, he had worked as an assistant at the Technical University in Aachen. This was a minor position, working for a relatively obscure physicist at a minor university. Unfortunately, little is known about Harig's time in the Soviet Union; we can only assume that his appointment at LFTI had more to do with party membership than with his skills as a scientist. For whatever reason, his employment at LFTI lasted less then a year and resulted in the publication of only one brief article in *Physikalische Zeitschrift*. That would be the last scholarly contribution he would ever publish in physics. Sometime in 1934, he became a historian of science. His first work, a study about Lenin and modern physics, appeared in a festschrift marking the tenth anniversary of Lenin's death. Harig left LFTI and joined the Institute for the History of Science and Technology in Leningrad; after this, he became a prolific writer on the history of science, and on science and Marxist theory.

The theoretical physicist Guido Beck's experience in the Soviet Union was also transformative, but perhaps not in the way the former assistant to Heisenberg would have anticipated. Hired to teach in Odessa, far from the intellectual centres of Moscow, Kharkiv, and Leningrad, Beck had little time or opportunity to continue his once promising research career. Instead he worked to establish a program for teaching theoretical physics at Odessa. He was soon asked to take on similar work at the University of Kiev. Neither university had subscriptions to prominent Western scientific journals, with the result that Beck soon found himself out of touch with the rapidly evolving world of theoretical physics. Even if he had maintained contact with the latest developments, he would have had little time to continue his research career. Beck found himself working eighteen-hour days, shuttling between the two universities. Soviet officials promised Beck money to hire a physicist at Kiev, but the funds to do so only arrived in 1937, by which time the situation in the Soviet Union was already so unstable that Beck found no takers for the job. Thus Beck was forced to give up his research career; however, he became a master teacher in the new science of quantum mechanics. He had a lively and charming personality and made an indelible impression on his students. One of his Odessa students recalls: "After the seminars, the students often walked with Beck all the way to his home, about an hour and a half's walk. Along the way, we discussed the latest developments in physics. These discussions invariably finished at his home. We sat on the floor, since he did not have enough chairs, and the discussions proceeded until very late. It was a wonderful time, and we all remember him with great affection." Beck always referred to his students affectionately as his "children."[34] He found the Soviet youth "good students" and considered teaching them a "very amazing experience." He explained his students' situation:

> All these boys from bourgeois families in Russia had just recovered their liberty to study, and they had some basis at least from home, some home education. And once they knew me better, these boys started complaining about the philosophy and the dialectical materialism they were taught at the university. They were afraid of telling it to anybody, but since they had confidence in me – once they knew me better – they started talking like that.

Beck also experienced some of the negative aspects of the Soviet academic system. Administrators placed tremendous pressure on him and others to inflate exam results "in order to get the statistics higher." Beck's friend Marcel Schein succumbed to the pressure and agreed to cheat a bit, but Beck was firmly against the practice. He warned that all one would achieve by cheating was that "in two years you can have a bad university." Beck avoided trouble, even though his students' test results were "a single point below the curve." One of Beck's students, however, caused a ruckus that drew the attention of the university's Communist Party organization. When the young physicist asked a chemistry student what exam grade he had received, the chemist replied: "'Otlichno,' that means the highest grade, exceptional." The chemist then asked Beck's student what grade he had received. He replied: "I got 'sufficient,' but an honest one." The conversation was reported to the party, and, Beck relates, it caused "lots of trouble and discussion. Finally they decided they wouldn't do anything, better not to do anything, so they just let go of it. That was the situation. Russia was a typical underdeveloped country." Beck was highly critical of the endemic corruption in the education system, and after his time in the Soviet Union he experienced similar problems in South American universities. Of the two places, he judged the Soviet Union to be far better, because "there was one thing the Communists didn't want: they didn't want to destroy work by not caring."[35] Beck did have one advantage over many of the other refugees: his first language was Czech, and he soon became fluent in Russian, another Slavic language.

Marcel Schein, Beck's close friend and fellow Czech, also prospered in the Soviet Union. In addition to his teaching duties in experimental physics research at the University of Odessa, he received funding to continue his research into cosmic rays. From 1935 to 1937, he led three summer expeditions to Mount Elbrus, the tallest mountain in Europe, to study cosmic ray showers. He established research camps as high as 5,300 metres to investigate these elusive stellar phenomena. The last of these expeditions took place while Schein was already planning his escape from the Soviet Union.[36]

Michael Sadowsky's first language was Russian, and this enabled him to carry a full load of teaching engineering mechanics to undergraduate students at Novocherkassk. Like Beck, he enjoyed teaching; the experience no doubt reminded him of his happy days in Minnesota educating keen young American students. However, he found no time to pursue research. He had been an active publishing scholar before being laid off in Minnesota, but he did not complete a single article during his time of unemployment nor while in the Soviet Union. This failure to publish would hamper both Beck and Sadowsky when it came time for them to find positions, yet again, after they were forced to leave the Soviet Union.[37]

Charlotte Schlesinger's work at the Kharkiv and Kiev Conservatories was also dominated by teaching. Although she had done a great deal of musical instruction in Germany, she found that teaching was all-consuming, and she had little time to compose music while in the Soviet Union.

Instead, she took on a multitude of tasks, many of which she would not have been allowed to undertake in Berlin. For instance, she had to study opera production in order to teach an opera class. She continued to educate music teachers and musicians, and, like many other Central European scholars, she found her students highly talented and enthusiastic but not sufficiently educated to take on advanced musical education. Despite not knowing Russian when she arrived in the Soviet Union, she was able not only to become an effective teacher but also to direct a series of national radio broadcasts of her students' performances. Looking back, she considered her time in Kiev the most exciting experience of her life.[38]

The language barrier was acutely felt by many of the academic refugees and their families. Edgar Lederer was more typical of the German scientists in the country. In addition to his position as director of the Vitamin Institute, he was a *Privatdozent* at the Faculty of Medicine in Leningrad. He explained: "I gave classes to the students, [although] I do not know if they understood my Russian."[39] At UFTI, the scientists and their families conversed at least as often in German or English as in Russian. The language barrier inflamed some of the tensions at UFTI. Certainly, all of the foreign scholars learned some Russian. Tisza, for instance, was able to teach courses at the polytechnic at Kharkiv and assisted Beck in Odessa with a month-long guest lectureship, where Tisza taught a course on electrodynamics.

While Beck and Sadowsky found themselves unable to continue active research careers, others were more fortunate. Kurt Zinnemann, assisted by his wife Irene, did extensive medical research on the problems of diphtheria at the Metchnikoff Institute in Kharkiv. Although Zinnemann was only able to publish one scientific paper while in the Soviet Union, he was able to exchange information with medical researchers in other countries; of particular importance was his correspondence with Professor J.W. McLeod at the medical school of the University of Leeds. Refugee scholars were not cut off from the outside world while in the Soviet Union, and for Zinnemann, as we shall see, his ability to carry out collaborations would turn out to be his family's salvation.[40]

In April 1936, Fritz Duschinsky arrived at the Optics Institute in Leningrad after his futile three-year quest for a position in Europe or North America from his temporary exile in Belgium. Duschinsky was the last of the refugee scholars to arrive in the Soviet Union, and his time in the country was a mere eighteen months before he fled to his native Czechoslovakia. The physicist was just twenty-nine years old in 1936 but was so highly regarded by Soviet officials that he was given the rank of university professor and allowed to undertake a completely independent experimental and theoretical research program. Duschinsky was able to publish at least four scientific articles during his brief time in the Soviet Union. He continued the work on spectral analysis he had begun in Germany. More than eighty years later, one of his articles, outlining the "Duschinsky Rotation Effect," "a simple and effective way to characterize the difference between the ground state and excited state potential energy surfaces," is still widely cited in the scientific literature.[41] Duschinsky also was assigned to give a lecture in the special fields of optics, and he worked as a scientific consultant as well. Within a month of taking up his post, Duschinsky was selected to be part of the official delegation sent by the Soviet Academy of Sciences to an international congress on luminescence in Warsaw. At the last minute, Soviet officials "banned the delegation from leaving the country." It was a worrying sign of things to come.[42]

Conclusion

In just a few years in the Soviet Union, the scholars from Germany and Central Europe made impressive contributions to the country's scientific and intellectual communities. In many ways, the successes of these scholars mirrored what their contemporaries were achieving in the United States and in Britain. Since they were building up the Soviet Union's scientific, educational, and industrial infrastructure, which was backward compared to that of Western countries, the work of these scholars had the potential to be even more transformative than the contributions of refugee academics elsewhere. Their notable achievements, like their contributions in Germany, would in the end cease to matter as they faced yet another, even more horrific round of persecution, this time under Stalin. Soon those refugee academics and their families were again forced to flee, if they were allowed. This time they were facing not just the loss of their livelihood but also the imminent risk of death. Some would never leave the Soviet Union; others would escape only to fall victim to the Holocaust. Most of the refugee scholars would endure, but their lives would be forever changed by their struggle to survive two of the greatest mass murders in history.

7 The Great Terror

The foreign academics who migrated to the Soviet Union were seeking a safe haven and an opportunity to continue their scholarly avocations. The Soviet Union, however, was never a stable or peaceful state, especially under Stalin's increasingly paranoiac rule. Waves of oppression and mass murder swept the country throughout the 1920s and 1930s, culminating in the Great Terror of 1937–38. By the end of 1938, all but one of the scholars had fled, been arrested, or been expelled from the Soviet Union. Those who were able to escape faced, once again, the challenges of finding refuge, but this time in a world made even more dangerous and in which many more displaced scholars were seeking safety in an ever-dwindling number of places that offered them refuge; safety had become far more important than a career. While most found refuge, many would find themselves in precarious circumstances.

The causes of the waves of repression and murder that occurred throughout Stalin's rule have been debated by historians for decades, and no amount of archival study will definitively resolve this question. As one recent study has pointed out, there can be no single person to blame nor any single cause of purges; "rather, it was a multi-faceted process, composed of separate but related political, social and ethnic dimensions, the origins and goals of which were differentiated, but which coalesced in the horrific mass repressions of 1937–8."[1] At various times during Stalin's rule, many groups became targets for extermination or other horrific treatment, including mass deportation. Those groups ranged from lowly peasants who resisted collectivization to members of the Politburo whom Stalin perceived as potential rivals or impediments to his policies. Historians continue to see Stalin as the "director general" of the purges and the Great Terror, but he was not the only perpetrator, nor were all of them based in Moscow. Differences are visible in the treatment of the foreign academics by various regional offices of the NKVD, government ministries, and educational institutions. Until the Great Terror swept away all of their protectors, some of the foreign scholars enjoyed immunity from the secret police and even from the Communist Party.

In 1932 the Soviet government received the first intelligence indicating that some political émigrés might be spies for hostile governments, including Poland and Germany. Foreign Communists had been allowed to enter the Soviet Union with little scrutiny. A verification process, requiring all foreigners to register, became law, but there was no clear way to identify espionage agents who might form a fifth column to disrupt the Soviet state.[2]

Expulsions of Refugee Scholars before the Purges

On 1 December 1934, Sergei Kirov, the First Secretary of the Leningrad Communist Party and a Politburo member, was assassinated by a disgruntled young Russian. Stalin suspected foreign involvement, and the police were able to verify this during the course of their interrogations of the assassin and the people close to him. It was long suspected that Stalin himself was behind the murder, but historians now generally agree that it is more than likely the assassin operated alone.[3] The evidence of foreign involvement was produced in order to identify a convenient scapegoat. As we shall see, NKVD interrogators were skilled at obtaining any confessions they desired; whether or not those confessions were genuine was seemingly irrelevant. As xenophobic fears of anti-Soviet foreign agents and spies gripped Stalin, the government, the NKVD, and the Comintern, all foreign Communists residing in the Soviet Union were targeted by the Great Terror. The fear of foreigners in part reflected the rising international tensions caused by German aggression, but Communists from all over the world also became targeted for elimination. Members of the Comintern were both victims and perpetrators of mass arrests and murders.[4]

As foreign specialists, the refugee academics were treated with suspicion, and, as outsiders with special privileges, they also attracted jealousy. The foreign specialists, however, were not yet specifically targeted for arrest and possible execution. However, those foreign scholars who were also affiliated with a targeted group, such as an international Communist Party, were far more likely to be arrested. As we shall see, the most vulnerable were those who had tried to integrate themselves and their families fully into Soviet society. One paradox of Stalinist-era paranoia is that the more a foreigner tried to be a loyal subject, the more suspicious the authorities grew.[5]

The first expulsions took place in 1935, when two physicists based in Leningrad, Herbert Fröhlich and Emanuel Wasser, were ordered out of the country. Both scientists were on the staff of the Leningrad Physical Technical Institute (LFTI), although for much his time in the Soviet Union, Wasser had been seconded to head the Photoelectric Laboratory of the Physico-Technical Institute of the Urals. According to Fröhlich's biographer, in the spring of 1935 the scientists became caught up in the anti-foreign hysteria that swept the city after the murder of Kirov in December 1934, when all non-Communist foreigners were expelled.[6] This explanation is not completely correct, since Herman Muntz, also a non-Communist, kept his position. Muntz may have been too well-established; in 1935 he held a very senior position at Leningrad State University. Another reason given by Muntz's biographers mentions that while he kept his German citizenship, he was "given a 'former foreigner' status within the Soviet Union." Whatever the reason, it appears that Muntz may have been able to prevent the expulsion of Stephan Cohn-Vossen and Michael Sadowsky, two mathematicians hired in late 1934 to teach in Leningrad, by arranging for them to be quickly transferred to positions elsewhere in the Soviet Union.[7]

Fröhlich and Wasser were bewildered by their sudden dismissal and by the cancellation of their Soviet visas. Little is known about Wasser's three years there, but we do know that in April 1935, he was the first foreign scholar thrown out of the Soviet Union. Wasser did not know why he was given three days' notice to leave the country, only that the orders had come not from his institute but from within the government. Later he explained to Walter Adams of the SPSL:

> I suppose that political reasons lie in the ground of my dismissal. I think it to be not necessary to assure You [sic] that I have never been active politically. It may be that any opinion of me, being not always positive, which I have expressed in the circle of my nearest colleagues or perhaps in a letter has given rise to such a treatment of me. I am unfortunately not able to give You some more accurate informations [sic], the said above being all I know about the affair.[8]

Fröhlich, without any explanation, was given just twenty-four hours' notice to leave the country. He had been well paid in Leningrad, but unlike some of the other foreign specialists, his salary was entirely in rubles, which could not be converted into other currencies. He used his rubles to buy a train ticket to Rome; he also purchased caviar, jewellery, and an ill-fitting British-made suit, the only foreign one he could find in Leningrad. He would sell all these items, even the unused portion of his train ticket, once he reached Western Europe, to provide some ready cash.[9] This period of persecution was relatively brief. Edgar Lederer arrived in Leningrad in October 1935 and saw nothing amiss. Fritz Duschinsky migrated to Leningrad in March 1936, and he too found foreign scientists warmly embraced.

Even before Duschinsky arrived in Leningrad, however, one other foreign scholar, Hermann Borchardt, found himself ordered to leave the Soviet Union. Borchardt left several somewhat contradictory accounts of his dismissal from his position at the teachers' college in Minsk, but the central feature of his story is that his struggle to introduce a modern and effective curriculum for teaching German to student teachers had led to his expulsion. He had been hired specifically for this purpose, but in his efforts to reform the system, he encountered "frightening resistance." The vast majority of the staff at the college were women, the wives and daughters of the educated class that had disappeared, having either fled, or been exiled to Siberia, or had died of deprivations. The women had an excellent command of the German language, but with few "exceptions they did not know how to teach." They "knew literally nothing about the new pedagogical ideas developed in the last fifty years." Even if the language instructors had this knowledge, "they were so scared, so beaten by life, that they lacked the will or the courage to depart from 'prescribed' methods."

Borchardt's new curriculum involved the use of simplified legends and fairy tales, recast to illustrate the basic vocabulary, verbs, and grammar used in the German language. His teaching methods gave very good results, and the local Cultural Commissariat offered him a few thousand rubles to write a new textbook, with the proviso that at least half the "reading matter, conversation pieces, and the like, must carry a Marxist moral." Borchardt believed that mixing propaganda with language instruction was a recipe for disaster, and he attempted to negotiate a compromise. When the commissariat refused to adjust their demands, Borchardt declined to proceed with the project. He claimed that he also emphatically rejected various other efforts to impose Marxist discipline on his academic work. He had observed the use of the wall paper or bulletin board, where a monthly newsletter subjected his colleagues to satirical attacks from the party leadership at the school. These attacks "might be a prelude to purge, expulsion, or even exile or prison." Borchardt, learning that an article was being written about him for posting on the board, went to the university director and warned him he would leave the Soviet Union if it appeared.

The director explained that the article would in fact praise him, but Borchardt would not agree to its appearance. He refused to back down, for he had observed that if an article praised one faculty member, it was usually at "some other teacher's expense."[10] In early January 1936, he was summoned to Moscow to meet with senior officials of the Education Department, which oversaw teacher training in the Soviet Union. He arrived at the meeting prepared to defend his teaching methods – specifically, his use of legends and fairy tales. He was blindsided, however, when the official he met, rather than discussing his use of stories, demanded to see his passport. It was then that Borchardt felt compelled to defend his right to retain his status as a foreign specialist; the official, meanwhile, pressed him to renounce his German nationality and become a Soviet citizen.

Upon his return to Minsk, both the university director and the chairman of the Worker Council assured Borchardt that he had little to worry about. Just two weeks later, however, the police commissioner in charge of foreigners in Minsk ordered Borchardt and his family to leave the Soviet Union within twenty-four hours. When Borchardt demanded an explanation, the police official simply told him, "You know yourself you are guilty, many times." The news devastated the family, but they decided to leave and return to Germany. They hoped that by living with Borchardt's mother in a small town in Pomerania, they would escape the scrutiny of the German government.

The problems of quickly leaving, however, proved too much. One logistical problem had to do with transporting the Borchardts' possessions to the train station. Now that he was no longer employed by the university, Borchardt had no access to the school's cart, and there were no private moving companies in the city. The Borchardts decided to sell their larger pieces of furniture, accepting any offer, no matter how low. Meanwhile, a small group of friends, risking arrest themselves, helped the family pack the rest of their possessions. NKVD agents arrived every half hour to check on their progress, and when it became clear that the family would not be ready to leave by the deadline, they granted the Borchardts a further twenty-four hours. Finally, on 24 January, the family was given a send-off by their friends at the restaurant in the Minsk train station, with Borchardt spending the last of his soon-to-be-worthless rubles on his friends. He worried for his friends' safety, for two NKVD agents sat nearby, carefully watching the activities.

Borchardt left the Soviet Union with a heavy heart, but he was hopeful that he might one day return. His students had sent a telegram to Stalin requesting a hearing, which they were certain would exonerate Borchardt. His wife and children had been happy in Minsk. Borchardt later discussed his own deep regrets about leaving: "I had not only been happy, but more useful to students than I could have been elsewhere in the world."[11] He would never return to Minsk, but at least he and his family were able to escape physically unharmed. Despite these early expulsions, foreign scholars continued to arrive in the Soviet Union. They seemed either oblivious to the dangers or too desperate to be frightened off by the uncertainty and potential hazards of their new refuge.

The Purges Begin

Borchardt was the only refugee scholar expelled in 1936. Even so, by the autumn of 1935, many others had seen and experienced ominous signs of growing state-sponsored intolerance, persecution, nationalism, and xenophobia. In October 1935, during his field research in Siberia, the Leningrad-based linguist and NKVD agent Wolfgang Steinitz encountered first-hand how quickly the Soviet government was shifting its policies. While in Siberia, Steinitz was interviewing students at a Soviet school for Ostyak students when a staff member accused him of being a counter-revolutionary spy. Specifically, he was accused of questioning the students about a local strike by kulaks, wealthy peasant farmers, who had been exiled to Siberia during the forced farm collectivization. The school banned Steinitz from conducting any further interviews. He appealed his banishment from the school to local party and NKVD officials; he may even have revealed that he was an agent for the secret police. The local officials interceded successfully on his behalf, which allowed him to continue his interviews; even so, increasing suspicions about his activities compelled him to return to Leningrad four weeks earlier than he had planned.[12]

Steinitz had one more nasty surprise before leaving Siberia. In the early 1920s, Lenin had authorized the use of Latin alphabet–based transcription to create a written language for the eighty or so pre-literate peoples living in the Soviet Union. Like the other linguistic experts at the Northern Peoples Institute, Steinitz worked to develop an alphabet using Latin letters and diacritical marks, which allowed for special pronunciation or for the correct stress to be placed on a syllable. For the Ostyaks, he developed an alphabet with twenty-five basic Latin letters and six diacritical marks. In late October, he attended a regional teachers' conference, where he was invited to present his new alphabet. As the conference began, a telegram from the People's Commissariat of Education was read to the audience, stating that Moscow now demanded the use of the Russian Cyrillic alphabet as the standard for all new written languages. Foreign influences, even the Latin alphabet, were to be rooted out of Soviet society. Steinitz was no doubt dumbfounded, for this decree undermined his own research. He told those at the conference that he approved of the decision to change to the Russian alphabet, but at the same time he defended the use of standardized spelling for all dialects, as well as the use of diacritical marks to reflect regional pronunciations. Under the new policy, all of his work on the alphabet and dictionary had been rendered useless. To no avail, he would continue to press for the adoption of his Latin-based alphabet until his expulsion from the Soviet Union in the autumn of 1937.[13]

In 1936, the situation in the country quickly deteriorated. The only extant detailed accounts of the lead-up to the arrests and expulsions concerns the scientists and their families at UFTI. The scientists there had hoped that, Davidovich having been removed from the directorship the previous November, the comparative freedom they had enjoyed in their professional and personal lives would be restored. Meanwhile, Moisej Korets, Landau's assistant, remained in prison while the scientists continued their efforts to have the trumped-up charges against him withdrawn. In June 1936 their efforts were rewarded when Korets was exonerated and allowed to return to the institute. Using their considerable political influence in Moscow, the scientists had won one last victory. Even before Korets's release, however, events well beyond Kharkiv would seal the fall of UFTI as a world centre of physics and a safe haven for the scientists and their families, be they Soviet citizens or foreigners.

The first incident happened in the Soviet capital. Early in the morning of 26 May 1936, Eva Striker was awakened by her mother Laura in their Moscow apartment. Laura informed Eva that there were people waiting to speak to her. Eva had separated from her husband Alex Weissberg and had left Kharkiv two years earlier. Since then, she had held several positions as a ceramics designer in Leningrad and Moscow. In the spring of 1936, she was the artistic director at the State China and Glass Trust, where she designed perfume bottles for the newly restructured cosmetics industry. The people waiting to speak to Eva were NKVD agents, who soon flooded into the apartment to look for evidence of criminal acts against the Soviet state. Eva was bewildered by this and had no idea what they were searching for. Without explanation, Eva was then arrested and driven off to be incarcerated in Butyrka prison. By early June, she had been transferred by prison train to Leningrad, where she was held in Bolshoi Dom prison.

Immediately after her daughter's arrest, Laura Striker contacted Alexander Weissberg to help her secure Eva's release. Striker and Weissberg launched a national and international campaign, soliciting letters attesting to Eva's good character. Arthur Koestler believed that Weissberg should have left the Soviet Union in 1936, while he could still secure the assistance of the Austrian embassy in obtaining an exit visa; instead, he had chosen to use his scientific and political connections in an effort to rescue Eva. Koestler commented on this: "In 1937 this would have been lunacy; in 1936 it was merely reckless."[14] It is not clear whether and when Laura Striker and Weissberg became aware that Eva had been charged with participating in a plot to assassinate Stalin, as well as a number of lesser offences. They must have understood, however, that the charges were serious. They asked people to sign letters in support of Eva, knowing that doing so might bring the signers unwelcome attention from the NKVD.

While Laura Striker and Weissberg worked to free Eva, the situation for everyone living in the Soviet Union grew more and more precarious. In June, the Stalin regime proposed that abortion be outlawed. The Communist Party secretary in Kharkiv organized a public meeting to discuss the proposal, an indication that public input was being solicited. Several of the foreign and Soviet scientists attended what proved to be a heated debate, one that eventually stretched over three nights. In the end, those attending the meeting voted in favour of maintaining legal abortions, even though the party had not authorized a vote. The following day, those in attendance were surprised to read in the newspapers that the vote had been overwhelmingly in support of Stalin's anti-abortion laws.

The show trials in Moscow in the summer of 1936 cast a further chill over the entire country. One summer evening, the Houtermans entertained Martin and Barbara Ruhemann and Alex Weissberg at their apartment. They all listened to the radio while Andrey Vishinsky, the chief prosecutor, presented his closing arguments in the infamous Zinoviev–Kamenev show trial, an event that some historians view as the opening salvo of the Great Terror. According to Weissberg, the absurdity of the state's case against the defendants gave him a headache, and he asked his host if they could turn off the radio. They then talked about the case. Also according to Weissberg, they spoke as if "we were perfectly convinced of the truth of the indictment, but then in perhaps a casual observation we remarked upon the senselessness with which the conspirators had gone to work." If one listened closely to the conversation, however, they could have detected that they were speaking "obliquely," in a way that they all understood as indicating their disbelief in the state's case against the conspirators. Weissberg explained the rationale behind their conversations:

> The growing terror instituted by the G.P.U. [NKVD] in recent years had forced such a secret language on us. We had good cause to fear G.P.U. spies everywhere. People became afraid to talk openly even to their friends. People were horrified at the thought of straying too far away from the official line. The gulf between public opinion, which was being suppressed into social subconsciousness, and reality grew wider and wider. The dictatorship

of the lie dominated the press, the school, the radio, the film, the factories, and the meetings of the Party.[15]

Weissberg and his friends in Kharkiv felt increasingly insecure in their adopted land. They would by no means be the only scientists to experience persecution as the purges evolved into the Great Terror. In 1936, foreign scholars were to some extent much safer than their Soviet colleagues. Moreover, it was not just in Kharkiv but throughout the Soviet Union, from Tomsk to Leningrad, that scholars came under attack. These attacks reflected what was happening throughout Soviet society. In July 1936, a public campaign in Moscow began criticizing leading Soviet mathematicians. The smear campaign was likely initiated by a group of younger mathematicians against the older generation of scholars, and particularly against Nikolai Luzin, a senior member of the Soviet Academy of Sciences. Though launched by younger scholars, the attacks against Academician Luzin and other senior scholars had been approved by higher political authorities. The newspaper *Pravda* published articles denouncing the mathematicians for behaving like the old nobility, holding "native language in contempt and … permanently cring[ing] before the west."[16] One article outlined the arguments against not just these mathematicians but also every scholar with international contacts:

Although reared by the Soviet land and by Soviet science, such scientists continue to cringe before every foreign country and every alien language. Perhaps, they do not possess a feeling of national pride in even the tiniest dose? … Perhaps, they do not experience Soviet patriotism at all? … Such a situation must not be tolerated any further. The Soviet Union is not Mexico or some kind of Uruguay, it is a great socialist power. Equally, Russian is the language of a mighty people, spoken by at least one hundred and fifty million human beings … It is time, it is really high time, to deride mercilessly and surround with disdain all those who show lackeylike servility at everything marked by a foreign stamp.[17]

Most of the mathematicians attacked in "Academician Luzin's Case" were able merely to acknowledge their guilt and continue their careers. Luzin also publicly acknowledged his crimes, but his punishment included removal from some his responsibilities at the academy, as well as the loss of his teaching position at the University of Moscow. Even so, he and the others got off comparatively lightly, as was still possible in the summer of 1936. The message of this particular case "to scientists was clear: be wary of publishing abroad, and stress nationalism over internationalism."[18]

The Luzin case served as the catalyst for a public campaign against Stefan Bergman and Fritz Noether, the German refugee mathematicians at the University of Tomsk, who were also associate members of the Siberian Physical Technical Institute (SFTI). On 15 September 1936, four Soviet members of SFTI published a newspaper article under the headline "Mercilessly expose and denounce concrete examples of servility before the bourgeois science," which singled out Bergman and Noether, as well as the Soviet physicist Petr Savvich Tartakovskii. In a later article, other Soviet physicists were attacked for having extensive contacts with foreign scientists. Vladimir Dmitrievich Kuznetsov, the director of SFTI, vigorously defended his scientists from these attacks. In the autumn of 1936, Kuznetsov's efforts to protect his staff were not yet suicidal, as they would become a year later. No one at SFTI was censored or arrested. Bergman, however, may have found the incident so unsettling that it led him to leave Tomsk and take a position at the University of Tbilisi.[19]

The Commencement of the Purges

For the time being, Noether was able to avoid arrest or expulsion. Other refugee scholars were not as lucky. In November 1936, Herbert Murawkin, the physicist and promoter of Esperanto, was the first of the foreign scholars to be arrested. His arrest seems to have been connected to the rise of nationalism and xenophobia within the Soviet state. Esperanto was international by nature, and since Lenin's time it had been considered a useful tool for helping promote revolution abroad. The Soviet Esperanto Union (SEU) was an informal club, not an officially recognized organization. Murawkin remained on the executive of the IPE, the German Communist Esperanto organization, which after 1933 operated in exile in the Soviet Union in cooperation with the SEU. Murawkin joined with the SEU in requesting official status. Without government recognition, the club had difficulty securing allocations of paper and other supplies necessary to print literature promoting the language in the Soviet Union and internationally.

In March 1936, frustrated by the lack of progress, Murawkin united with a leader of the Soviet organization and wrote directly to Stalin, asking him to order that Esperanto receive its long-coveted official recognition. They warned the Soviet leader of the consequences of further inaction: "It seems to us – for its utility, particularly in the defense of the USSR – that SEU should find itself in a position if not better, then at least normal, in comparison with the working conditions in capitalist countries. But the situation is such that our enemies abroad are beginning to compare the conditions of the USSR Esperantists with those in fascist Germany."[20]

As we saw with Borchardt's expulsion, prior to the Great Terror, letters to Stalin were a common form of appeal in the Soviet Union, though they could be very dangerous if the political winds in the Kremlin suddenly shifted direction. Instead of winning recognition for Esperanto, Murawkin's letter made him a casualty of Stalin's fickle decision to promote all things Russian and to fear all potential foreign influences. Instead of receiving official recognition, the Esperantists were one of many groups that suddenly found themselves singled out for arrest. Murawkin was one of the first of the leaders detained, on 26 November 1936, accused of being a foreign spy and agitator who had promoted Esperanto plots against the Soviet Union.[21]

By the end of 1936, the warning signs that the Soviet Union would not remain a place of refuge for displaced academics were becoming increasingly clear. At some point after the arrest of Eva Striker that spring, Charlotte Schlesinger arrived at the Kharkiv Conservatory and learned that her mentor, Grigory Maksimovich Weller, had been arrested by the local NKVD. Weller was charged with conspiring to assassinate Stanislaw Kosier, the deputy prime minister of the Soviet Union and architect of the policy of collectivization and the Great Famine. Soon after that, Schlesinger met with Alexander Weissberg, who apologized to her for not having foreseen what would happen in the Soviet Union and for arranging her position in Kharkiv. He counselled Schlesinger to immediately seek an exit visa. Schlesinger, fearing that she herself would be arrested, fled to Kiev, where she stayed with a friend from Vienna whose husband represented the Siemens company in Ukraine.[22]

At the end of 1936, Ernst and Mia Emsheimer were summoned to the German consulate in Leningrad, where they were ushered into the office of a diplomat, who presented them with a stark choice. He made a considerable show of opening a file and placing it on his desk in clear view of the couple. He then apologized for needing to leave the office for a short while. Ernst and Mia quickly read the first page of the file, which stated that as Communists and Jews, their German citizenship was to be revoked, and that if caught by the German government, they would immediately be arrested and transported to an unspecified location. With the long-term validity of their passports now threatened, the Emsheimers could either accept Soviet citizenship or leave immediately for a third country. They were deeply troubled that their private conversations in their apartment were being spied on, and also, no doubt, by the growing persecution in the Soviet Union. They decided to leave as quickly as possible. By early January 1937, the couple had found refuge in Sweden. This decision likely saved their lives, given the fate of the foreigners who opted for Soviet nationality.[23]

The Destruction of the "Oasis of Freedom in the Desert of Stalin's Despotism"

The destruction of the physics community at UFTI must be viewed in the broader context of the Great Terror that swept across Soviet society in 1937. No one, however powerful or well-connected, could be certain they would survive the terror unscathed. Most of the senior officers of the Red Army, and many of the junior ones, were arrested, and large numbers were executed. Senior party officials, including Sergo Ordzhonikidze, head of Narkomtjazhprom (People's Commissariat of Heavy Industry) and the man directly in control of research centres like UFTI, were subjected to arbitrary arrest, trial, and likely execution. Fearing the worst, Ordzhonikidze committed suicide on 18 February 1937, the day after NKVD officers searched his apartment. The Soviet press reported that Ordzhonikidze died of a heart attack. The weakening and eventual removal of UFTI's protectors in Moscow cleared the way for local party and NKVD officials to go after the scientists, who had all too often flaunted their immunity from authority.

At UFTI, at the end of 1936, the situation went from bad to worse. Landau, who had travelled extensively in the West, had always been critical of the Communist Party's official "religion" of dialectical materialism, especially as it affected theoretical physics. In private social gatherings, Landau and some of the others at UTFI turned to dark humour and satire to critique the regime. Landau and other Soviet scientists had part-time teaching appointments at the University of Kharkiv. This sort of double appointment was common in the Soviet Union and allowed for the better use of talent, which was in critically short supply. Landau, however, was especially contemptuous of the university faculty and students, who slavishly followed party doctrine; even worse, he viewed them as incapable of understanding the latest theoretical developments. In late 1936, the university administration, tired of Landau's impertinence, fired him. When the rest of the UTFI scientists went on strike in protest, the Ukraine Ministry of Education declared that the strike "was illegal in the socialist state; you were not supposed to 'strike against yourself.'" The UFTI people soon returned to their teaching duties.[24]

At a UFTI staff meeting in early 1937, Landau was publicly denounced by the senior administration as a counter-revolutionary. Tired of being harassed and fearing arrest, Landau turned for assistance to Pyotr Kapitsa, the brilliant experimental physicist. Kapitsa invited Landau to join his newly founded Institute of Physical Problems in Moscow. Hoping to leave his tormenters behind in Kharkiv, Landau accepted the offer.[25] With Landau's departure, UTFI's brief period as a world-leading research facility came to an abrupt end. The remaining foreign scholars were soon attacked in local newspapers, which accused them of being German spies.

One uncorroborated account of the events at the research centre links the targeting of foreigners to an incident at a social gathering at the Ruhemanns'. George Placzek, a Czech physicist known for his often ill-judged sense of humour, had been a regular visitor to UFTI. During his 1936 trip to the Soviet Union, he was offered a professorship at Kiev, to relieve Guido Beck of his double duty. Despite the temporary nature of his current position at the Bohr Institute in Denmark, Placzek had turned down the job offer. At the Ruhemanns', Placzek was asked by someone to explain why he had rejected the offer of a senior-level research position in the Soviet Union. Placzek gave a tongue-in-cheek reply, explaining that Soviet officials would not meet his demands, which included that "Chasain" resign. Chasain was a Russian word meaning boss and a slang term for Stalin. Most of the attendees laughed, but Barbara Ruhemann saw nothing funny about this insult to the Soviet leader. According to some accounts, she immediately reported the incident to the local Communist Party office, and soon after, the newspaper attacks began.[26]

After Landau fled to Moscow, the next targets of the NKVD were Alexander Weissberg and his associates. Already, Weissberg's close friend Konrad Weisselberg had been subjected to intense pressure, apparently intended to force him to leave the Soviet Union. Unfortunately, Weisselberg was determined to remain. He had married Galia Mykalo, whom he had met while staying at Weissberg's apartment during his first visit to the Soviet Union in 1933. Galia and Konrad intended to start a family. Galia was unwilling to migrate and probably unable to secure permission to do so in any case. In early 1936, Konrad began a legal process to become a Soviet citizen, hoping that this would ensure his right to stay permanently.[27]

In July 1936, Weisselberg was abruptly dismissed from his position as a chemist at the Coal and Chemistry Institute in Kharkiv, despite having been just weeks earlier singled out as a model worker. The head of Weisselberg's laboratory was aghast that he was losing his best chemist, but the only explanation he could offer Weisselberg was that the facility was being forced to downsize its workforce. Over the following weeks, as Weisselberg searched for another position, it became clear that the secret police had targeted him. Every time Weisselberg seemed to have won a competition for a job, the offer would be withdrawn at the last moment or he would hear that the position had been cancelled. The problem stemmed from the last step required before one could be hired in the Soviet Union – getting clearance from the NKVD. Either a false allegation had been made against him, or he had been targeted as a foreigner, or he was being victimized because of his close association with Weissberg. The relationship between the two men was especially close; after Weiselberg's dismissal by the Coal Institute, Weissberg had allowed his friend and Galia to share his UFTI apartment. To keep Weis-selberg employed, Weissberg found him part-time employment at UFTI. As a contract worker, Weisselberg did not appear on the list of institute employees, which meant he did not require vetting by the NKVD.

On 25 December 1936, Galia gave birth to a son, Alexander, named for the parents' great friend. Weisselberg remained confident in the Soviet Union and in his ability to stay with his family. In January 1937, he became a Soviet citizen, renouncing his Austrian nationality. He urged Weissberg to emulate him and become a citizen. Weisselberg was naively confident that the madness gripping the country would soon end and that the more he integrated himself, the sooner "he would enjoy complete equality in Soviet society." Weissberg did not share his friend's confidence, but he also felt that as a fellow Communist, he could not advise against becoming a Soviet citizen. Weissberg, however, "knew perfectly well that by abandoning his last shred of protection … he would get no thanks for his devotion."[28]

By January 1937, shaken by Eva Striker's continued imprisonment, the growing number of arrests on clearly false charges, and the forced departure of Landau, Weissberg was preparing to find a way out of the country. He doubted, however, that the NKVD would grant him an exit visa. He began to think of volunteering to fight the fascists in Spain, a last ditch route of escape from the Soviet Union, and one taken by many foreigners. On the morning of Sunday, 24 January 1937, Weissberg received a phone call ordering him to report to the Kharkiv NKVD at 11 a.m. He was warned not to inform anyone he had been summoned. Weissberg's summons appears to have been sparked by his recent visit to the Leningrad military prosecutor and NKVD office to deliver the letters of support for Eva. Weissberg's account of this first interrogation, which began with a discussion of his activities in Leningrad, reveals the impossibility of anyone proving their innocence to the NKVD:

> "Well, I went to the Military Prosecutor and to the N.K.V.D. on behalf of my former wife."
>
> "Your wife was arrested as an enemy of the people and yet you intervened on her behalf? So you support an enemy of the people, eh?"

"I'm certain she's quite innocent."
"So now you're saying we arrest innocent people."[29]

Throughout February, Weissberg was repeatedly ordered to return to the NKVD for increasingly intense and detailed questioning about all aspects of his life, and concerning a vast array of people with whom he had associated even before he joined the Austrian Communist Party. Each time, Weissberg went to the headquarters not knowing whether he would be released or arrested. The physical and psychological strain on him was intense. At one point, despite NKVD orders to the contrary, he consulted Leipunski, asking the UFTI director to help him. Leipunski promised to talk to the head of the NKVD in Kharkiv, to vouch for Weissberg's devotion to the party and the Soviet Union. But he also warned Weissberg that he would report this unauthorized discussion. The director explained to him that spies were everywhere, even in his own office, and if he did not report to the secret police, Lejpunski himself would fall under suspicion.[30] Despite Leipunski's efforts, on 1 March, Weissberg was arrested. Three days later, Weisselberg too was detained.

Shorty after these arrests, the NKVD held a meeting at UFTI for all employees, to announce that Weissberg had organized attacks on Soviet leaders, had committed counter-revolutionary acts of sabotage, and had spied for an unspecified foreign power. Then, "in accordance with Soviet rituals in such matters," all but one of Weissberg's former colleagues "solemnly arose one after the other and damned the enemy of the people who had once been their comrade." The one exception was Martin Ruhemann, who, when it came his turn to speak, defended his close friend and mentor:

> Personally I can't believe that Weissberg was a counter-revolutionary. You all know me and I think you trust me. When I first came here the Soviet cause was not particularly dear to me, and if there was one man more than any other who won me over to it was Weissberg. He did his utmost to show me what was going on here and to make me understand its historical significance. Why should he have done that if he had been a counter-revolutionary all the time? He didn't try to win me for the counter-revolution but for the cause of socialism.[31]

Ruhemann's remarks infuriated the NKVD agents and led the institute's party members to incessantly bully him to publicly renounce his statement. Two days later, Ruhemann agreed to publish a "feeble declaration" on the wall newspaper, where he "mildly criticized his attitude at the meeting."[32]

The arrests marked the end of any hope for most of the few remaining foreigners at UFTI that they could remain in the Soviet Union. Fritz Houtermans sent Charlotte to Western Europe to seek assistance in finding a new position. Charlotte found it impossible to convince their mainly socialist and Communist friends that the situation in the Soviet Union had degenerated so suddenly. Before Charlotte could find a position for her husband, the NKVD began questioning Fritz about what she was doing and when she would return. Fritz was forced to ask his wife to return to Ukraine posthaste.

The Great Terror and the Ensnared

The arrests of Weissberg and Weisselberg, however, did not begin the wholesale departure of foreign scholars from the rest of the country, or their detention. These early arrests were initiated by the local NKVD. The only other foreign scholars arrested before the autumn were also victims of the Kharkiv secret police: the medical scientists Kurt and Irene Zinnemann, who were detained for espionage in August. Only two foreign scholars left the Soviet Union during the spring. Lazlo Tisza decided to end his association at UFTI after Landau's abrupt move to Moscow. Without Landau, Tisza had no reason to remain in Kharkiv. Tisza tried in vain to secure an academic position elsewhere in the Soviet Union; he visited Guido Beck in Odessa, who could not offer him a job. It appears that the long vacant post at Kiev was no longer available either, or perhaps it was not open to a foreign scientist. The two men took a pleasant weekend boat trip to Sebastopol in the Crimea, followed by a taxi ride up over the mountains to the subtropical forests of Yalta. It was, recollected Tisza, a "nice closing episode" to his stay in Ukraine. After his trip with Beck, Tisza decided it was time to leave the Soviet Union. Tisza, however, encountered difficulties securing an exit visa, since it required the approval by the NKVD. German citizens were further hampered by their inability to seek assistance from German diplomats, whereas non-German scholars were able to turn to their embassies for help. Tisza, despite his prison term for his left-wing political activities, was assisted by the Hungarian delegation to secure permission to leave, but this still took several months. He returned to Budapest in June 1937.[33]

Michael Sadowsky's exit from the Soviet Union was far less pleasant. In late April or early May, he was dismissed from his position as professor of engineering mechanics and department head at the Engineering College in Novocherkassk. Like Borchardt, Sadowsky had "refused to introduce Communistic ideas" into his lectures. He had also resisted intense pressure from the NKVD to renounce his "German citizenship and become a Soviet subject." Sadowsky was given just two days' notice to depart the country: the NKVD had "tried to convert me about two years, but after they saw their efforts were gone lost [*sic*] in vain they lost their temper and I have to leave." By 17 May, Sadowsky and his Jewish wife and child found themselves back in Germany. In Novocherkassk, as in Kharkiv, it was the local NKVD, not a national policy, that led to Sadowsky's dismissal.[34]

Tisza, Beck, and Sadowsky were among the fortunate ones who were able to leave the Soviet Union in the spring of 1937. Tisza and Beck were helped to get exit visas by their embassies, and Sadowsky's expulsion came with a visa. Not everyone was so fortunate. German citizens had no one to turn to for assistance. Charlotte Schlesinger, after fleeing Kharkiv for Kiev in late 1936, eked out a living teaching piano to the children of Viennese business people in the city and occasionally acting as an accompanist at the Kiev Conservatory. Heeding Weissberg's advice, Schlesinger gathered together all the necessary documents and applied to Moscow for an exit visa. In Kiev, fear was omnipresent – a universal dread had gripped everyone. She recalled: "On the surface, life went on as usual; people fell in love[,] children were born. But almost every week one or another musician disappeared. They were never mentioned any more." After many months waiting in terror for word about her visa, Schlesinger decided in the autumn of 1937 to travel to Moscow to try to advocate for the required documents. Before his arrest, Weissberg had provided Schlesinger with the phone number of Friedel Cohn-Vossen, wife of the German mathematician who had died in Moscow of pneumonia the year before. Cohn-Vossen had been a member of the same left-wing social circle that had often met in Eva Striker's Berlin apartment years before. Despite their never having been close friends, Cohn-Vossen agreed to shelter Schlesinger and help her. After her husband's death, Cohn-Vossen had become involved with Alfred Kurella, a leading member of the KPD, whom Schleslinger believed had good connections with the NKVD. Despite Kurella's help, Schlesinger found herself trapped in Moscow, unable to leave the country as the Great Terror continued to escalate out of control. In October, Kurella's brother Heinrich became one of the many German Communists in Moscow arrested by the NKVD and executed.[35]

In the first half of 1937 it was the regional NKVD that directed the arrests and expulsions of refugee academics. It was not until July that the first dismissals took place in Moscow, and not until October in Leningrad. By the summer, however, foreigners found themselves subjected to intense pressure to leave. Most of their employers feared their own arrest if they retained their foreign specialists. Thus, as the foreign scholars' contracts expired, they were not renewed.

In July, Victor Weisskopf reported to the SPSL in London on his recent trip to Odessa, where he had observed the deteriorating situation facing Marcel Schein. Schein had initially had great success building up the experimental physics program at the university, a success made possible by the unqualified support of the school's senior administrators. Eventually, his work was undermined by jealous Russian faculty, who used every means at their disposal, including "political insinuation," to make his life "unbearable." In the climate that now existed in the Soviet Union, Schein as a foreigner had no means of defence. He was desperate to remove himself and his family from the Soviet Union, and he remained in Odessa only because he had nowhere else to go.[36] Soon after Weisskopf's visit, the situation became even more frightening. Everywhere the family went, they were trailed by NKVD agents. In September, Schein fled to Prague; his dread of arrest by the NKVD had overtaken his fear of unemployment. The family left in considerable haste, leaving most of their possessions behind, including Edgar's beloved model trains and steam engine. Edgar's parents promised to compensate him for his loss with a Schwinn bicycle, if and when the family managed to move to the United States. Edgar has no idea how he knew about Schwinn bicycles, but at the time it seemed like "reasonable compensation" for the loss of his precious toys.[37]

A similar situation compelled the physiologist Ernst Simonson and his family to flee from Kharkiv, also for Prague. The exact date of his departure is unknown, but if he did not flee before his friend Kurt Zinnemann was arrested, it would have been soon after. From the temporary safe haven of the Czech capital, Simonson described the deteriorating situation to his British colleague, A.V. Hill: "The last months in Russia were terrible for us, we were, as foreigners, in every way isolated, even physicians – personally good friends of mine – refused to visit us in case of disease fearing to get accused of connections with foreigners. Every day we had to fear arrest."[38]

On 9 August, Kurt and Ena Zinnemann were arrested by the Kharkiv NKVD. No longer were the Kharkiv NKVD politely summoning their victims to visit their headquarters. Instead, NVKD officers were driving around the city in the middle of the night in a thinly disguised van to drag their victims off to jail. Kurt and Ena's daughter Pamela, born in 1945 long after her parents were deported from the Soviet Union, wrote a poem vividly recounting her parents' story of their arrest:

Arrest

Every night we lie awake
sweating in the August heat.
One night: 4.00am,
Kurt says,
"Listen!"
A muffled engine sound;
the quiet click:
the van doors shutting
before it comes, the dreaded, half-expected knock.

I find us some clothes
and see the van known as Black Crow
from the window;
and the Crow's disguise:
letters denoting "Meat Van"
everyone can recognize.

They take us down together.
I have no time to say "good-bye."
I watch his straight back bend as they shove him in;
I see how he bows his head before they push me too.

There is no time to say good-bye.[39]

The couple were charged with being German agents engaging in espionage, having been ordered by the Gestapo to carry out diversionary activities in the Soviet Union by spreading bacterial materials in order to cause illness among the Soviet population. NKVD searches of the couple's laboratory found diphtheria cultures – not surprising, given their scientific research into the disease. To bolster their case against the couple, the NKVD claimed that during their search of the Zinnemanns' apartment they had discovered portraits of Hitler and Göring, something that loyal Nazi spies carried with them everywhere they went.[40]

Three days prior to the Zinnemanns' arrest, two of the Soviet scientists at UFTI, Lev Shubnikov, head of the low-temperature physics group, and Lev Rozenkevich, were also arrested by the Kharkiv NKVD. The Kharkiv NKVD had been given direct approval for their actions by V.I. Mezhlauk, who after Ordzhonikidze's suicide had taken over as head of Narkomt-jazhprom in Moscow. By early November, both scientists had been forced to make false confessions, given mock trials, and executed.[41]

The arrest of the Soviet scientists at UFTI that summer left the Houtermans in a state of panic. Prior to his arrest in March, Alexander Weissberg visited Fritz in his laboratory, where he found Fritz unperturbed by the growing danger faced by those working at UFTI. After Weissberg's arrest, the Houtermans came to realize that they needed to leave the Soviet Union; this led to Charlotte's aborted trip to Western Europe. Since her return to Kharkiv, they had tried without success to secure an exit visa from the local NKVD. Each arrest or departure of scientists they knew heightened the couple's anxiety over their fate and that of their children. Charlotte's recollections of the events of that summer, written years afterwards, are somewhat muddled, an understandable result of the psychological terror the couple endured.

The couple's mental breaking point came when Valentin Fomin, Fritz's research assistant, tried in an especially gruesome manner to kill himself to avoid arrest, certain torture, and likely execution. Fomin's elder brother had already been incarcerated by the NKVD, and Valentin now worked himself into a state of intense anxiety, fearing he would be next. When he received a summons to visit the local headquarters of the secret police, he rushed into a UFTI laboratory and drank sulphuric acid. The acid did not kill him but left him in agonizing pain; likely out of his mind, he raced home and threw himself out of a third floor bathroom window. He survived the fall. Rather than escaping his fate, Fomin was hospitalized. When he was well enough to be released from hospital in early October, he was arrested. After a sham trial, Fomin was executed by firing squad in early December. The news of Fomin's botched suicide attempts, Charlotte later explained, left her paralysed with fear; she "seemed to freeze inside, like holding my breath waiting for the next blow to fall." Fritz, for his part, had a near total mental collapse. He walked around his room, chain smoking, talking out loud to himself about ways and means to escape or to be rescued from the Soviet Union. His schemes became "more and more fantastic," and he kept on repeating himself, "like a patient in a delirium."

Charlotte, at a loss for how to help her husband, went to see Sasha Leipunski, who had recruited them to UFTI. Since March, Leipunski, like most of the Russian scientists who feared they would be found guilty by association, had avoided socializing with any of the remaining foreign specialists. Charlotte, however, "bluntly and practically forced" Leipunski to visit Fritz and reassure him that he would be safe and was in no danger. Leipunski did Charlotte's bidding, even though they both knew that any reassurances were lies. The lie did calm Fritz down, although nothing could allay the couple's fears, which continued to grow as more and more people were arrested.

As the confusion and terror in the Soviet Union escalated in the autumn, the pace of voluntary departures and expulsions of the refugee scholars increased. Some, like Edgar Lederer, remained oblivious to events outside their laboratories. Lederer's wife noticed the degrading political situation in Leningrad long before he did. Lederer later mused that his wife was perhaps more aware of what was happening because she did not share his "communist faith!" He explains that his wife made him aware of the "purges, trials and the rest": "In short, after two years spent in the USSR, it was necessary that I return to France. The director of my Institute had suggested to me to go to Moscow to request from Foreign Affairs to obtain permission to stay one more year, but the atmosphere became really oppressive. None of the Soviets dared to have any contact with us."[42] He and his family arrived back in Paris at the end of December 1937.

By October of that year, the Great Terror had spread, from a series of local actions to a national campaign of arbitrary arrests, show trials, imprisonments, and executions. None of the foreign scholars in Leningrad were safe. Even Hermann Muntz, who had survived the expulsions after Kirov's assassination, was dismissed from his post as professor of mathematics at Leningrad University, a position he had held for nearly a decade.

The almost complete breakdown of social order in the Soviet Union that developed in the last half of 1937 is best exemplified by the curious case of Wolfgang and Inge Steinitz. Both were members of the KPD, but their party status had been kept hidden from everyone in Leningrad except for their NKVD handler and the director of the Institute for the Peoples of the North. Steinitz's contract at his institute was up for renewal in October, and he feared, like the other foreign academics, that he would be kicked out of the Soviet Union. At about the same time, the Steinitzes' value as secret agents came to an abrupt end when German diplomats at the city's consulate were replaced with devout Nazis. No longer would Jews be allowed in the consulate; moreover, the German diplomats were aware that since June, Steinitz had been on a Gestapo list for immediate arrest. There would be no more invitations to social events at the consulate. Wolfgang, now free of his secret duties, made desperate efforts to reveal himself as a party member and NKVD operative. This proved impossible. His handler had himself been arrested, and the director of the institute had been dismissed for political reasons. Letters from the Steinitzes to the Comintern went unanswered. The Comintern had been decimated by arrests, and chaos reigned at the organization's headquarters in Moscow.

In September, a series of meetings were held at the Northern Peoples Institute to discuss the use of Cyrillic letters for the Ostyak language. At one meeting that went late into the night, Steinitz, supported by two Soviet linguists, defended his previous work using Latin letters. The next morning, Steinitz was attacked on the institute's wall newspaper, "accused of deliberately introducing complicated spelling to make it harder for Ostyak to learn the written language." Steinitz was officially dismissed from the institute on 2 October and ordered to leave the Soviet Union by 1 November. Inge and their two children were permitted to stay for two more months while she packed the family's possessions and Wolfgang's scientific papers. Inge also worked diligently to have Wolfgang's dismissal reversed, even writing a personal appeal to Stalin. Looking back, she realized how foolish she had been not to be frightened. She later wrote: "I felt so safe, but it was more that the sword of Damocles [was] hanging over us."

Later, Wolfgang and Inge came to believe that their downfall had been caused by fellow Communist Gerhard Harig, the physicist turned historian of science. The couple never provided any proof for these allegations, but it is likely they knew Harig, and knew of his close affiliation with the NKVD.[43] Whoever was responsible, it is more than likely that no one at the institute believed Steinitz to be a Communist and NKVD spy. His expulsion very likely saved his and Inge's lives, since they could easily have shared the fate of so many German party members during the terror. Having been forced into a second exile, the couple's greatest fear was of being captured by the Gestapo. Instead, relatives of the couple rescued them by arranging for tourist visas to Sweden. It would be a long holiday – they would not return to Germany until 1948.

If Gerhard Harig indeed denounced the Steinitzes, it may have been to try to avoid his own arrest by the NKVD. In 1937, Nikolai Bukharin, the Marxist philosopher and former political ally of Stalin, was the director of the Institute for the History of Science and Technology at the Academy of Sciences of the USSR in Leningrad, where Harig worked. In February, Bukharin was arrested and the institute dissolved, with some of its remaining employees moving to Moscow. Harig was left behind in Leningrad to work as an external member of the institute; he was also given a low-paying position as a research assistant at the Leningrad State Public Library. During the transfer of the institute to Moscow, Harig's pay ceased, and his wife Katharina was forced to work long hours, even though she was still breast-feeding one of their children. Despite Katharina's efforts, her earnings were insufficient to feed the family, and they were heavily in debt. One can only wonder whether the death of their infant daughter was a result of their increasing poverty.

Katharina, without her husband's knowledge, wrote to a Moscow-based official of the institute, pleading with him to pay what was owed to her husband. She outlined their plight: "The emotional pressure from which my husband, and with him, the whole family suffers is terrible. He work and works and there is no recognition. I see how he becomes sadder and sadder every day without admitting it to me." The situation did not improve, and on the night of 20–21 October 1937, Harig was arrested on suspicion of espionage. Whether Harig was arrested and, like the others, faced torture and possible execution, or whether the arrest was staged by the NKVD to create a cover story for Harig's return to Germany, remains unknown. Harig's account suggests that he first believed the charges were real, but he was soon informed by the NKVD that they were fake. In early December. Harig was officially exonerated. By the end of the month, the NKVD had asked him to return to Germany. Whether Harig was a willing volunteer or felt coerced remains another mystery. In early 1938, he was given espionage training by the NKVD, which included instructions in operating a shortwave radio. In April, Harig said goodbye to Katharina and his son Georg. He was then "officially" deported by the Soviet government and placed on board a ship transporting German citizens who had *actually* been kicked out of the Soviet Union. Harig did not know it at the time, but he would not see his family again for more than a decade.[44]

The majority of the academic refugees were able to escape or were forced to leave the Soviet Union by early 1938. Among the last were Martin and Barbara Ruhemann, who departed for Britain in December 1937. The Ruhemanns had been informed in the summer that their contracts would not be renewed. Whether their departure was caused by Martin's defence of Weissberg, or by the general hostility in the Soviet Union toward foreigners, or was a front to hide the fact they were agents working for the NKVD, remains unclear. Certainly, the Ruhemanns were not the only ones suspected of being Soviet spies; in Sweden the Emsheimers and Steinitz were watched carefully by counter-intelligence agents. Harig had been trained to spy for the Soviets in Germany, but his almost immediate arrest by the Gestapo there put a swift end to his career in espionage.

Perhaps the last refugee scientist forced out of the Soviet Union was Walter Zheden. The Moscow-based physicist's contract had not been renewed when it expired in the autumn of 1937, but he remained behind in the Soviet capital. Although little is known about his life in the Soviet Union, he likely lingered in Moscow either to try to stay with his wife and daughter or to get permission for the entire family to leave the country. He was unsuccessful in his quest to keep his family together. In January 1938, when his Soviet visa expired, he was forced out of the country.[45]

Other Arrests during the Great Terror

Five other academic refugees – Fritz Houtermans, Siegfried Gilde, Fritz Noether, Julius Schaxel, and Hans Hellmann – would join Murawkin, Weissberg, Weisselberg, and the Zinnemanns in NKVD prisons. On 15 September 1937, Fritz Houtermans, without warning or reason, was fired from his position at UFTI. Immediately after this, the Houtermans left Kharkiv for Moscow, believing that the secret police in the Soviet capital would be more amenable to issuing an exit visa. This might have been true in the spring, but by the autumn it was doubtful. Still, the Houtermans had many Soviet friends in Moscow, including Landau and Kapitsa, who would help them secure the long-coveted visas.

The family finally received their exit visas on 17 November, but they were denied permission to take their books and papers. Fritz summoned Charlotte and the children to Moscow, while he lobbied Soviet officials for permission to take all of their property out of the country. Charlotte arrived in the city on the last day of November and found Fritz still unwilling to leave. The following day, Fritz, accompanied by Charlotte Schlesinger, visited the Moscow Customs House with an inventory of books he was hoping to take with the family to the West. But while he was delivering his list, he was arrested by waiting NKVD officers and taken to Lubyanka Prison.

After learning of Fritz's arrest from Schlesinger, Charlotte Houtermans was left in a desperate situation. The Houtermans had been staying in the home of Soviet physicist Pyotr Kapitsa, but now Kapitsa, fearing his own arrest, ordered Charlotte and her children to leave, only relenting to allow them to stay one night. The following day, Charlotte went to Lubyanka Prison to retrieve her passport and whatever money Fritz had with him when he was arrested. Charlotte then checked herself and the children into a hotel, unsure of what to do next. The next morning, the Houtermans were joined by Schlesinger and Freidel Cohn-Vossen. The three women discussed what Houtermans should do next. Houtermans was reluctant to leave, in the wishful hope that Fritz would soon be released. Cohn-Vossen was adamant that Houtermans had to leave the Soviet Union immediately. With her husband under arrest, no one was going to hire her, and her German passport would be expiring in three weeks. Reluctantly, Charlotte agreed that she had little choice, and she and her two children undertook a perilous train journey to Riga. At the border the Houtermans were removed from the train and held for eight days while local NKVD agents tried to determine whether she could proceed. Finally allowed to cross the border into Lithuania, Charlotte and her children were temporarily safe, although she was destitute and her passport would expire in just a few days.

In early December, Charlotte Schlesinger remained confined to Moscow, in an increasingly tense situation in Cohn-Vossen's apartment. Alfred Kurella was furious that Schlesinger had involved them with Houtermans, as any association with an enemy agent could bring the NKVD to your door. Kurella refused to speak to Schlesinger, except to ask her to pass the salt at dinner. One week later, Kurella suddenly broke his silence to announce that Schlesinger's long-awaited exit visa would arrive in the next two days. Schlesinger, determined not to repeat Fritz's mistake, decided to abandon all of her possessions in Moscow, and as soon as she received her passport with the coveted visa, she raced to the train station and purchased the last available ticket to Prague.[46]

Charlotte Houtermans and Charlotte Schlesinger were fortunate to be allowed to leave the Soviet Union. Siegfried Gilde, the medical researcher, had not been as lucky when he and his family reached the very same border station two weeks before the Houtermans. On 11 November 1937, Gilde, like so many other refugee scientists, had been, without any notice, expelled from the country. The Gildes were ordered to depart within four days, but Siegfried managed to persuade officials to delay the date they had to be out of the country until 21 November. Tragically, the change of date was not noted on their passports, though "the official promised to notify the border." At the Soviet border control for Lithuania, Siegfried left the train to clear Soviet customs; Ruth and the baby were allowed to stay on board. Ruth reported what happened next: "I did not see my husband again. The luggage came back. As the train was already moving, an officer brought me my passport and said, 'Do not worry, your husband is coming on the next train, the visa is not in order, we have to check with Moscow.'"

Six days later, from Riga, Ruth wrote to their family friend Professor Karl Weissenberg about her now desperate situation: "Frequently, I think he has to come every day, and then again I have the horrible idea I will never see him again." Ruth told him she had found a place to live, though with great difficulty, since no one wanted to rent to a young woman with a child. She had done everything possible for her husband, even visiting the Soviet consul, who could offer no assistance. She was afraid to approach the German embassy. With nothing else to do, she sat in her room "waiting half mad with fear for my husband." She and her husband had hoped to go to America, and by herself she felt helpless. She asked Weissenberg to contact Roland de Margerie, the French diplomat who had tried to help them two years earlier, for assistance in getting her a visa to travel to Paris. After waiting a few more weeks for Siegfried, she decided there was nothing more she could do, and she and the baby moved to the French capital. It is now known that Siegfried Gilde was arrested and sent back to Moscow. He would never see his family again.[47]

Fritz Noether had survived the attacks made against him by his colleagues in 1936, but one year later, no one could or would protect a foreigner. On 22 November, he was arrested by the Tomsk NKVD on charges of espionage and sabotage for Germany.[48]

Very little is known about the time, place, or cause of Julius Schaxel's arrest, except that it likely occurred sometime in 1937. It is possible that Schaxel, who had been stripped of his German nationality the year before, had no way of leaving the Soviet Union.

Hans Hellmann was arrested by the NKVD at his Moscow apartment on the night of 9–10 March 1938. He had known for some time that his position was becoming increasingly dangerous. On New Year's Day 1938, he wrote to his mother: "Dear mother, it is certainly not neglect on my part that I don't write more often. You do have some idea of the present international situation. The mood seems almost pre-war, so I am really afraid to engage in too much correspondence with other countries. In fact, the wall between us gets higher every day."[49] Hellmann, however, had no path allowing him to leave the Soviet Union; he had become a Soviet citizen. Like Weisselberg and, perhaps, Schaxel, Hellmann discovered too late that rather than ensuring his integration into Russian society, his new nationality had doomed him.

A few days after Hans's arrest, Viktoria Hellman visited the Karpov Institute in search of her husband's final paycheque. Viktoria saw that Hans had been denounced in an article posted on the institute's bulletin board. The article was signed by two of Hans's colleagues, A.A. Zhukhovitskij and Michail Tjomkin (or Temkin); the former had until recently been the party secretary at the institute, and the latter, his successor. Zhukhovitskij benefited personally from Hellmann's arrest, since it enabled him to take over the German scientist's theory group and his graduate students. As we have seen, it was not uncommon in the Soviet Union for people to advance their careers by denouncing their colleagues. By this time, it was "open season" on all foreigners, and as leading members of the Communist Party at the institute, perhaps the men felt compelled to turn Hellmann over to the NKVD in order not to face arrest themselves.

The Mysterious Survival of Fritz Lange

After Hellmann's arrest, all but one of the refugee scholars either fled the country or faced the tender mercies of the NKVD. Only physicist Fritz

Lange remained unmolested, even though he worked at UFTI, where the terror had first reached the foreign scholars. The arrests and persecutions at UFTI continued well into 1938. Alexander Leipunski was stripped of his directorship of the institute and expelled from the Communist Party before being arrested on 14 July. Very little is known about Lange's activities at UFTI; three of his friends from Berlin – Charlotte and Fritz Houtermans, and Weissberg – have left accounts of their time in Kharkiv, but they make almost no mention of Lange at UFTI. The Houtermans, before migrating to the Soviet Union, even shared their house in England with Lange, who lived on the upper floor.

In Weissberg's book, Lange is mentioned only twice, despite Weissberg stating that at UFTI he and Lange were close friends who often discussed technical and economic questions. Weissberg's silence on Lange in *The Accused* may have been an attempt to protect his old friend, who remained a defence scientist in the Soviet Union during the early Cold War. If so, Lange was not the only person Weissberg attempted to shield in his account of the purges. In the late 1940s, the fate of Konrad Weisselberg and his family remained unknown in the West, so Weissberg gave them code names. Konrad, for instance, became Marcel. The first time Weissberg mentioned Lange in his book was just before Weissberg was arrested, at a time when he was making plans – futile, as it turned out – to escape the country. Weissberg states that up to that point, he had attempted to keep Lange out of controversies at UFTI. Weissberg was not always successful in protecting Lange. During the Davidovich affair, Lange learned from other sources what was happening and came out in open support of Weissberg. When Weissberg informed Lange of his intention to flee to Sweden, Lange suggested that he move to Siberia instead, since in the remote regions they were always looking for talented people. Weissberg rejected this idea, but he would later reflect that it was actually a good one, and that it might have saved him, since the Kharkiv NKVD had no power beyond Ukraine. Weissberg realized that Lange had an astute understanding of the Soviet system under Stalin. When Weissberg insisted that he wanted to leave the Soviet Union, Lange offered him money he had set aside in a Danish bank, and considerably more funds he had in the United States. When Weissberg protested that he did not want to use his friend's nest egg, Lange replied: "Promise me rather that you'll use it the moment you need it. I don't need it here. I've applied for Soviet citizenship, which means I shall never leave the country again, and I don't know how better I could use my money abroad than in helping friends."[50]

If Lange did renounce his German citizenship and become a Soviet national, then his survival at UFTI is all the more inexplicable, something that Weissberg later discussed with a jailed Soviet colleague whom he met by chance while being transferred between prisons.[51] This was the second and last occasion that Lange appears in Weissberg's account. Weissberg provides one of the only clues concerning Lange's unique survival, explaining that Lange was an excellent experimental physicist who was also a first-class technician capable of building his own equipment. These practical skills, by themselves, were not enough for Walter Zehden to remain in the Soviet Union; he was dismissed from his position in Moscow in the autumn of 1937 and left the country after his residency visa expired in January 1938.

The only other information related to Lange's ability to remain in the Soviet Union is found in the history of UFTI published in Ukraine in the late 1990s. According to an unverified story, Lange was briefly detained by the Kharkiv NKVD. When the secret police examined Lange's internal Soviet passport, they discovered that it had been signed by Stalin himself. Lange was immediately released and was not further molested by the NKVD. No explanation is provided for why Lange's papers had been personally signed by Stalin.[52] The key to Lange's survival, then, may have been his work on high-voltage electrical systems, work that it was believed would have direct military applications. Lange's absence from the Houtermans's accounts is perhaps related to the former working within a top secret military laboratory, from which most of the other foreign scientists were excluded. Misha Shifman, the Russian American physicist and historian who has written extensively about UFTI, confesses that he "simply does not know why" the 1937 disaster at UFTI did not touch Lange.[53]

Conclusion

Hans Hellmann's arrest in March 1938, marked the end of the story of German and Central European academics working in the Soviet Union. Of the thirty-six scholars who migrated voluntarily before 1933 or who fled Germany for the Soviet Union later on, one had died of natural causes, nine were incarcerated, and the rest fled or were expelled; only Lange remained unscathed. The scholars and their families able to escape the Soviet Union were desperate to find another haven, hopefully a more stable one. They would have to escape Hitler's Germany at least one more time if they were to survive. Not all of those who escaped Stalin's madness would survive their second brush with Hitler's murderous regime. Before tracing the stories of the scholars who were able to leave the Soviet Union, let us first examine the experiences of those who faced the horrors of Stalin's prisons and court system.

8   Into Stalin's Frying Pan

The fate of the nine scholars left behind in NKVD prisons was determined by the brutal, idiosyncratic, surrealistic, and arbitrary world of what passed for a justice system in the Soviet Union. All of those who entered the Soviet penal system were subjected to brutal psychological and physical torture, designed to extract false confessions or, at the very least, to coerce a signature on documents outlining fantastic, sinister plots against the Soviet Union. These plots were a reflection of the delusional fantasies of Stalin, a despot who believed in intricate international conspiracies that had no basis in truth.

A great deal is known about the treatment of the scientific prisoners. Alexander Weissberg and Fritz Houtermans left detailed accounts of their time in the Soviet prison system. More than a decade after his initial arrest, Weissberg wrote his memoirs without any supporting documents. During that time, he survived three years in Soviet prisons, followed by five years in Nazi-occupied Poland. In May 1945, just days after the end of the war in Europe, Houtermans wrote an account of his imprisonment; this account has been translated and published in Misha Shifman's *Physics in a Mad World*.[1] Early in the 1950s, Houtermans co-wrote with K.F. Shteppa, formerly professor of history at Kiev University, *Russian Purge and the Extraction of Confession*. The two men had shared a cell for much of Houtermans's incarceration. Rather than simply recounting their own experiences in prison, they focused on explaining how the NKVD had coerced prisoners to admit to involvement in absurd, extraordinarily elaborate conspiracies against Stalin and the Soviet state. Fearing a Soviet backlash against their work, they wrote under the pseudonyms F. Beck and W. Godwin. It is impossible to identify specifically which of the events described in this book actually occurred to Houtermans himself, but the book does confirm much of Weissberg's account. Supplementing these accounts are the recollections of Kurt and Ena Zinnemann, which they recounted to their daughter Pamela. Pamela subsequently wrote a book of verse in which she retold her parents' stories. Other less extensive testimonies survive in Kurt Zinnemann's SPSL file and in the memories of the family of Konrad Weisselberg. Eva Striker, Weissberg's wife, left her own story of her experiences in the Soviet prison and judicial systems.

Since the 1990s, a number of NKVD files of imprisoned academics have been published, although at best they are useful only for helping us grasp the nature of the conspiracies that NKVD agents created from their own warped imaginations, and their success in coercing their victims to sign documents attesting to their veracity. The secret police files do not contain information on the means used to force prisoners to betray their friends and their families; none of them were guilty of any crime, except perhaps for occasional verbal criticism of the regime. Confessions under torture were technically illegal under Soviet law, and the thin veneer of legality behind the mass murder of so many thousands was such that detailed records were not kept regarding the means used to extract those confessions.

Torture and Confessions

Alexander Weissberg described in considerable detail the torture he endured, a technique known as "the conveyor." As tortures go, the conveyer seems relatively benign: the prisoner was required to sit in a chair without moving while a team of three interrogators took four-hour shifts interrogating him round-the-clock. Occasional, very brief food and bathroom breaks were allowed, but always, the prisoner was forced to return to his chair. Some of the examiners were occasionally friendly, sometimes even telling jokes, but one of them, named Shalit, took particular pleasure in hurling abuse and threats at the prisoner. "The 'conveyer,'" Weissberg explained, "worked automatically and silently. After a few days a prisoner's limbs began to swell. The muscles in the neighborhood of the groin became extremely painful and it was an agony to move." The threats and abuses of torturers like Shalit were unnecessary. "All the examiners had to do was to wait patiently. Time was their ally." Everyone, Weissberg discovered, had their breaking point under the relentless torment of the conveyer. At first, Weissberg attempted to refuse to admit any part in conspiracies against the state. After a few days, he was willing to admit his own guilt in a fictional plot to murder Stalin and the head of the NKVD, but he refused to implicate others.

After six days, Weissberg could stand the pain of sitting on the stool no longer; he feared that he would soon be willing to confess to even worse conspiracies implicating even more innocent people. When Captain

Tornuyev, the friendliest of the interrogators, began his shift, Weissberg decided to try to end the torture:

"Citizen Examiner, I am at the end of my tether. It is physically impossible for me to go on. What you accuse me of is fantastic, but I am prepared to admit it." "A confession with such reservations is no confession," he replied. "If it's untrue you have no right to confess it. That would be provoking the Soviet power." "But I can't stand this any longer. What am I to do? You force me to confess by torture and then you don't want to accept my confession when I'm prepared to make it. If I were not under pressure I would neither confess nor provoke the Soviet power." "We are conducting our investigations in a civilized fashion. We are using no physical violence and all we ask you to do is to tell the truth. You are to confess what you really have done against the Soviet power. We certainly aren't prepared to accept conditional confessions which you reject in advance."

Eventually, after seven days of round-the-clock interrogation on the conveyor, Weissberg's spirit was broken and he realized there was "nothing left for me but capitulation and 'confession.'"[2] Weissberg states in his memoirs that while he confessed to his involvement in an imaginary anti-Soviet conspiracy, when his interrogators demanded the names of his co-conspirators, he refused. "On this point," Weissberg wrote, "I was determined not to budge, whatever happened: I would not denounce them."[3]

In the 1990s, however, Ukrainian scholars researching the history of UFTI in the 1930s found a copy of Weissberg's confession in Weisselberg's NKVD file.[4] J.L. Reinders, who translated part of the confession into English, believes that the document proves that Weissberg "left out a lot that might be damaging for the heroic picture he paints of himself."[5] Weissberg could not resist the pressure to denounce his colleagues and friends, and, understandably ten years later, he was unable to publicly admit that he, like everyone else, succumbed to the relentless torture of the NKVD.

Weissberg's confession need not be explored in detail, but in summary, the NKVD forced him to state that there were two groups who conspired to sabotage military scientific research at UFTI. The first group consisted of foreign right-wing extremists led by Weissberg and including the Houtermans and the Ruhemanns. The second was made up of Trotskyite sympathizers led by Landau; this group included many of the top Soviet scientists at the institute. According to the confession, every incident at UFTI, from the dismissal of Davidovich to the civilian scientists' opposition to secret military projects being foisted upon the institute, had been designed by the conspirators to sabotage, delay, or spy on vital Soviet defence research.[6]

Kurt Zinnemann suffered the conveyer in much the same way as Weissberg. Fritz Houtermans, whose imprisonment began nine months later, suffered a slightly different but equally effective variation on the torture. Beginning on 11 January 1938 at the Kharkiv NKVD headquarters, he was interrogated for eleven straight days by NKVD agents. As with Weissberg, three agents interrogated him round-the-clock. On the first day he was given a five-hour break, on the second about two hours; thereafter, interrogation was continuous. For the first two days Houtermans was allowed to sit on a chair, for the next two he was only allowed to sit on the edge, then he was forced to stand; when he fainted, cold water was thrown over him. His feet became so swollen that his shoes had to be cut off. Houtermans was occasionally slapped and punched, but no weapons were used. After the NKVD threatened to arrest his family, Houtermans confessed to being a German spy, though he of course had no way of knowing that Charlotte and the children had by then escaped the Soviet Union. Once Houtermans broke down and agreed to sign a brief confession, he was fed a luxurious meal, with tea, and allowed by his interrogators to return to his cell. There he slept for thirty-six hours. Once refreshed, Houtermans was returned to the interrogation room and ordered to write a full confession. He explained:

> I had to write about espionage, sabotage and counterrevolutionary agitation, and I was absolutely free to invent anything I liked, no corroboration by facts or by evidence being needed. I made nuclear physics the theme of my espionage, though at that time no technical applications of nuclear physics were known, since fission was not yet discovered, but I wrote a lot of phrases that nuclear energy is existent and that it needed only the right way to start a chain reaction as described in popular novels on this matter.[7]

The conveyer was an effective way to break the will of prisoners, but it required a great deal of time and resources to be effective. Less subtle and more overtly brutal means soon became the norm. Weissberg recollected:

> In the years 1936 and 1937 the G.P.U. [NKVD] rarely used direct physical torture. As I learned later, physical violence toward an accused had first to be sanctioned from above. When the mass arrests began at the end of August, 1937, physical maltreatment became general. Up to that period the "conveyer" was the utmost physical pressure they were allowed to apply. It was enough. It was as painful as any torture. But later on it had one grave

disadvantage – it took up a lot of time. Until I experienced it myself I did not believe that a strong-willed man could be made to capitulate by mere interrogation alone.

Weissberg stated that the switch to using violence to extract confessions began on either 17 or 18 August, a date attested to by a number of people he met later in prison, including former NKVD officers. It is impossible to confirm the date of this distinct shift in interrogation techniques, but it coincides with the dramatic increase in the pace of arrests late in the summer of 1937. Weissberg recounts that after he recanted his original confession made under the threat of the conveyor, he suffered an unprovoked, violent assault by a plain-clothes NKVD agent he had never seen before. When the officer asked Weissberg if he would confess, Weissberg barely began to profess his innocence when the agent punched him "full in the face." The agent then knocked Weissberg to the ground and kicked him in the head several times, before forcing him to stand again, whereupon he pummelled his victim with a series of further punches. All Weissberg could do was try to shield himself from the blows as best he could, and to scream in a horrible way in desperate effort to persuade his attacker to stop his assault.

Konrad Weisselberg was subjected to physical torture even before this August shift in policy. His family relates that after Konrad's imprisonment, his wife Galia was initially allowed to visit the prison to bring food and to collect his clothing to wash. His underwear was frequently "completely soaked" with blood. One day, ten months after her husband's arrest, when Galia arrived at the prison in Kharkiv, she was told that neither the food nor the clean clothing could be accepted, as they were no longer necessary.

Living Conditions in Soviet Prisons

When not subjected to interrogation, the scientists were forced to endure prison conditions that varied only in the intensity of the horrors and deprivations they suffered. Conditions might change over time, between different prisons, depending on whether the inmate was undergoing active interrogation. As far as can be determined, none of the scientific and academic prisoners were sent to the work camps in Siberia – the Gulag – where conditions were generally much worse than in the prisons. In the Gulag, many tens of thousands of people died of exposure, overwork, or starvation; in the prisons, every effort was made to keep inmates alive, at least until they were sentenced.

Prison officials followed "a practice ironically referred to by prisoners as Stalin's 'care for the living person.'" The NKVD prison officers took every precaution to prevent suicides, including stringing nets in stairwells to prevent prisoners from "throwing themselves over the banisters."[8] Inmates received relatively good medical and dental care. Prison staff did everything possible to keep the prisoners in their charge alive. The guards rarely subjected their prisoners to physical assault. The prison staff and the interrogators worked for separate branches of the NKVD. According to Houtermans and Shteppa, "this explained the paradox that in the same cell you would find prisoners suffering severely from the effects of interrogation, about whom nobody bothered, while every conceivable medicine for the prevention and cure of coughs and colds, and headaches was regularly distributed."[9] While physical abuse was rare, jail staff engaged in humiliating their prisoners; Weissberg and Ena Zinnemann both recall being forced to strip and parade naked past the guards before proceeding to the showers. This was a particularly upsetting experience for the women. Weissberg notes that their clothing was steam cleaned while they washed in order to prevent epidemics.[10]

Guards were particularly worried that prisoners would engage in hunger strikes, which they deemed to be an anti-Soviet activity. Houtermans went on an eight-day hunger strike after he was denied the pencil and paper he needed to work on some scientific ideas he was developing while in solitary confinement. When his health began to fail, the prison officials capitulated to Houtermans's demands. He had his paper, but by that time "I was very much weakened, since I had been in a bad state when I started."[11] Weissberg became involved in several such strikes, all of which led to some sort of negotiations with prison officials to improve conditions. Sometimes prisoners voluntarily resumed eating after some concessions were made by the NKVD; other times, the strikes collapsed, and Weissberg witnessed prisoners being removed to the hospital to be force-fed.

One of the principal factors that provoked group hunger strikes was overcrowding in the cells. Weissberg and Houtermans experienced every type of conditions in their various cells. Solitary confinement was common, especially during the initial stages of interrogation at a local NKVD headquarters. Eva Striker was kept segregated from other prisoners for twelve of her sixteen months of imprisonment. For Weissberg and Houtermans, the main period of seclusion ended when they were transferred to prison. Both men would spend the majority of their time in Kharkiv's notorious Kholodnaya Gora Prison, although they were also transferred back and forth from Kiev.

Weissberg was initially placed in a small 4 by 2 metre cell, designed in Czarist times to hold one inmate, but now, ominously, crowded with three small cots. Weissberg was soon joined by two cellmates. The men were confined to the cell for twenty-three hours a day, with a brief period of mandatory outside exercise in the jail yard. Weissberg was regularly removed from his cell and transferred to Kharkiv NKVD headquarters for further interrogation. Initially, he was confined by himself to a small holding cell, where he would wait hours or sometimes several days for the interrogation to start. As more and more people were arrested, however, there was no place to isolate him, and the holding cells were often packed like "vertical sardine tins," so tight that the warders had difficulty closing the door after adding another prisoner.[12]

By early 1938, the same cell designed for one, and which held three when Weissberg was first transferred to Kholodnaya Gora, now was crammed with as many as fourteen prisoners. Many of the newcomers were peasants, who were bewildered by their situation. There was no more room for beds, and many were forced to sleep without a mattress on the concrete floor. In the summer of 1938, Weissberg was put into a large cell holding 130 other prisoners. The overcrowding was so bad that unless a number of men were removed for interrogation during the night, there was not enough floor space for everyone to lie down. Many of the prisoners were forced to lie on their sides on the hard floor, their bodies pressed so tightly together that they could only move their position in unison with their cellmates. By the end of the year, Weissberg had been moved to another large cell, this one containing around 260 prisoners in even more appalling conditions. Ena Zinnemann confirms Weissberg's account of the appalling conditions in the Kharkiv prison. She was held with 110 women in a cell designed for 25.

The prisoners in the large cells were a microcosm of Soviet society, revealing the full scope of the Great Purge. In one of the cells he occupied, Weissberg counted men from twenty-two different ethnic groups; minorities in Russian towns had been particularly targeted by the NKVD. Peasants were the largest single group in the prisons, but they were generally dealt with quickly and sent off to their fate in the Gulag. There were many political prisoners, among them former national and local officials of the Communist Party. By the end of 1938 this group included a large number of ex-NKVD agents, including some of Weissberg's original interrogators and accusers. Criminals were intermingled with the victims of Stalin's madness and became an important part of the social structure that developed in the larger cells. Weissberg soon discovered that establishing good relations with those criminals would be key to his survival.

The main job of the prison staff was to keep their charges alive for interrogation and to prevent communication between them. Human ingenuity prevailed: when kept in solitary, prisoners developed a tapping code to allow some discussion between them. All inmates were aware that the person in the next cell, or even a cellmate, could be a spy. Conversation was often guarded. Houtermans, fearing *agents provocateurs*, refused to use tapping, which prevented Weissberg from communicating with him even when they were briefly in the same cell block. Whenever possible, Weissberg communicated in German, believing that an NKVD plant would be less likely to understand the language.

When Ena Zinnemann was removed from her cell and placed in a wooden cubicle to await her turn for interrogation, she tried to get word about her husband by tapping a message to the prisoner in the next cubicle. When she tapped out hello and asked who was there, Ena noted that the answer came slowly. This was a good sign that she was not communicating with a stool pigeon, who would have been "too quick" to respond. They exchanged information on the people in their cells, but her neighbour had no word on Kurt. Using a contraband pencil stub she had hidden in her hair, Ena scratched into the door a message: "Kurt, I love you, Ena." A few days late Ena was thrown into the same cubicle. To her amazement she found a reply: "Ena, I love you, Kurt." It was the only word the couple had of each other while they were imprisoned.

As the prison population expanded exponentially, it became impossible to fully isolate prisoners. Every new transfer into a cell was quizzed about whom they had met while in prison. Debates about why they had been arrested were frequent. In larger cells with well-educated inmates, some prisoners organized lectures to educate and entertain one another. Others told stories, often summarizing a favourite novel or play for their often illiterate fellow prisoners. Another favourite activity was playing chess using pieces made by kneading bits of bread. Both Houtermans and Weissberg spent a great deal of time working out complex mathematical problems to pass the time.

Food provided to the prisoners was barely adequate for survival. Inmates received 500 to 600 grams of coarse black bread, 20 grams of sugar, and, twice daily, a thin cabbage soup of little nutritional value. After a few months, inmates who relied on the prison food alone were perpetually hungry, eventually "reduced almost to a skeleton, but once having reached a certain absolute minimum weight they remained at it and lived on." These ghost-like men could survive, but only if they exerted themselves as little as possible.[13]

With little to keep themselves occupied, the women in Ena Zinnemann's cell exchanged recipes for the dishes they dreamed of preparing if they were ever released. Ena used her pencil stub and cigarette papers to record a recipe from each woman. Her daughter Pam recounted the scene:

> In the cramped space
> on the stone floor
> my mother and her
> cell-mates live on
> black bread and cabbage;
> they call up
>
> the normal kitchen chat,
> the swapping of recipes
> out of their cavernous hunger,
> making feasts.

When Ena was released from prison, she somehow managed to keep the recipes. Her daughter only found them after Ena's death, tucked away in the attic of their house in Leeds, hidden in an old suitcase, inside a small envelope. On the front of the envelope Ena had written, "Russische Kochrezepte, nicht wegwerfen" (Russian recipes, don't throw away). These flimsy pieces of paper are the only remaining physical evidence of the horrors suffered by the Ensnared in Stalin's prisons. The recipes reveal that, like the men in Weissberg's cell, the women in Ena's came from all over the Soviet Union, representing different nationalities, races, and religions. There were traditional recipes from Latvia, Georgia, Ukraine, and Russia, as well as Jewish recipes. The recipes, Pamela Zinneman-Hope remarks, consisted of "my only map of who was with her."[14]

In 1937, families and friends were allowed to bring prisoners packages with clean clothes, food, and tobacco. Konrad Weisselberg's wife Galia provided packages to her husband and to Weissberg. Likely before the end of the year, prison authorities stopped allowing packages to be delivered. Instead, they established a canteen system where prisoners could purchase goods one or twice a month. Weissberg was one of the fortunate prisoners: shortly after his first arrest, he had been able to arrange for his friends to deposit money from his bank into a prison account. Houtermans was less fortunate, but while he lived in the same cell as Shteppa, he was able to share in his purchases. Houtermans promised he would return the favour, which he in fact did during the brutal German occupation of the Soviet Union during the war. Those with money in an account were permitted to spend fifty rubles at a time to purchase additional

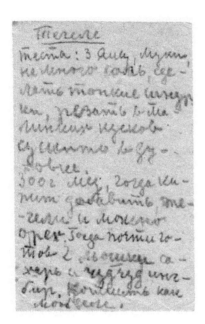

Image 12. One of the prison recipes written in Russian on cigarette paper carefully preserved by Ena Zinneman (reproduced with permission of Pamela Zinneman-Hope).

food, tobacco, and a few "luxuries" from the canteen. In the cells, prisoners generally agreed to share some of their canteen purchases with those who lacked an outside source of funds. Weissberg was able to make canteen purchases for the first year or so of his captivity. Once his funds ran out, however, his physical condition began to deteriorate. Weissberg repeatedly requested that the jail's authorities arrange for some of his money in a Kharkiv bank to be transferred into his prison account. By the late autumn of 1939, Weissberg's health had so deteriorated that in a desperate effort to force his jailers to accede to his demands, he launched his final hunger strike.[15] As he would soon discover, this was a pointless act.

Executions

Weissberg, Houtermans, and most of the other arrested scholars remained in remand facilities, left in limbo by the NKVD and by what passed as a justice system in the Soviet Union. It was perhaps their foreign nationality that saved them. It is no coincidence that two of the three foreign scholars sentenced to death during the Great Terror had renounced their nationality and become Soviet citizens. The one exception, Herbert Murawkin, had Russian citizenship through his parents;

even though he was born in Berlin, the Soviet authorities considered him a citizen.

Herbert Murawkin, who was researching death rays, was the first foreign scientist to be arrested and the first to be executed. As with so much of his life before prison, very little is known about his time in captivity. There can be little doubt that he was tortured, but the methods used are, as in other NKVD files, not recorded. All that can be ascertained from his file was that he eventually confessed to the imaginary crimes of becoming a "member of the Trotskyist terrorist and espionage-sabotage organization, on whose instructions he carried out sabotage in scientific institutes." Moreover, he admitted that he was a German intelligence agent, and he implicated many other supporters of Esperanto.[16] Murawkin's fate was sealed because he had a Soviet passport, was a member of the KPD, and was almost completely unknown in the West. No one outside the Soviet Union came to his defence. Only very recently have historians, writing about the Soviet purge of Esperantists, revealed his melancholy fate. After a summary hearing, without a defence lawyer being present, Murawkin pleaded guilty of espionage, spreading anti-Soviet propaganda, and agitation. He was sentenced to be shot. He was executed on 11 December 1937, just five days before Stalin's next victim was executed.[17]

Konrad Weisselberg's love of Galia and of Communism sealed his fate. On 16 December 1996, his son Alex was finally permitted to read and make a copy of his father's NKVD file, exactly fifty-six years after he was murdered. Konrad was interrogated four times in March 1937, the month of his arrest, and three more times in the subsequent three months. There is nothing in the file to show how his interrogators treated him, nor is there evidence of torture. Only the family story of Konrad's bloody underwear informs us that something much more sinister happened to him than is revealed in the official record. The file is markedly different from those of most other foreign scholars in the NKVD archives; nowhere does Konrad accuse anyone else of criminal activities, nor does he confess to any crimes against the Soviet state. The worst offences to which he admitted were that he, Weissberg, and some of the other foreign scientists at UFTI had not reported visiting scientist George Placzek's critical remarks about the Soviet political and judicial system. The transcripts of Konrad's interrogation reveal that far from being an enemy of the Soviet state, he was an idealist who truly believed in Communism and in the great Soviet social experiment, but that he was also remarkably naive about the world. He had renounced his Austrian citizenship and his father's wealth to live in Kharkiv with Galia. His answers to the interrogators' questions about his past political and personal life were truthful, and devoid of any guile. Running through his responses was a strong sense of his complete bewilderment at the mere thought that he could be considered a danger to Stalin or to the Soviet state.

Except for the transcripts of his interrogations, the official summaries of the charges, and the approval of his death sentence, the only other materials in Weisselberg's file are extracts of confessions made by Weissberg and two other prisoners, Valentin Fomin and Rudolf Anders. Fomin was Houtermans's former research assistant, who had tried and failed to kill himself prior to his arrest. Erwin Anders, also known as Rudolph Anders and Ervin Kohn, was a German economist and Communist Party member who had moved to the Soviet Union around 1929. Prior to his arrest he had been in charge of rebuilding two large Soviet iron and steel manufacturing plants. Fomin and Anders had both been coerced into making confessions, which were used primarily against Weissberg. Fomin accused Weissberg of leading a Gestapo organization at UFTI and claimed that Weisselberg was also a member of an espionage ring that spied on the activities of the Coal Chemistry Institute. Anders, by contrast, made no specific claims against Weisselberg. In fact, Anders's only mention of Weisselberg was that Weissberg, during a meeting with Anders in Moscow in December 1936, had used his friend's dismissal from the Coal Chemistry Institute as an example of the Soviet mistreatment of foreigners.

According to NKVD investigators, Weisselberg had "only admitted to taking part in anti-Soviet discussions with the physicist and Trotskyist Placzek." The "testimony of Anders, and partly that of Weissberg, exposed him as a member of a counter-revolutionary Trotskyist group," while Fomin's statement revealed him "to be a German spy."[18] These statements against Weisselberg, like the coerced confessions of Weissberg, Fomin, and Anders, are dubious at best. Not a single piece of evidence linked Weisselberg to foreign intelligence services, and his connection to the groups "conspiring to sabotage" UFTI was almost exclusively social in nature, a direct result of his living in Weissberg's apartment. Despite this complete lack of proof, Weisselberg's case file was sent by the Kharkiv NKVD to Moscow for an administrative ruling on the Austrian scientist's fate. There would be no trial. On 11 August 1937, Nikolai Yezhov, People's Commissar for Internal Affairs (head of the NKVD) and Andrey Vyshinsky, the Attorney General of the USSR, heard the "evidence" against the accused. On 28 October they issued their decision: "WEISSELBERG K.B.: TO BE SHOT." On 16 December, Weisselberg was executed. His family would not learn of his fate until twenty-one years later.

Hans Hellmann, the German theoretical chemist renowned for the Hellmann–Feynman theorem, was the only other foreign scientist to share the fate of Murawkin and Weisselberg. The cases of Hellmann and Weisselberg are eerily similar. Like Weisselberg, Hellmann had renounced his citizenship and become a Soviet national. Without any way of escaping from the Soviet Union, he was forced to continue living in Moscow; almost all the other foreign scholars had either fled the country or been arrested. Many of his Soviet colleagues had also been imprisoned, then sentenced to long prison terms or executed. The Great Terror came later to scientists in Moscow than it did elsewhere in the Soviet Union. On the night of 9–10 March, just over a year after Weissberg and Weisselberg's detention by the NKVD in Kharkiv, "men in dark coats" arrived at the Hellmanns' apartment and arrested him.[19] Hans Hellmann Junior recalls that horrible night: "I was eight and a half years old then, and I can still remember this event well. I was awakened and [they] searched my bed for anti-Soviet writings and evidence of espionage activity."[20]

Hellmann's NKVD file, like Weisselberg's, does not contain any proof of his guilt, nor is there any evidence of the means used to extract his confession. There is only a brief handwritten statement, not written by Hellmann himself, where the German scientist admits to being a German spy. According to the statement, Hellmann had been recruited by Hans Gussmann, his assistant at Hannover and a known "Nazi." Hellmann was alleged to have provided information on the Karpov Institute's production of poison gas and other secret military work to Germany through a Polish intermediary. The confession concluded with Hellmann's signature, signed in "a faint hand." Again, there was no formal trial or hearing where Hellmann might have offered a defence. On 17 May 1938, Andrey Vyshinsky signed the order authorizing Hellmann's execution. He was shot twelve days later.[21]

The one remaining scholar incarcerated by the NKVD and actually sentenced was the mathematician Fritz Noether, who was arrested in Tomsk in November 1937, charged with spying for Germany and sabotaging Soviet armament manufacturing. Until 1988, almost nothing was known about Noether's time in prison or about his fate. Finally, fifty years after his arrest, Noether's sons were informed by the Soviet government: "On October 23, 1938 Professor Noether was found guilty of allegedly spying for Germany and committing acts of sabotage and was sentenced in Novosibirsk to 25 years of imprisonment. He served time in different prisons."

Why Noether was singled out for a lengthy sentence remains unclear. His trial, if there was one, was held while the Great Terror was finally winding down. Nikolai Yezhov was still nominally in charge of the NKVD, but his power was being undercut by his deputy and eventual successor, Lavrenty Beria. Death sentences were less common than just a few months earlier, which raises a question: Why was Noether sentenced at all, rather than left, like the others, to languish in perpetual remand? Only one clue exists to link Noether's case to those of Hellmann and Weisselberg: in May 1938 the German government had stripped Noether of his citizenship.[22] Deprived of the protection of his foreign nationality, Noether was vulnerable to the full power of the Soviet police state. Loss of German citizenship was not, in itself, a guarantee of being condemned to death or to a lengthy prison sentence. Julius Schaxel's citizenship was revoked in March 1937, and while he too was arrested during the terror, his fate, as we shall see, was quite different from Noether's.

International Campaigns and Releases from Soviet Prison

Prior to the Second World War, most of the remaining refugee scientists who had been arrested continued to languish in jails, with no resolution of their cases ever being reached. Even so, it was possible to be released from NKVD custody, as happened to three people associated with our story. Eva Striker, whose arrest in May 1936 had led Weissberg to doubt his own future in the Soviet Union, was released and deported to Poland in September 1937. As with so many events in Stalin's Soviet Union, it is impossible to know exactly why Eva survived. The charges against her were no less dubious than the accusations against other foreigners who had been arrested. Furthermore, she was accused of plotting to murder Stalin, a crime far more serious than the conspiracies imputed to the scientists, such as her husband. Various theories have been put forward to explain her release, including international pressure on the Soviets through the letter-writing campaign that Weissberg and Laura Striker had launched to rescue her. Other accounts focus on the role of the Austrian ambassador, who pressured the Soviets to release Eva. Eva's own account confuses the situation still further; she claims that prior to her arrest, she had a love affair with a high-ranking NKVD official, who may have used his influence to sway opinion in her favour.

Eva's case became even more muddled after her NKVD file was made available to her. Karen Kettering, an art historian who assisted with the publication of Eva's memoirs of her imprisonment, wrote: "Rather than providing a neat explanation for the reasoning behind the decision to free Zeisel, [the documents in the NKVD file] were arcane and opaque, written in the stilted and formulaic language of official Soviet discourse of the mid-1930s."[23] Kettering suggests that, at least in part, Striker's case became part of a bureaucratic tug-of-war between the NKVD and the attorney general or prosecutors, a conflict over who had the jurisdiction to determine her fate. Kettering goes so far as to speculate that the final determination may have been made by Stalin himself, although she admits there is no evidence for this. Stalin, if he did become involved in a specific case, usually rendered his decisions verbally. The multitude of explanations for Striker's release reflect the chaos that overwhelmed Soviet society as the purges morphed into the Great Terror. All that can be said for certain is that Striker was deported from the Soviet Union. Among the many people accused in this particular imaginary plot to assassinate Stalin, only Eva and a German colleague were not executed.

Even less is known about the April 1938 release of Kurt and Irene Zinnemann. Their NKVD file shows that they were deported along with a group of eight other Germans and one Czech national. Once released, the couple was subjected to one final indignity; all of Kurt's scientific papers and research notes were destroyed. Like Striker, the couple was deported across the Polish frontier.[24]

Foreign citizenship undoubtedly kept most of the other refugee scientists alive, but they may also have been shielded by a letter-writing campaign by Western European scientists, who tried to assist their colleagues and friends, particularly Weissberg and Fritz Houtermans. In December 1937, as Fritz's prison ordeal was just beginning, Charlotte Houtermans, trapped in Riga, turned to the couple's wide circle of left-wing scientific friends for help. Harald and Niels Bohr arranged for temporary travel documents to be provided by the Danish government. Charlotte quickly left Copenhagen for London, where the SPSL had made provisions to support her and the children. The SPSL had also arranged with the Foreign Office to provide Fritz with papers to facilitate his travel to Britain, if and when he was released. Fritz was to receive a one-year stipend once he reached Britain.

Even before Charlotte's arrival in Copenhagen in late 1937, the Bohrs commenced an international campaign requesting socialist and Communist scientists and scholars to lobby Soviet authorities for Fritz Houtermans's release. Efforts to free him spread quickly throughout the West. Charlotte informed Elsa Houtermans of her son's arrest while she was still in Riga. Elsa had escaped from Vienna by accepting a low-paying teaching position at a private school in New England. Elsa approached James Franck, Fritz's doctoral supervisor, who was then teaching at Johns Hopkins University, asking him to appeal to Albert Einstein to write a letter to Soviet officials in support of Fritz. On 5 January 1938, Einstein, always willing to put his name behind a good cause, wrote a letter to the Soviet ambassador to the United States, attesting to Fritz's excellent character and outstanding scientific abilities. Charlotte arrived in London early in the New Year and tirelessly lobbied on her husband's behalf. In February 1938, Charlotte convinced Patrick Blackett, the renowned British physicist and Fabian socialist, to ask Ivan Maisky, the Soviet ambassador to Britain, for information on the imprisoned scientist and to ascertain when he could be released.[25]

By the spring of 1938, Charlotte's efforts were being paralleled by those of Eva Striker. As we have seen, Weissberg's and Laura Striker's efforts to solicit support for Eva from Soviet and European scholars may have been a factor in her eventual release. Upon her return to Western Europe, Eva decided to do all she could to aid Weissberg and the other scientists still detained in the Soviet Union. She worked to free Alex, even though she had divorced him soon after returning to Vienna. After her release from prison, her mother Laura had visited Kholodnaya Gora Prison, where she persuaded prison officials to pass to Weissberg a document giving his brother, a lawyer in Vienna, permission to act on his behalf for the legal dissolution of his marriage. Weissberg agreed, hoping this would free Eva and give her a chance to have children. Eva began a relationship with Hans Zeisel, a sociologist who had been a regularly attendee at her Berlin parties before she left for the Soviet Union in 1932.

Eva asked her uncle, Michael Polanyi, to help with the international campaign to free the scientists who had been arrested in the Soviet Union. Polanyi wrote to physicists throughout the world, including Albert Einstein, in Princeton, asking them "to vouch for Weissberg and Houtermans' innocence and to ask for their release."[26] Polanyi's efforts led Einstein on 18 May 1938 to write a second letter to Stalin regarding the arrests of foreign scientists, Weissberg in particular. Einstein carefully crafted his letter so as to avoid criticizing either the Soviet dictator or his country. Instead Einstein suggested that unfathomable mistakes had been made: "I understand how easily suspicion may fall, in times of crisis and excitement, on innocent and valuable men." Einstein suggested that Weissberg was innocent and that he had made invaluable contributions to the Soviet Union. He requested that Stalin take a personal interest in Weissberg's case.

While in Vienna, Eva also reached out to Robert Millikan, the Nobel Prize–winning physicist at the California Institute of Technology, asking him to request US government intervention on behalf of Weissberg. Not much is known about who made the approach to Millikan, but it may have been Felix Bloch, an Austrian physicist and family friend of the Polanyis, who had moved to Stanford University in 1934. Millikan was perhaps the single most politically influential physicist in the United States, and his request to the US government carried considerable weight. Sometime in the spring of 1938, the US embassy in Moscow requested information on Weissberg from the Soviet government. When the Soviets replied that Weissberg was a member of the Communist Party, the Americans "immediately withdrew from the matter," and Weissberg's fate was "immediately rubber-stamped as a matter internal to the USSR." Neither Bloch nor Millikan had been informed by Eva that Weissberg was a party member.[27]

Just before the German annexation of Austria in March 1938, with the assistance of their old family friend and Eva's former lover, Arthur Koestler, Eva and Laura were whisked out of Vienna to London. There, Eva had an opportunity to talk to Koestler about the horrors that had taken over Soviet society and engulfed her and his good friend Weissberg. Eva's information profoundly transformed Koestler's political views (something that will be discussed later in this book); its immediate effect was to galvanize him to do whatever he could to save his friend and the other imprisoned scientists. In June, he travelled to Paris and met with Georg Placzek regarding the fate of their friends in Kharkiv. Together, they composed a letter to Vishinsky, and Placzek arranged to have this letter signed by three Nobel laureate physicists who were prominent in Parisian left-wing intellectual circles: Irène and Frédéric Joliot-Curie and Jean Perrin.

Nothing resulted from any of these efforts. By the late spring of 1938 a growing threat to the safety of Landau, Kapitsa, and other Soviet scientists compelled a number of Houtermans's most active supporters to rethink any escalation of international pressure for the German physicist's release.[28] In April, Landau, along with other Soviet physicists, was arrested in Moscow. He spent a year in prison before Kapitsa made a successful personal appeal to Stalin for his release. Landau was scarred for life by the experience. He never again risked becoming involved in anything political. He was one of the lucky ones; many others were far less fortunate.[29]

In October, Eva, now Zeisel, and her new husband departed for New York. While she would do what she could to assist Weissberg from America, she now focused on starting her new life in the United States. Charlotte remained in London, where she continued to lobby on her husband's behalf, refusing to allow Fritz to be forgotten like the other foreign scholars imprisoned in the Soviet Union. In February 1939, in desperation, she wrote directly to Lavrenty Beria, the new chief of the NKVD. She pleaded: "I am confident that his case will be justly dealt with and I shall be very thankful for every effort made to ensure his release for which I my children and his aged mother are longing from day to day."[30] Beria did not reply. With no hope left, Charlotte received one further grant from the SPSL to help pay for her and the children's passage to New York. Charlotte and the children's emigration to the United States was sponsored by her old romantic flame and classmate Robert Oppenheimer. In Britain, only Blackett continued the effort to free Fritz, sending periodic letters to Maisky requesting his release.

In April, one further effort was made by the SPSL to persuade the Soviet government to free Fritz. At the suggestion of the British Foreign Office, the SPSL decided to appeal to leading left-wing scholars, labour leaders, and politicians to once again lobby Soviet officials on Houtermans's behalf.[31] The SPSL approached the Society for Intellectual Liberty, a left-wing group of "intellectual workers concerned with the active defence of peace, liberty and culture," to take the lead in one last effort to free Fritz.[32] The society's chairwoman was Margaret Gardiner, a well-known patron of socialist, Marxist, and artistic causes. Gardiner was also the lover of J.D. Bernal, professor of physics at Birkbeck College, University of London, and a pioneer in the fields of molecular biology and the sociology and history of science. Bernal was a committed Communist and an unwavering supporter of the Soviet Union and of Stalin.

It took Gardiner until early June to persuade five leading left-wing intellectuals to sign a letter to Maisky in support of Houtermans, since it was a "rather tricky matter" to get such people to agree to support "such a question." The signatories were Bernal, the economists Beatrice and Sydney Webb, the poet and writer Cecil Day Lewis, and the politician, Marxist-Leninist theorist, and anti-fascist organizer John Strachey.[33] Bernal hand-carried the letter to a meeting with Maisky, who informed him that he had made inquiries as a result of previous letters and that "he could state that Dr. Houtermans was being held on very serious charges which could not, in the present international situation, be made public. He hoped to be able to make a statement about it at a later date." Bernal did not press the Soviet ambassador on the matter. Houtermans was still in prison when Germany and the Soviet Union invaded Poland less than three months later.[34] Houtermans was, of course, not alone; Alexander Weissberg, Siegfried Gilde, Fritz Noether, and, possibly, Julius Schaxel also remained in detention in September 1939.

Julius Schaxel was eventually released. Little is known about why he survived or when he was released. When war with Germany began in June 1941, Schaxel and several other prominent anti-Nazi German survivors of the Great Terror founded the National Committee to Free Germany. The committee disseminated propaganda that endeavoured to persuade Germans to form an Anti-Nazi Popular Front. On 15 March 1943 he died under mysterious circumstances in a sanatorium near Moscow; he was sixty-six years old.[35]

Conclusion

With the notable exception of Fritz Lange, the story of academic refugees in the Soviet Union came to an end. What had started with such
high hopes had become an unimaginable horror perpetrated by Stalin and the Soviet state. Still caught in Stalin's web were the families of
Weisselberg, Anna and their young son Alex, and of Hellmann, Viktoria
and Hans Junior; the wife and daughter of Walter Zehden; and Friedel
Cohn-Vossen, wife of Stephen. Additionally, the sons of Fritz Noether
were orphaned, but they were allowed to escape to Sweden in 1938. The
suffering of the families of the victims of the Great Terror was only just
beginning. The tales of their perseverance and ultimate triumph will be
a fitting end to this study. While most of the Ensnared managed to escape
the Soviet Union, before we trace their continuing search for another
place of refuge, let us first consider the fate of those left behind in the
Soviet prison system. Their ordeal was only just beginning.

9 From Stalin's Frying Pan into Hitler's Fire

Siegfried Gilde, Alexander Weissberg, and Fritz Houtermans had the horrific experience of being transferred from their NKVD prisons directly into the hands of the Gestapo during the Soviet–German prisoner exchanges of 1939 and 1940. Nominally allies after signing the Non-Aggression Pact in late August 1939, the two former ideological foes implemented a system of prisoner exchanges as part of the secret protocols of the agreement. Far more German prisoners of the NKVD were handed over to the Gestapo than Soviet prisoners were returned to the Soviet Union.

The Transfer from the NKVD to the Gestapo

The eyewitness accounts of Weissberg, Houtermans, and others attest to the events preceding the exchange. Weissberg and Houtermans were both transferred to Moscow at the end of September 1939. At first, neither man knew why they were in the Soviet capital. Both were interrogated, but the questioning was quite different in tone than previously. The NKVD agents seemed willing to accept that they were innocent and hinted that they would soon be released. Both men pleaded not to be sent back to Germany.

Despite the nature of the interrogations, conditions did not immediately improve. In desperation, Weissberg, already in terrible physical condition from the lack of food, began his final hunger strike, demanding access to his money for use in the prison canteen. An NKVD officer pleaded with Weissberg to be patient and to abandon his strike, since efforts were being made to finally resolve his case. Weissberg was warned by a senior NKVD interrogator that the secret police could not be seen to respond to his threats. Weissberg ended his strike only after the interrogator personally promised him that his demands would be addressed. When

Weissberg was returned to his cell, to his astonishment, the manager of the canteen arrived and asked him to write down a list of what he wanted. When Weissberg explained he had no money, the manager brushed aside his concerns, telling him everything would be provided. Weissberg's list included fantastic items he had not seen since his arrest, including his favourite chocolate bonbons. To his astonishment, within thirty minutes the manager returned with everything on Weissberg's list, including the bonbons! The miracles continued; that evening, the warder brought not thin soup, but a meal of roast meat and potatoes.

Houtermans's experience was somewhat better. He was locked in a good cell by himself and given "special food in quite sufficient quantity and a package of cigarettes every day." Paper and pencils were provided to Houtermans so that he could continue his mathematical work. In December, Houtermans and Weissberg were given new clothes and transferred to large but not overcrowded cells in Butyrka Prison. The other prisoners in their cells were fellow Germans or Austrians. The prisoners were fed large portions of excellent food, including unheard of luxuries, such as cocoa and eggs for breakfast. It was in these cells that Houtermans had a brief encounter with Fritz Noether.

Before their departure from Moscow, the prisoners were interviewed one final time by a troika of NKVD officers, whom the Germans had dubbed the "Three Magi." The Magi asked some of the prisoners if they would be willing, once they were released, to spy for the Soviet Union on the Germans. While the request appeared voluntary, rumours soon circulated in the cells that refusing the request could have dire consequences. Carola Neher, a famous German actress, in a cell for women one floor down from Weissberg's, "refused the request with great indignation." As a result, Weissberg heard through the prison grapevine, she was left behind when the other female prisoners were handed over to the Germans. Weissberg's account of Neher's exclusion from the exchange, while accurately reporting the rumours circulating among the German prisoners, was not true. Historians have subsequently discovered that Neher was rejected, not by the Soviets, but by the Germans, who refused to accept the actress because of her very public opposition to the Nazi state. Weissberg never knew Neher's fate, but modern researchers have uncovered that she died of typhus in 1942 in a Soviet prison camp.[1]

Neher's fate may hold the best clues about what happened to Fritz Noether after his espionage conviction. He was to have joined the other three scientists in the prisoner exchange. Houtermans had met Noether in Moscow's Butyrka Prison in December 1939, in the cell where he and other prisoners were being fattened up before their transfer to the Germans. Shortly after Houtermans's arrival, without any explanation,

Noether was removed from the cell. It is impossible to know with certainty why Noether was excluded from the exchange. Carola Tischler has shown that the lists of prisoners to be exchanged were rather arbitrary; often specific individuals requested by the Germans had died, or simply disappeared. Local NKVD offices had not been asked to provide particular people to Moscow to be processed for return to Germany; rather, a specific quantity of German nationals had been requested.

There are three possible reasons why Noether had been excluded from the prisoner exchange. First, since he was no longer a German citizen, Soviet authorities deemed him ineligible. A second possible reason may have stemmed from his refusal to spy on Germany for the NKVD. The only evidence to support this suggestion is contained in Weissberg's and Houtermans's accounts, but as we have seen, their observations were based more on prison gossip than on actual Soviet policy. The third and more likely scenario is that the German government refused to accept him; after all, they had stripped him of his citizenship. The Germans tried as much as possible to exclude Jews from the prisoner exchange. If this last theory is true, however, it does not explain the inclusion of Gilde or Weissberg, both Jews and the latter a well-known Communist. Moreover, Gilde was also on the list of those whose German citizenship had been revoked.[2] Whatever the reason for his exclusion, Noether did not benefit from avoiding the Gestapo. In 1941 he was transferred to another NKVD jail in Orel, southwest of the Soviet capital. In late summer, with the German army approaching the city, the Red Army decided to prevent the Wehrmacht from liberating any Nazi "spies." On 8 September 1941, the Military Collegium of the Soviet Supreme Court ordered Noether's execution because he was "engaging in anti-Soviet agitation." Two days later he was shot; "his burial place remains unknown."[3] He was one of 157 German prisoners shot by a special NKVD squad.

Weissberg was never asked to spy, possibly, he speculated, because the NKVD knew what his answer would be and simply preferred to be rid of him. They only required him to sign a document attesting that he was aware why he was being expelled from the Soviet Union. A few months after Weissberg's departure, Houtermans signed a similar document in which he promised to engage in espionage and not to reveal anything about his treatment by the NKVD. Houtermans had learned from other prisoners of the potential consequences of not signing, so he acceded to the NKVD's final demands, even though he had no intention of helping the Soviet Union ever again.[4]

On the night of 12–13 December 1939, Gilde was one of thirteen men and women dispatched from Moscow to Brest-Litovsk, where a bridge over the Bug River demarked the German–Soviet frontier after their division of Poland. On 31 December, Weissberg followed, one of a group of twelve. Gilde and Weissberg were both immediately arrested by the Gestapo and imprisoned in Kraków and Lublin. The prisoner exchanges continued once or twice a week until mid-January 1940, when the exchanges were temporarily suspended, supposedly for technical reasons, until May. On 2 May the exchanges resumed, and Houtermans was included in the first group. Unlike what had been done with the earlier groups, soon after arriving in Poland all thirty-seven prisoners were transferred to Germany for processing.[5]

Gilde and Weissberg in Poland

The initial experiences of Weissberg and Gilde in German-occupied Poland were very similar. After a period of imprisonment by the Gestapo, both were released but were denied permission to return to Germany because they were Jews. They were required to wear armbands bearing the Star of David, marking them as Jews. It is unknown how long Gilde remained in the hands of the Gestapo, but Weissberg was imprisoned for four months in Warsaw and Lublin before his release to live with relatives in Kraków. Remarkably, both men were given the opportunity to write to family and friends in the West, to request assistance in securing visas that might allow them to emigrate from Poland.

In late January 1940, Ruth Gilde, living in Paris for the last two years, received a telegram informing her that her husband was now residing in Warsaw. This was the first word Ruth had received from Siegfried since that fateful day in November 1937 when he had left the train to clear customs on the Lithuanian frontier. Siegfried also wrote to his sister Mascha Gilde-Berent, who had escaped Germany but was, in early 1940, trapped in Copenhagen and prevented by the outbreak of war from reaching safety in Australia. Siegfried begged his sister to help him reunite with his family. Time was of the essence, he explained – he was in great danger in Warsaw. Mascha did all she could in Denmark to secure a visa for her brother; when she failed, she turned in desperation to Professor Karl Weissenberg, Siegfried's friend and mentor. Mascha asked Weissenberg, "Is there no possibility to get a permit for him to another country on account of his scientific abilities?"[6] Weissenberg once again pleaded with the SPSL to assist Siegfried, knowing that it was likely a forlorn hope. Esther Simpson understood the gravity of the situation immediately, but she was at a loss as to how she could aid Gilde. She turned to the Society of Friends (Quakers): "[Do] you know of any group that would be able to assist in an unusual case of this kind – a German refugee imprisoned in Soviet Union, and now in Warsaw, *out of the frying pan into the fire?*"[7]

On 24 February Bertha Bracey, the secretary of the Quakers, explained to Simpson that they could do nothing to aid Gilde: "Our Friends in the neutral countries can only get neutral permits if a visa for another country is guaranteed; while we cannot approach the British authorities, even in thoroughly good cases, so long as they are in German controlled territory. It is an appalling vicious circle, which, so far, we have failed to break."[8] This Catch-22 doomed Gilde. He never was able to contact his family again, and as far as they could learn, he died in 1940 in Warsaw. In fact, as we shall see, Gilde survived for three more years.

By late May, Weissberg had managed to get word to Eva Zeisel in New York, telling her he was trapped in Poland and in desperate need of a visa to escape the Gestapo. Zeisel sprang into action in an attempt to mobilize support for her former husband, to secure permission for him to come to the United States. Zeisel and her husband Hans concocted a scheme to have Weissberg offered a one-year teaching fellowship at St Lawrence University in Canton, New York. The university's president was happy to extend an offer to Weissberg, but it was conditional on Weissberg's friends posting a $600 bond, which would be returned when and if Weissberg arrived there. The purpose of this bond is not explained in the documents, but the federal government probably required it before it would approve the visa.

Raising the money proved an insurmountable task for Hans and Eva Zeisel. Having arrived only recently as penniless refugees themselves, the couple was deeply in debt after having "squeezed money" out of all of their friends to post a bond to fund Hans's mother's escape from Europe. All of the family's valuables had been pawned or sold to help other relatives reach America. As well, Eva was expecting a child, and she would need money to pay for a doctor and baby supplies. The very most the family was able to provide was $200. For the remaining $400, Eva turned to Victor Weisskopf and George Placzek, old physicist friends from Berlin and regular visitors to Kharkiv. Both men had managed to find faculty positions at universities in New York state, and Eva believed they should jump at the opportunity to assist Weissberg.

Eva was shocked when Weiskopf and Placzek not only initially refused to provide money but also declined to have their names in any way associated with Weissberg. Fearing for their jobs and their immigration status, the men would not agree to be linked to a well-known Communist. Placzek went so far as to demand proof that the president of St Lawrence University was aware of Weissberg's party membership. Eva was especially incensed by Placzek's rebuff. Through the letters Eva had received from Weissberg, and eyewitness accounts relayed to her by Charlotte Houtermans and other survivors from UFTI, Eva believed that Weissberg's arrest by the NKVD had been the result of Placzek's ill-considered barbs about Stalin and his other criticisms of the Soviet Union. She was not afraid to tell Placzek as much, and demanded that he do something to redress the wrong he had done Weissberg. On 21 July, Placzek replied to Eva, promising to do what he could but that it would not be enough. He explained that he had just learned that his own family needed rescue, and he required whatever resources he could get to help them arrange US visas. Placzek warned Eva that is was "little use to our cause if we start to heap mutual recriminations considering Alex's fate over each other." He believed that Eva was still suffering trauma from her own imprisonment: "I can well understand that, behind Russian jail walls, you gradually start to take the abominably wretched parody of a legal process that they play there seriously in every detail and to see the characters of the torturers' tragicomedy as actual reasons or causes for the charge. But, really, by now you know as well as I do that this is not the reality."[9] Placzek was, of course, quite right. Eva's views were at best a gross oversimplification of the events that had led to Weissberg's detention by the NKVD. In her despair, and perhaps out of her own sense of guilt, Eva was not thinking quite rationally. Eva even told Charlotte Houtermans that Weissberg was not a Communist, something both women knew to be false. Despite Eva's frantic efforts, Weissberg, like Gilde, remained trapped in Poland. It is unclear whether a visa would have afforded either scientist a chance of escape.

Until 2018, Gilde's time in Poland remained a mystery. His family believed he had perished in Warsaw in 1940. Michal Palacz and Paul Weindling at Oxford Brookes University were examining the lingering legacy of Nazi medical experiments conducted during the Holocaust when they discovered at a Berlin hospital a set of pathological slides of brain tissue made by doctors at a Jewish hospital, Czyste Hospital in Warsaw during the Second World War. Further research revealed that among the staff of the pathology department at the Czyste Hospital was Siegfried Gilde. Palacz and Weindling knew nothing about Gilde prior to 1940, and the author of this book had no idea what fate awaited Gilde after he was trapped in Warsaw. In October 2018 the annual Powell–Heller Conference for the Holocaust, "First Do No Harm: Medical Science, Ethics and the Holocaust," was held at Pacific Lutheran University in Tacoma, Washington. At that conference, the author of this book gave a talk that outlined Gilde's tragic journey to Warsaw. In the question-and-answer session that followed, Weindling revealed his team's findings on Gilde. For the first time, Gilde's story could now be told in full.

Gilde's employment at the Czyste Hospital had begun by 22 August 1940, when identification papers were issued to him.[10] Gilde was not a pathologist but apparently had been forced to take whatever work he could find. The head of the pathology department at the time was Dr Josef Stein; Gilde and Dr Henry (Henryk) Fenigstein carried out the bulk of the laboratory work. Of the three doctors, only Fenigstein survived the Shoah. Fenigstein described Gilde as "a quiet bespectacled man" who had been placed in charge of microscopic examinations.

After an underground Jewish medical school was founded in April 1941, Gilde involved himself in educating students. According to Fenigstein, Gilde was not well-liked by the students, who considered him "an autocratic German who resented being treated like a Jew." Fenigstein believed that much of the problem was linguistic: the students spoke poor German, and Gilde's Polish was also not very good, "so communication was difficult." Fenigstein, however, spoke good German, and the two men "got along well."[11] Prior to the war, the hospital had been located in the heart of Warsaw, outside the area the Germans later set aside for the Jewish ghetto. When the Warsaw Ghetto was established in October 1940, the hospital remained outside it. In early 1941, however, it was forced to abandon its campus and move into the ghetto. Just before the move, Stein's department began a clinical study of "the pathological anatomy of typhus, typhoid fever and dysentery."[12] It is the slides made for this study that survive today in Berlin.

In February 1942, as conditions worsened in the ghetto, the doctors in the two Jewish hospitals, the Czyste and the Bauman and Berson Children's Hospital, began a scientific study of "the changes in human organs caused by starvation." They performed physical examinations of live subjects and autopsies of the dead. They kept meticulous records on the quantity and quality of food consumed. Gilde prepared and examined slides of the organs; certainly there was no shortage of bodies to be examined. The work was done to a high standard; according to one account, the statistical material "could not have been bettered in the best university of Europe."[13] While scientific purpose guided their work, the doctors and medical staff were also documenting the sadistic mass murder of the Jews of Poland. Their research was performed in secret, since the Germans would not have wanted their crimes recorded with such painstaking detail. Gilde, in the most deplorable conditions imaginable, had this one opportunity to be involved in a major scientific research project. On 22 July 1942, the first forced deportations from the ghetto brought a halt to all research projects at the hospital. It was forced to move once again, deeper into the shrinking ghetto; somehow the staff "saved all of the specimens, slides and charts for the Hunger Disease project."[14]

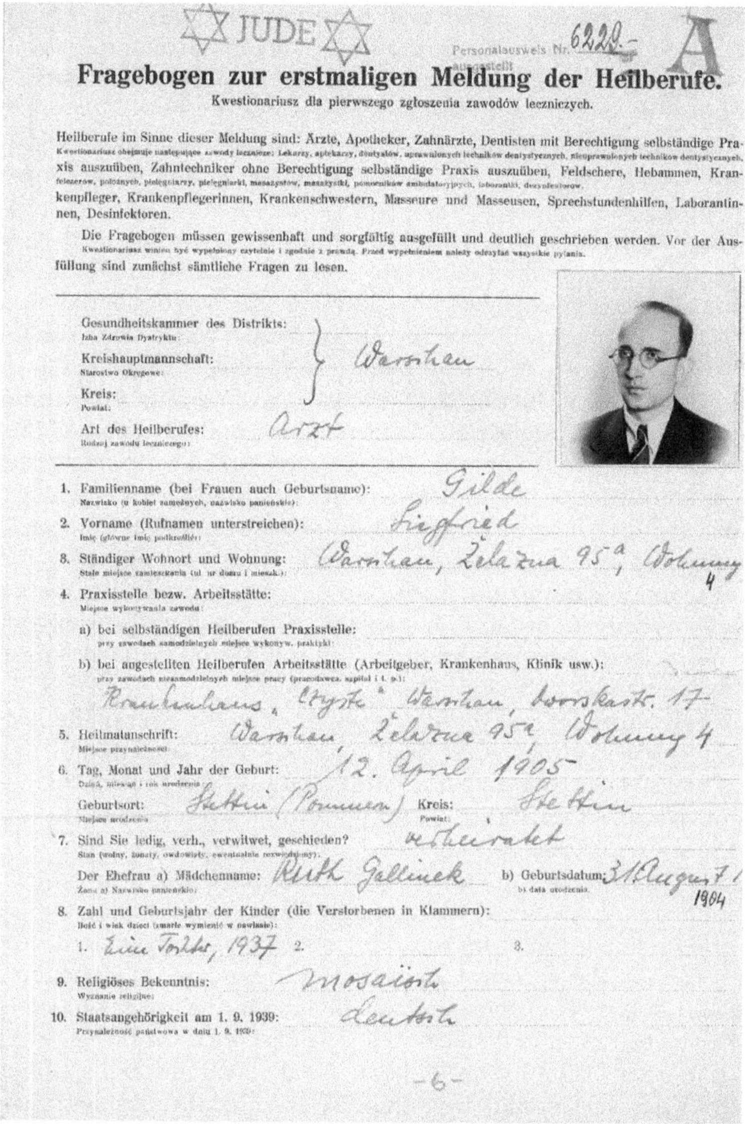

JUDE

Personalausweis Nr. 6229.-
ausgestellt

A

Fragebogen zur erstmaligen Meldung der Heilberufe.

Kwestionariusz dla pierwszego zgłoszenia zawodów leczniczych.

Heilberufe im Sinne dieser Meldung sind: Ärzte, Apotheker, Zahnärzte, Dentisten mit Berechtigung selbständige Praxis auszuüben, Zahntechniker ohne Berechtigung selbständige Praxis auszuüben, Feldschere, Hebammen, Krankenpfleger, Krankenpflegerinnen, Krankenschwestern, Masseure und Masseusen, Sprechstundenhilfen, Laborantinnen, Desinfektoren.

Kwestionariusz obejmuje następujące zawody lecznicze: Lekarzy, aptekarzy, dentystów, uprawnionych techników dentystycznych, nieuprawnionych techników dentystycznych, felczerów, położnych, pielęgniarzy, pielęgniarki, masażystów, masażystki, pomocników ambulatoryjnych, laborantki, dezynfektorów.

Die Fragebogen müssen gewissenhaft und sorgfältig ausgefüllt und deutlich geschrieben werden. Vor der Ausfüllung sind zunächst sämtliche Fragen zu lesen.

Kwestionariusz winien być wypełniony czytelnie i zgodnie z prawdą. Przed wypełnieniem należy odczytać wszystkie pytania.

Gesundheitskammer des Distrikts: / Izba Zdrowia Dystryktu:
Kreishauptmannschaft: / Starostwo Okręgowe: } Warschau
Kreis: / Powiat:

Art des Heilberufes: / Rodzaj zawodu leczniczego: Arzt

1. Familienname (bei Frauen auch Geburtsname): / Nazwisko (u kobiet zamężnych, nazwisko panieńskie): Gilde
2. Vorname (Rufnamen unterstreichen): / Imię (główne imię podkreślić): Siegfried
3. Ständiger Wohnort und Wohnung: / Stałe miejsce zamieszkania (ul. nr domu i mieszk.): Warschau, Żelazna 95a, Wohnung 4
4. Praxisstelle bezw. Arbeitsstätte: / Miejsce wykonywania zawodu:
   a) bei selbständigen Heilberufen Praxisstelle: / przy zawodach samodzielnych miejsce wykonyw. praktyki:
   b) bei angestellten Heilberufen Arbeitsstätte (Arbeitgeber, Krankenhaus, Klinik usw.): / przy zawodach niesamodzielnych miejsce pracy (pracodawca, szpital i t. p.):
   Krankenhaus „Czyste" Warschau, Dworskastr. 17
5. Heimatanschrift: / Miejsce przynależności: Warschau, Żelazna 95a, Wohnung 4
6. Tag, Monat und Jahr der Geburt: / Dzień, miesiąc i rok urodzenia: 12. April 1905
   Geburtsort: / Miejsce urodzenia: Stettin (Pommern) Kreis: / Powiat: Stettin
7. Sind Sie ledig, verh., verwitwet, geschieden? / Stan (wolny, żonaty, owdowiały, ewentualnie rozwiedziony): verheiratet
   Der Ehefrau a) Mädchenname: / Żona a) Nazwisko panieńskie: Ruth Gallinek b) Geburtsdatum: / b) data urodzenia: 31. August 1904
8. Zahl und Geburtsjahr der Kinder (die Verstorbenen in Klammern): / Ilość i wiek dzieci (zmarłe wymienić w nawiasie):
   1. Eine Tochter, 1937 2. 3.
9. Religiöses Bekenntnis: / Wyznanie religijne: mosaisch
10. Staatsangehörigkeit am 1. 9. 1939: / Przynależność państwowa w dniu 1. 9. 1939: deutsch

-6-

Image 13. Siegfried Gilde's questionnaire for the registration of health professionals in occupied Warsaw, Poland. Gilde is already working at the Czyste Hospital, which is still located outside of the ghetto. There is no date, but the hospital was forced to move no later than late 1940. This is the first page of a three-page document.* (Main Medical Library in Warsaw, Documents of the Warsaw-Białystok Chamber of Physicians, "Gilde, Siegfried" provided by Dr. Michal Palacz.)

The liquidation of the ghetto continued. Gilde's fate is unknown, though he certainly perished there or in one of the extermination camps sometime between September 1942 and May 1943. We do know that on 6 September 1942, all the remaining people in the ghetto, including the Czyste staff, were ordered to assemble in a small plaza for selection. Dr Stein, who was also medical director of the hospital, was given two hundred right-to-live cards. Stein had to select who would survive among the five hundred employees of the hospital.[15] Those without cards were rounded up onto cattle cars and sent to Treblinka for extermination. Further liquidations and deportations followed. Those who survived each event endured ever increasing brutality from the SS, as well as the Jewish Police and others who worked for the Germans. Disease, already endemic, spread like wildfire through the ghetto. Heroically, surviving medical staff continued to compile their research on starvation. The completed manuscript of the study was smuggled out of the ghetto just before the final liquidation on 19 April 1943.[16]

After the war, that study, titled *Hunger Disease*, was published in Polish and French. A partial summary of the work in English appeared only

---

* The page displayed in Image 13 reads as follows: Questionnaire for the first registration of the health professions. No. Identity card issued 6229.

Medical professions within the meaning of this notification are: Doctors, pharmacists, dentists, dentists with authorization to practice independent practice, dental technicians without authorization to practice independently, midwives, nurses, masseurs and masseuses, receptionists, laboratory assistants, disinfectors. The questionnaires must be filled out conscientiously and carefully and written clearly. Before filling out the form, you must first read all of the questions.

District Chamber of Health and District Authority: Warsaw.

Type of medical profession: Doctor.

1. Surname (for women also birth name): Gilde.

2. First name (underline first name): Siegfried.

3. Permanent place of residence and apartment: Warsaw, Lela Lanc 95a, Warsaw 4.

4. Place of practice or place of work: a) for self-employed medical professions, place of practice: [blank] ; b) in the case of salaried medical professions, place of work (employer, hospital, clinic, etc.): Hospital "Czyste," Warsaw, 17 Dworska Street.

5. Home Address: Warsaw, Lala Lane Warsaw 4.

6. Day, month and year of birth: 12 April 1905. District: Steffin (Pomerania). Place of birth: Steffin.

7. Are you single, married, widowed, divorced? Married. The wife a) Maiden Name: Ruth Gallinek; b) Birthdate: 31 August 1904.

8. Number and year of birth of children (the deceased in brackets): 1. one daughter, 1937; 2. [blank]; 3. [blank].

9. Religion affiliation: Moses.

10. Citizenship on 1/9/1939: German.

in 1967; the full English translation was not published until 1979. Long before the first English translation, however, the research conducted at the Czyste Hospital was being used by medical researchers examining the effects of famine on the human body. Unfortunately, somewhere in the shifting of documents, loss of life, and editing for eventual publication, Gilde's important role in the pathology component of the study was forgotten. Henry Fenigstein specifically states that Gilde certainly should have been included as one of the major contributors to the project.[17] It is hoped that any further editions of *Hunger Disease* will include Siegfried Gilde among the list of researchers, since his name, which Stalin and Hitler did so much to erase, should not be allowed to be forgotten.

Siegfried's wife Ruth and their child Marina, then just five years old, were also murdered by the Nazis. Nothing is known about their lives in France after Ruth's desperate efforts to rescue her husband from Warsaw failed. Ruth and Marina were detained, likely by French police, and held in the notorious Drancy transit camp. On the morning of Tuesday, 9 February 1943, Ruth and Marina were among 1,000 Jewish deportees "transported from the Drancy camp to the Le Bourget Drancy railway station in buses with the gendarmes standing guard. At 11:09 the same day, a train designated 901 [transport 46] departed Le Bourget Drancy for Auschwitz." The train arrived at the extermination camp two days later. Seventy-six men and ninety women were selected for labour. The other deportees, including almost certainly the Gildes, were murdered in the gas chambers as soon as they reached the camp. Of the 1,000 deportees, only fifteen men and seven women survived the war.[18]

Alexander Weissberg's account of his experiences in occupied Poland is a remarkable story of survival, rivalling any fictional tale of defying the odds. After his release by the Gestapo in April 1940, he lived with relatives in Kraków for six months to recover from his ordeals at the hands of the Soviet and German secret police. Before the end of the year, as German persecution of Jews escalated, Weissberg was determined to take steps to improve his chances of survival. Kraków had been chosen by the Germans to be their capital, and they wanted the city to be free of Jews. In contrast to what was done in other Polish cities, instead of bringing in Jews and concentrating them in severely overcrowded ghettos, most Jewish residents of Kraków were expelled into smaller towns nearby. To avoid deportation, and to raise funds for his survival, Weissberg "founded factories, organized undercover companies, broke all the German laws, and bribed and double crossed them." Weissberg soon discovered that, now that he was no longer "inhibited" by his "political principles," he was a superb capitalist businessman. In March 1941, by the time the small

Kraków Ghetto was walled off from the rest of the city by the Germans, Weissberg was a leading employer of Jews.

For a short time, Weissberg was able to pay his workers enough that they were able to eat better than the non-Jewish citizens living outside the ghetto. In March 1942, however, the Germans began liquidating the ghetto, beginning with the murder of fifty leading members of the Jewish community, who were dragged from their beds in the middle of the night and, according to Weissberg, shot. It is now believed that the fifty were not shot, but rather transferred to Auschwitz, where they were murdered. Weissberg was on the list, but "luckily" the Germans did not find him. For the next few months he shifted his address every few days, avoiding arrest by the Gestapo. He continued to run his companies, making a small fortune, which he used over the following three years "supporting comrades and friends, buying out those who had been arrested, organizing passports, apartments, and weapons."

In May 1942, the Germans continued the liquidation of the Kraków Ghetto. Weissberg organized an operation to try to save as many as possible of his friends and colleagues in the anti-Nazi resistance movement. Through contacts with the Polish underground, he managed to secure identification papers showing his friends to be Aryans. He organized the smuggling of people across the Carpathians into Hungary. His boldest operations involved establishing a network to smuggle Polish Jews to safety across Austria into Hungary. Somehow, while hiding in Kraków, Weissberg established contact with Doctors Kurt and Ella Lingens, anti-Nazi Catholics who had been acquaintances of his prior to his move to Kharkiv. The Lingens were already doing what they could to assist Jews, hiding some in their Vienna apartment and caring for the valuables of others so that they would not be looted. The Lingens agreed to help Weissberg.

Soon two couples, Bernhard and Jakob Goldstein, who were brothers, and their wives Helene and Pepi arrived at the Lingens' apartment from Kraków. One of the Jews the Lingens had assisted, a former stage actor named Rudolf Klinger, agreed to guide the couples across the border. Unbeknownst to the Lingens, Klinger was in fact an informer for the Gestapo. At the border, the Gestapo swooped in and arrested the Goldsteins. On 13 October 1942, the Gestapo arrested the Lingens. Kurt was sentenced to an army punishment unit and sent to the Russian Front. Ella was sent to Auschwitz, where she was forced to work as a camp doctor. While there she managed to save several Jews from the gas chambers. Miraculously, both survived the war, though Kurt was badly wounded at the front. In 1945, Ella endured the death march of camp inmates from Auschwitz to Dachau. They were both recognized by Yad Vashem, the

Holocaust memorial centre in Jerusalem, as being among the "Righteous Among the Nations." In 1943, after the Gestapo determined he was no longer useful as an informer, Klinger was sent to his death in Auschwitz.[19]

Weissberg's struggle to survive and fight the Germans continued. During his time in Kraków, he fell in love with and married Janka Skarowa, an assimilated Polish Jew who was so completely absorbed into Polish society that she was not identified by the Germans as a Jew. Despite that good fortune, Skarowa volunteered to enter the ghetto in Kraków, where she "worked day and night as a secretary for the Jewish community in order to alleviate the fate of the poorest and most oppressed." On 22 August 1942, as the Germans continued to deport the ghetto's inhabitants, Weissberg persuaded a reluctant Skarowa to escape and go into hiding in Warsaw. Just two weeks later, however, Skarowa and her mother were arrested during a routine street check by the Gestapo. Weissberg moved heaven and earth to free the women, offering half his fortune to the Gestapo as a bribe. The Gestapo took the money but arrested the messenger who carried it. On 22 October, Janka and her mother were shot. Weissberg was devastated: "I could think of nothing else for years. She was torn from me at the heyday of our great love. And great love is so rare. I think she was the only true love of my life."

After Janka's death, Weissberg escaped from the Kraków Ghetto, moving thirty-two times in the next five months in an attempt to avoid the large-scale manhunt the Germans had launched to capture him. His last place of refuge was the home of Countess Zofia Cybulski. There he met and fell in love with the countess's daughter. They were married at some point afterwards. On 4 March 1943, Weissberg was finally arrested at Cybulski's home. He was certain he was about to die. He somehow managed to manipulate paperwork so that rather than being shot immediately, he was included in a group of resistance fighters sent to a concentration camp. Weissberg arrived at the camp on a Tuesday, and the next day he was able to get word to his friends in Warsaw; by Saturday he had escaped and returned to Warsaw.

Weissberg worked actively with the growing Polish underground for the remainder of the war, mainly in Warsaw. With the last of his diminishing funds, he supported Jews passing as Aryans, organized the forging of documents, and was able to pass himself off as a German because of his excellent command of the language. When the Warsaw Uprising began on 1 August 1944, he joined the fight until the final surrender to the Germans on 2 October. Afterwards, he hid in the basement of a convent, knowing that arrest would bring his immediate execution. He managed to persuade some Austrian soldiers to smuggle him across the barbed-wire lines the Germans had established to catch any remaining Polish fighters. Weissberg settled in a suburb with his new wife and awaited the arrival of Soviet troops. On 17 January 1945, the Red Army liberated Warsaw, but for Weissberg, concerns lingered that the Soviets might arrest him once again. He had survived, but at great personal cost. He had lost his entire extended family, his parents and siblings, and his second wife and her mother; all had been killed by the Germans.

Fritz Houtermans in Germany

Fritz Houtermans also survived his transfer to the Gestapo by the NKVD. Unlike Weissberg's heroic tale of resistance, Houtermans's story is much more controversial. Houtermans has been viewed by some as an immoral war criminal and adulterer; others consider him someone who had done what was necessary to survive while doing as much as he could to thwart the Nazis. His transfer to the German government was quite different from those of Gilde and Weissberg. He was in the first group of prisoners transferred to Germany, with only an initial period of interrogation in Poland. The German government decided that Poland was too unstable to process the mainly German prisoners. Instead, after a brief stay in Lublin Castle, Houtermans and most of the other prisoners in his group were transferred to Berlin, where they were incarcerated in a Gestapo prison on 24 May 1940. Nine days later, Houtermans was transferred to the jail at the Central Berlin police station.

Professor Max von Laue, winner of the 1914 Nobel Prize in Physics, soon learned of Houtermans's plight and rushed to help his close friend. As a Nobel laureate, von Laue had substantial political power and influence in the German scientific community. He used his prestige to oppose, as much as possible, the efforts of Hitler regime's to Nazify German physics. He arranged for Houtermans to have money and additional food and cigarettes. He even managed to inform Houtermans's wife Charlotte, and his mother Elsa, by then both living in the United States, that Fritz was alive and reasonably well. He strenuously lobbied officials to have Houtermans released from prison. Houtermans was only one-quarter Jewish, which, under Nazi racial laws, meant he would not automatically be sent to a concentration camp. The Gestapo had a thick file on Houtermans's Communist activities before his departure to England in 1933, and of his time in the Soviet Union. Yet despite that file, von Laue managed to convince the secret police that Houtermans could be released, albeit under close watch. Houtermans might be free from prison, but he was still considered *Mischlinge* (mixed race), in the racist language of the Nazis, and therefore he was ineligible to return to his university position. In January 1941, after Houtermans had spent a few months working for a German scientific journal writing abstracts of physics articles written in Russian, von Laue persuaded Manfred von Ardenne, a wealthy inventor and physicist, to hire Houtermans to work in his private scientific laboratory. When Houtermans arrived, Ardenne's laboratory was researching the practical applications of the latest breakthroughs in atomic physics, a project funded by the German Post Office.

In the years that Houtermans had languished in prison in the Soviet Union, other physicists in Germany and elsewhere had made major strides in understanding nuclear energy. In 1938, two German scientists, Otto Hahn and Fritz Strassman, conducted a series of brilliant experiments. One result of bombarding uranium with slowed neutrons, they concluded, was the creation of small amounts of a radioactive barium isotope with a much smaller atomic weight than uranium. In December, several weeks before these results were published, Hahn wrote to Lise Meitner outlining his discoveries. Meitner was a sixty-year-old Austrian Jewish physicist who had been forced to flee from the Nazis to Denmark in 1938. During December, Meitner discussed the findings with her nephew, physicist Otto Frisch. What had taken place, they concluded, was the collision of a neutron with the nucleus of a uranium atom; the impact had triggered violent internal motions in the uranium atom, causing it to split into two approximately equal fragments. This was a new type of nuclear disintegration – radically more powerful than any observed before. Because it seemed to be similar to the division of a biological cell, the physicists decided to name the new process "fission."[20]

Before fleeing Germany in 1933, Houtermans had publicly discussed the possibility of neutrons causing a nuclear chain reaction. In Kharkiv, he had researched slow neutron absorption, one of a number of important avenues of scientific work that led to the discovery of atomic fission. At Ardenne's laboratory, Houtermans investigated the latest theoretical research on atomic fission, summarizing his findings in August 1941 in a thirty-three-page report, "On the Question of Unleashing Chain Nuclear Reactions." The report "discusses fast-neutron chain reactions, neutron slowing, uranium 235 and the magnitude of the critical mass." Houtermans paid special attention to element 94, which "the Americans had already secretly christened 'plutonium.'" He suggested that it would be easier to separate chemically quantities of plutonium, rather than separating U-235, the highly fissionable isotope of uranium.[21] Houtermans's report summarized the then current state of nuclear research, using the latest scientific results from Germany, France, Great Britain, the Soviet Union, Denmark, and the United States. It was only in mid-1941, as theoretical and experimental research began to show that building nuclear reactors, and perhaps, an atomic bomb, might just be possible, that governments began to classify work in the field. The German government considered Houtermans's report a state secret, one that mirrored research being undertaken for the German military by a team of physicists led by Werner Heisenberg. Houtermans was well aware of the potential military applications of nuclear fission, even if the means to build a nuclear bomb remained unknown. While researching his report, Houtermans took the highly risky step of trying to inform colleagues in the West that Germany might dramatically accelerate its military research into atomic energy. After the war, Houtermans explained his daring actions: "When faced with a totalitarian regime, every man should have the courage to commit high treason."

In the early spring of 1941, German-Jewish physicist Friederich Reiche received a visa to migrate to the United States. The day before his departure, Houtermans arrived at Reiche's apartment for a meeting arranged by von Laue. Reiche had assumed that the purpose of the meeting was so that Houtermans could send word to his family in the United States. To Reiche's surprise, Houtermans asked him to memorize a message to be passed on to physicists in the United States. Reich later recalled Houtermans's words:

> "Please remember if you come over, to tell the interested people the following thing. We are trying here hard, including Heisenberg, to hinder the idea of making the bomb. But the pressure from above" – the whole thing is very funny – this in parentheses – because it seems now to me that the thing just opposite. But at least he told me this – "Please say all this; that Heisenberg will not be able to withstand longer the pressure from the government to go very earnestly and seriously into the making of the bomb. And say to them, say they should accelerate, if they have already begun the thing." This was, by the way, February – no, March, '41. "They should accelerate the thing."[22]

In April in New York, Reiche was taken to a party by his friend Rudolf Ladenburg, attended by a group of expatriate German physicists, including Hans Bethe, Eugene Wigner, Wolfgang Pauli, and John von Neumann. Reiche delivered Houtermans's message. Reiche recollected: "I saw that they listened attentively and took it. They didn't say anything but were grateful."[23]

Shortly after the party, Ladenburg took further action, writing directly to Lyman Briggs, chair of the US government's Uranium Committee, outlining Houtermans's message but not actually naming the sender. It is impossible to assess the significance of Houtermans's message. Of the expatriate physicists at the party, all but Pauli would later play important roles in the Manhattan Project. Years later, however, none of the physicists present at the party had any strong memory of this dramatic event. Nor did Ladenburg's note to Briggs have any influence on shaping American nuclear research. In the spring of 1941, Briggs and other senior American scientific officials remained convinced that there was little risk of anyone developing a practical nuclear weapon anytime in the near future.[24] The Manhattan Project was months away from being authorized by President Roosevelt, and the origins of the atomic bomb program had little to do with refugee physicists living in the United States.[25] Houtermans may not have influenced the establishment of the Manhattan Project, but perhaps his efforts played a role in motivating the refugees, and, through them, the American scientists working on the bomb. The fear that the Germans, under the brilliant Heisenberg, might be ahead certainly spurred completion of the bomb in 1945.

In 1942 and 1943, Houtermans continued to work at Ardenne's laboratory conducting research into slow neutron capture and photonuclear reactions. Houtermans was not associated with the main investigations into atomic energy sponsored by the German military, but he does appear to have been kept informed of the progress of Heisenberg's team and other research groups. Houtermans risked his life once more to get word out to scientists in the West about the seemingly rapid progress being made in Germany. In 1943 he arranged with a friend who had been given special permission to travel to neutral Switzerland, to send from there a brief telegram to Eugene Wigner in the United States. The telegram read "Hurry up. We are on track." Wigner had no doubt what Houterman meant, but again, it is doubtful this telegram did much to spur the Americans' atomic bomb project.

Houtermans left Ardenne's laboratory in 1944, moving to the Imperial Physical Technical Institute (Physikalische-Technische Reichsanstalt or PTR), located first in Berlin and later, as a result of Allied bombing raids, in Ronnenburg. Houtermans continued his basic research in neutron physics. He was fired in early 1945 after the Director of PTR discovered Houtermans's scheme to illegally acquire tobacco by requisitioning several kilograms for fake military scientific experiments. Houtermans's friends, including Heisenberg, quickly arranged for Houtermans to be employed at the University of Gottingen, the same university where he had earned his doctorate and met Charlotte more than fifteen years earlier. The rush to find employment for Houtermans was spurred by fears that if he was not in a protected position, he might be conscripted to serve in the military in the dying days of the Third Reich.

Houtermans survived the war, but his reputation was sullied by events in his personal and professional life. After his release from prison in

1940, he was able to exchange correspondence with Charlotte and his mother in the United States. Watched by the Gestapo, however, he could not leave Germany. After Germany declared war against the United States in December 1941, correspondence with his family ceased. In 1942–43, Ilse Bartz, a young and beautiful chemical engineer working in Ardenne's laboratory, began collaborating with Houtermans on his neutron research. They published several scientific papers together. Fritz Houtermans had not seen Charlotte and his children since that fateful day in December 1937 at the Moscow Customs House. Working in close proximity, Ilse and Fritz Houtermans fell in love.

In 1944, without informing Charlotte, Fritz decided to marry Ilse. Ilse was likely already pregnant with the first of their three children. Fritz used a Nazi era law that allowed a spouse separated from their partner for more than five years to obtain a quick divorce without either informing the other party or having their consent. Houtermans's decision was particularly unethical; the divorce laws he used were designed to allow Aryans in mixed marriages with Jews to easily terminate their then illegal unions. Fritz did not even bother trying to inform Charlotte, although he could have sent a message in an emergency to her via the International Red Cross. Indeed, Charlotte received news of her divorce through the Red Cross, but the sender was her sister, not her former husband. Charlotte recalled her reaction: "To say that I was ruined is to say nothing. My love, my plans for the future, my dreams, my years of struggle for Fritz's liberation, my desperation and hope, everything was destroyed in a blink."[26] The worst part for Charlotte was having to tell their two children what their father had done.

Houtermans's other controversial action was his return to Kiev and Kharkiv at the behest of the German government shortly after the occupation of those cities in October 1941. Much mystery surrounds this trip. Houtermans would later claim that he felt that he had little choice but to agree to return to the Ukraine, since his refusal might have led to his rearrest. Houtermans also claimed that while the purpose of his mission was to assess any useful equipment and personnel for the German war effort, his personal goal was to assist any former colleagues remaining in the occupied city. The historical writings on Houtermans's mission are generally contradictory: Soviet sources claim the mission was under the authority of the Gestapo and that Houtermans wore its uniform, whereas German documents indicate that the mission was authorized by the Luftwaffe and that Houtermans wore a German air force uniform. In 1947, Houtermans insisted that he wore civilian clothes throughout his mission. Further sources suggest that his mission had been authorized by the German navy.[27]

There is much evidence that Houtermans did what he could to assist the surviving staff at UFTI and other people known to him. While in Kharkiv, he encountered his former cellmate Konstantin Shteppa, who had decided not to evacuate the city before the German occupation and to follow the Red Army eastward. After his imprisonment, Shteppa loathed everything about the Soviet system, and like many Ukrainians, he harboured the misplaced hope that the Nazis would be better masters. Houtermans arranged for regular food packages to be sent to Shteppa's apartment. Houtermans's report to the German government about his findings in Ukraine was carefully worded, indicating that among the remaining staff at UFTI there were no specialists important enough to be conscripted to work in Germany. Moreover, he did not recommend the looting of the institute's equipment, even though this included the large Van der Graaff accelerator, which was significantly more powerful than any available in Germany.

After the war, however, Houtermans was branded a war criminal by the Soviet government for his actions in Ukraine. During a postwar meeting of physicists held in Moscow in 1945, Peytor Kapitza told Max Born that Houtermans was "a traitor, and said that the Russians would hang him if they found him." Kapitza went on to explain to Born that in Kharkiv, Houtermans had denounced his former colleagues to the Gestapo. On his return to Scotland, Born made sure to write to Houtermans, warning him of the grave risk he would run if he dared enter Soviet-controlled Germany. While Houtermans was in Gottingen in the British occupation zone, some of Houtermans's friends feared the British might turn him over to the Soviets if they requested his arrest.[28]

On 22 November 1945, Marie Rausch von Traubenberg, widow of the physicist Heinrich Rausch von Traubenberg, wrote to the SPSL requesting they intervene with the British government to pre-empt a Soviet request for Houtermans's arrest. Heinrich had been a professor of physics at the University of Kiel until he was dismissed in 1937 because he refused to end his marriage to Marie, who was Jewish. Marie wrote a stirring defence of Houtermans's activities in the Ukraine and in Germany during the war; she painted him as a hero who had done all that he could to assist Jews and to oppose the Nazis. Marie was arrested by the Gestapo in 1944 for racial reasons; within hours of watching his wife's arrest, Heinrich died of a heart attack. Von Traubenberg credited Houtermans, along with Professor Otto Hahn and other German physicists, with saving her and her mother from certain death by arranging for their release from the Theresienstadt concentration camp and for their return to Germany.

Houtermans was also instrumental in rescuing the German-Jewish theoretical physicist Richard Gans from hard physical labour, which came close to killing the sixty-three-year-old scientist. According to Trauben-berg, Houtermans was assisted by Theodor Cammann, a friendly SS offi-cer whom Fritz met while on his mission to Ukraine. Traubenberg may have misidentified this officer; his name does not appear on lists of SS officers, but there was a Luftwaffe officer named Theodor Cammann. If the Luftwaffe officer was the one who assisted Houtermans, it tends to support the view that his mission to Ukraine was at the behest of the Ger-man air force.[29] As well, Traubenberg reported that Houtermans "very often hid Jewish persons, whom he scarcely knew, in his lodgings." "All this was done," she explained to the SPSL, "with great personal risk, so that my husband and I often were much afraid the Nazis would catch him."[30]

Many of Von Traubenberg's claims about Houtermans have been veri-fied by historians. Edgar Swinne wrote a biography of Gans in which he traced the extraordinary efforts by Houtermans and other physicists to rescue him. Houtermans and Heinz Schmellenmeier, a close friend and fellow Communist who ran his own small private scientific research company, concocted a scheme to convince Nazi authorities that Gans was vital to a new project to build a small electron accelerator, the "Rheotron" (later renamed the Betatron). The device was presented as the key com-ponent for a death ray, designed to shoot down aircraft at long range. Houtermans and Schmellenmeier knew such a device was impracticable; the idea simply needed to appear plausible enough to convince officials to fund the project and authorize Gans's transfer to Schmellenmeier's laboratory. The ruse succeeded, and Gans survived the war.

Soviet accusations that Houtermans had actively assisted the Gestapo continued to haunt the physicist well after the war. In 1947, Houtermans felt compelled to defend his actions in a six-page letter that was delivered to Born by Bruno Tousckek, a young Austrian physicist working in Born's laboratory in Edinburgh. Writing to the SPSL, Born expressed concerns about Houtermans's character as they related to his treatment of Char-lotte but believed his "explanations quite creditable." Born viewed the Soviet assertions as far less plausible; after all, since the end of the war, "they seem to be little better than Nazis."[31] Houtermans's reputation was damaged, but he had survived more than a dozen years of exiles, impris-onments, and life in police states.

Conclusion

Houtermans, Weissberg, and Gilde had experienced the shared horrors of being handed directly over to the Gestapo by the NKVD. Houtermans's story shows how much more likely it was for Germans of mixed heritage to survive in Germany than Jews, even if they were known members of the Communist Party. Houtermans's survival, of course, was also in large measure predicated on the support he received from the German physics community. Weissberg stayed alive as a result of his extraordinary survival skills, perseverance, and luck. Gilde shared the fate of many Jews in Poland but left an important legacy that is revealed here for the first time. As the fate of Gilde's family illustrates, even those who managed to escape from the Soviet Union during the Great Terror were by no means safe from being swept up in the Holocaust. Some of the scholars who managed to leave the Soviet Union would not survive the mass murder of Europe's Jews.

10 From the Great Terror to the Shoah

Those who were not arrested by the NKVD and who were able to escape from the Soviet Union once again faced the challenge of finding refuge, this time in a world even more dangerous than when they had first migrated. There were now many more displaced scholars seeking to migrate to an ever dwindling number of places that might offer them safety. Pursuit of a career had become secondary. Remarkably, most of those who fled Stalin found another place of refuge, although many lived in precarious circumstances for years after their flight from the Soviet Union. The Germans' annexation of Austria and Czechoslovakia just before the Second World War, and their conquest of much of the rest of Europe between September 1939 and December 1941, once again placed those survivors who had not found a route off the continent in a terrifying predicament – having to flee or hide in order to survive.

The success or failure of the scholars who fled the Soviet Union to find new places of refuge depended on much the same factors as had determined their survival when they first fled Germany. The scholars' discipline and specialty, their status in the international community of scholars, and their ability to link up with a personal network of friends and colleagues all helped determine whether they survived. Fortuna was always present in determining the outcome of their quest for safety and security for themselves and their families, as was their ability to persevere even in the most trying circumstances. Many of the scientists were caught up directly in the Holocaust. Some of them and/or their families did not survive it, while others made remarkable escapes.

The Fate of the Early Deportees from the Soviet Union: Herbert Fröhlich and Emanuel Wasser

For the first two scholars expelled from the Soviet Union, physicists Herbert Fröhlich and Emanuel Wasser, all of the above factors were in play in their search for a second refuge, leading to polar opposite outcomes. Both men were forced to leave Leningrad in 1935 in the aftermath of Kirov's assassination. Fröhlich's charmed life continued after he departed by train from Leningrad with his caviar, jewellery, and ill-fitting English-made suit. When he arrived in the Netherlands, he cashed in the remaining portion of his train ticket to Rome. Shortly after his arrival there, he learned that the University of Bristol's physics department was holding a conference on electrons in metal, a subject on which Fröhlich had focused his research in the Soviet Union. He wrote to Professor A.M. Tyndall, head of Bristol's physics department, asking for funds to attend the conference. Bristol was one of the few physics departments in Britain with considerable financial resources, thanks to a large private endowment, and Tyndall gave Fröhlich a small grant so that he might attend.

Going to Bristol was a serendipitous decision for Fröhlich, since it was one of the few places in the world likely to have an interest in hiring him as well as the means. Bristol already had a sizable contingent of refugee physicists, including W. Heitler, H. London, K. Fuchs, and K. Hoselitz. At the conference, Fröhlich impressed Tyndall with his knowledge of electrons in metals, and he was offered a still rare, highly coveted one-year Academic Assistance Council fellowship. The AAC had assurances from Tyndall that if Fröhlich lived up to his expectations, there was a good likelihood that the young German scientist would be offered a long-term position. Tyndall kept his promise; after Fröhlich's AAC grant expired, he extended him a continuing position at Bristol.[1] Fröhlich was able to find a safe haven and a secure university position after being cast out of the Soviet Union. By contrast, Emmanuel Wasser's work as a physicist ended once he left Leningrad, and he would spend the remaining eight years of his life in a never-ending and ultimately futile struggle for survival for himself and his family.

After his flight from the Soviet Union in 1935, Wasser and his wife and child lived with his parents in Łódź, Poland. As an Austrian citizen, Wasser could only stay there for two months. In May, like many others who would follow, he applied to the SPSL for assistance in finding employment. For Walter Adams, executive secretary of the SPSL, Wasser's case was not straightforward. By 1935, the society had been inundated with hundreds of desperate pleas for help from German scholars who had lost their positions. Wasser had voluntarily left gainful employment in Vienna for Leningrad and in 1935 could still return to the relative safety of Austria. Moreover, the SPSL had been pressured to broaden its mandate to include other groups of displaced scholars, specifically Russians who had lost their academic positions after the revolution and civil war between 1917 and 1921.

Another factor was that Adams had already been briefed by Max Born, one of the most important of the refugee scientists in Britain. Born regularly advised the society about physicists seeking assistance. Born's comments about Wasser are unknown, but clearly he did not rate the Austrian physicist highly enough to warrant extraordinary action by the society. Adams wrote Wasser a polite, carefully worded rejection letter, informing Wasser that the society's terms of reference were "to give assistance to University teachers who are prevented, on grounds of race, religion or political opinions, from continuing their work *in their own country*." Adams concluded: "I should warn you that we are finding it extremely difficult to discover means of assistance for displaced scholars and that, therefore, although we shall, if possible, do everything to discover a new position for you, it would be unwise to depend upon assistance from this Council."[2]

Four months later, a desperate Wasser again wrote to Adams, expressing his surprise for not having heard from him about employment opportunities. Wasser asked him if there was any possibility of finding anything for the upcoming school year and, if not, "if You would help me in finding a *unpaid* occasion [*sic*] to continue my scientific researches in any Physics Institute in England." Wasser feared that if he remained unemployed he would lose contact with the world of science.[3] Adams could offer little comfort to Wasser, only some details on the remote possibility of winning a competition for a position at the University of Cape Town. As for seeking an unpaid position, here too Adam could offer nothing but discouragement. Britain was already inundated with German scholars seeking temporary employment. British professors were reluctant to take on anyone who did not have financial means, and the government was unlikely to allow a destitute scholar without prospects to enter the country. After this rejection, nothing more was heard from Wasser for three years.[4]

In late 1935, Wasser returned to Vienna. There he applied for numerous positions, but he was unable to find employment as a physicist. In April 1938, after the German annexation of Austria, Wasser and his family fled back to Poland, and once again he appealed to the SPSL for help. He hoped that now that Jewish academics were being forced to flee Austria, the society would look more favourably on his application for assistance.[5] Unfortunately, as Esther Simpson explained to Wasser, the society had made efforts to find him a position in the intervening three years, but to no avail. Whether or not this was true, Simpson was forthright in explaining that the employment situation was now even worse than it had been a few years before. Moreover, the British government was being "much more severe" in its policies on admitting academic refugees who did not have some sort of support already committed. All she could suggest to Wasser was that he try to find employment in the United States by corresponding with colleagues who had already immigrated there.[6]

In September, Wasser appealed for help from the SPSL one last time. He had taken Simpson's advice and written to colleagues around the world. His efforts to find "some position or at least some possibility to enter any democratic country, where I could live with my family," were unsuccessful. The only suggestion had come from Professor Samuel Sambursky, a physicist at Hebrew University, that he ask the SPSL for help in acquiring a migration certificate for Palestine.[7] Again, the society could not help Wasser either to obtain a certificate to enter Palestine or to find employment. As David Cleghorn Thomson, the new general secretary of the SPSL, said, "we would be only too glad to help you if we could, we must frankly admit that the opportunities are exceedingly limited."[8] Wasser and his family were left stranded, unable to stay in Poland and under no illusion about what would happen to Jews who dared to return to Vienna.

Flight from the Soviet Union to Germany

Hermann Borchardt was the next scholar to be expelled from the Soviet Union when he was dismissed from his position teaching German at the University of Minsk in January 1936. Borchardt took a very different approach; instead of trying to stay out of the reach of the Nazis, he was one of only two academic refugees to try to return to Germany after being forced to leave the Soviet Union. He had reason to believe he might return to Germany without being molested by the Nazis, despite being a Jew and having just returned from the Soviet Union. During his second trip to the Ministry of Education in Moscow, when he rejected demands to become a Soviet citizen, he realized he might be expelled. As a precaution, he visited the German embassy and spoke to the ambassador, Count von der Schulenburg, who assured Borchardt that he could safely return home, suggesting he seek employment in one of the new segregated schools for Jewish children. Borchardt's son Frank explained: "It can be assumed that the good count honestly believed that Borchardt could return to Germany without losing his freedom or even losing his life."[9] Like so many of the old Prussian nobility, Schulenburg believed that Hitler would soon stop the persecution of the Jews; he could not accept that Hitler meant to carry through his clearly stated policy of eradicating Jews from German society.

Over the objections of his wife Thea, Borchardt decided to accept the count's assurances and brought the family home to Germany. He did find employment teaching at a Jewish school in Berlin. The pay was not sufficient to support the family, however, and the Borchardts were forced to rely on handouts from friends for their survival. At first he had some hope that his decision was the correct one: "In 1936 when I returned to my native Germany, after a two-year sojourn in Stalin's country, the contrast seemed so striking that I thought I was in a relatively free land – until the brutal Nazi state police taught me the sad truth."[10] The intensifying Nazi persecution of Jews, and the hopelessness of their situation, led Borchardt's wife Thea to propose to Hermann that the family commit suicide. Borchardt refused to consider the idea because he would not in any way hurt his children.[11]

On 28 July 1936, the Gestapo arrested Borchardt, just six months after his return to Germany. He was charged under a new law that had made it illegal for expatriate Germans to return home. He was first interned at Esterwegen concentration camp, where he was brutally beaten by other prisoners, some of them Communists, and by camp guards. In one particularly brutal encounter, an *Oberscharführer* (sergeant) for fun asked Borchardt what he wanted. When Borchardt replied sarcastically he wished to die, the *Oberscharführer* ordered him to dig his own grave. The entire time he dug, Borchardt must have felt that he was about to be killed. Only when the top of the grave reached his lips was he ordered out of the hole. As a result of his treatment at Esterwegen, Borchardt lost his hearing, twelve teeth, and the middle finger on his right hand. He was then transferred to Dachau, where he was forced to help build the camp.

Meanwhile Thea, who as a non-political non-Jew remained unmolested, worked to liberate her husband. She persuaded Otto Warburg, the Nobel laureate medical researcher, and Henri Jourdan, the cultural attaché at the French embassy, to negotiate with German government officials for Borchardt's freedom. Warburg and Jourdan succeeded. In April 1937, Borchardt was transferred to the Sachsenhausen concentration camp, near Berlin. On 11 May he was finally set free, but the Gestapo informed him that his release was on the condition that he leave Germany – otherwise, he faced rearrest. The Gestapo also warned him to not reveal any information about his treatment, since even in America, they had "a fine ear and a long arm."

Borchardt turned to his old friend George Grosz, the renowned artist, who had migrated to the United States just as the Nazis came to power in 1933. Grosz was well enough established in New York to sponsor the Borchardts' immigration. Additional support was received from Bertolt Brecht and the Jewish-German banker Max Warburg. When the family arrived destitute in New York, Hermann was a broken man; he was neither physically nor emotionally fit enough to resume an academic career.

He had to rely on the financial assistance of friends in the German expatriate community to support his family.

Borchardt's return to Germany had almost killed him and had scarred him for life. The only other refugee scholar to return home after his deportation from the Soviet Union was no more successful. Michael Sadowsky, the Russian-born mathematician, was already four times a refugee. After his expulsion from Novocherkassk in May 1937, he believed that he and his family had no choice but to return to Germany. His entry visa to the United States had long since expired, and with no prospects anywhere else, Sadowsky hoped to be allowed to use his contacts in Germany to find employment. He had few illusions that he could stay for long in Germany. Upon arrival there, he immediately wrote to the SPSL in London asking them to renew his file, but the society had been unable to help him in 1934, and it was even less likely that three years later they could find him work. Esther Simpson at the SPSL realized that Sadowsky was now in a precarious situation. She wrote to the American Emergency Committee, imploring their assistance and emphasizing that Sadowsky had spent two successful years at the University of Minnesota. Neither organization, however, could find any immediate opening for Sadowsky, though his name was proposed for positions as far afield as Venezuela.[12]

Sadowsky did not last out the summer in Germany. Upon his arrival, Nazi officials and Gestapo agents informed him that he would not be given permission to work on account of his "non-Aryan" marriage. This rendered him unemployable; even personal friends who offered Sadowsky work were forced to withdraw their offers after a visit from the Gestapo. The Labor Exchange refused to register him. The German officials twice demanded that he divorce his wife Hilda and separate from her and their infant son. He refused to take up this repugnant suggestion, despite their threats of arrest and incarceration in a concentration camp.

Only one option for escape remained for Sadowsky and his family. Sadowsky's wife's parents and brothers had migrated to Palestine, and they were able to sponsor the family to go there. They arrived some time before 4 September. Sadowsky was the only one of the scholars in the study who managed to escape to Palestine, despite the pleas of several other scientists to the SPSL for assistance in migrating there. Ironically, while Sadowsky was the only one to reach Palestine, he was also one of the few who had no prospects of benefiting from the move, except for the raw fact of surviving. He soon found that he could not find employment in the Jewish communities of the region: he was not a Jew, and perhaps even worse, he was in a mixed marriage. He reported that the economy in Palestine was in a depression, "which makes things worse than they should be and creates hostile feelings among the population who consider themselves at home here towards people they think should be strangers."[13] Safe from the Soviets and the Nazis, but facing destitution, Sadowsky turned his attention to finding a way out of Palestine. His efforts to do so will be examined in the next chapter.

Gerhard Harig also voluntarily returned to Germany, though in his case not as a refugee from the Soviet Union, but as an NKVD spy. In early March 1938, shortly after his arrival from the Soviet Union, he was arrested by the Gestapo in the Baltic city of Stettin (now Szczecin, Poland). It does not appear that the German authorities were aware that Harig was an NKVD agent; instead, he was arrested as an anti-fascist agitator and a member of the KPD. After five months in a Leipzig prison, Harig was transferred to the Buchenwald concentration camp. Buchenwald had been established a year earlier, primarily to house criminals and political prisoners. Harig's Buchenwald number, Political Prisoner 173, attests to his being one of the camp's first inmates. Despite his early arrival at the camp, Harig survived the war, in large measure because he was not Jewish, but also because he was a Communist.[14]

In Buchenwald, Harig became enmeshed in what Evelyn Zegenhagen describes as the "bitter struggle" between the criminal, so-called green prisoners (for the colour of badge they wore) and the political or "red" prisoners, over positions in the camp's prisoner administration. The strongest and best organized groups among the political prisoners were the Communists, both German and international. By 1943 the Communists had control of most of the important camp positions. The SS found the Communists crucial for running the camp, which enhanced their chances of survival. The Communists had to collaborate with the SS, but to a certain degree they could manipulate the SS's terror. The degree to which the Communists were able to save other inmates, particularly Jews, is a matter subject to considerable historic debate, but there is no doubt that party members were central to the underground resistance in the camp.[15] Although Harig's camp file has survived, it contains little information about his specific experiences. In April 1939 he was allowed to fill out a form for the German embassy in Moscow, attesting to his desire for his son Georg to be repatriated back to Germany to live with his grandmother in Leipzig. No mention is made of Harig's wife Katherina. The application was not successful, and it is known that Harig's family remained in the Soviet Union until after the war.[16] Harig's camp records do attest to his having a series of increasingly important administrative positions. In 1943 he was transferred to the camp administration's political department to work as a translator. He spoke German, Russian, French, and English and was certainly qualified for the job. He worked in the political department until the camp was liberated in 1945.[17] Borchardt, Sadowsky, and Harig had faced the Gestapo directly in Germany, and the fact that all three survived says something about the tremendous survival skills of the academic refugees.

Including Sadowsky, twenty-two scholars either fled or were expelled from the Soviet Union in 1937–38. Most of them avoided Hitler's Germany at all costs; a few, however, made very brief stopovers there after departing the Soviet Union before leaving as quickly as possible. It is noteworthy that no one was *un*able to leave Germany and that all found some form of secondary refuge. Fourteen found places of sanctuary outside the path of the German army's conquests in 1940–41. But before tracing their escapes, let us examine the fate of those ensnared by Hitler's conquests.

Escape and Death in France, 1937–1945

The remaining scholars who had survived the Great Terror and sought refuge in Western Europe would be forced to confront the Holocaust first-hand, having been swept up in Hitler's conquests in the first year of the war. Most of the refugee scholars would survive this second entanglement with the Nazis and their murderous hatred of Jews. Their survival might seem miraculous, but rather than some divine intervention, they lived because of their now finely honed survival skills and their ability to connect with the international community of scholars.

Prior to the spring of 1940, France again proved a popular place of refuge. Seven of the fleeing scientists fled there – Guido Beck, Stefan Bergman, Fritz Duschinsky, Edger Lederer, Helmuth Simons, Laszlo Tisza, and Emanuel Wasser. Bergman migrated to the United States in the summer of 1939. Of the six remaining scientists who were in France when the Germans invaded in the spring of 1940, four survived the war. What appears to have been crucial in determining survival was the forging of strong connections with the French scientific community and civil society, either to find safety within France or to provide time and opportunity to escape. Those who remained foreign refugee Jews perished in the Holocaust.

By 1937, when the Ensnared began to arrive in France from the Soviet Union, the situation for academic refugees in the country had been transformed. Between 1933 and 1936 there were almost no prospects of permanent employment in France because the restrictive civil service laws that governed universities required faculty to be French citizens. In 1936, Louis Rapkine, a brilliant biologist, organized a new French Committee for the Reception and the Organization of the Work of Foreign

Scientists (Comité français pour l'accueil et l'organisation du travail des savants étrangers). Modelled closely on the SPSL, it recruited leading left-leaning French scientists to the Reception Committees board, including Georges Urbain, Jean Perrin, Frédéric Joliot, Paul Langevin, and Irène Joliot-Curie; as we have seen, all of them had been assisting refugee scholars since 1933. Aided by an initial government endowment, Rapkine raised considerable funds to create one-year fellowships for displaced scholars. At first, like the SPSL, the board focused on established scientists, but by 1938 the emphasis of the grants program had shifted to assisting younger scholars. As well, Jean Perrin, a Nobel laureate physicist and Under Secretary of State for Scientific Research in the Popular Front government in 1936, founded a precursor agency of the National Center for Scientific Research (CNRS). Under Perrin's leadership, the CNRS began offering foreign scholars contracts as researchers after their fellowships with the board ended. The CNRS operated outside of the restrictive civil service laws of universities, thus making room for the long-term employment of refugee scientists.[18]

After fleeing the Soviet Union, Laszlo Tisza returned to Budapest in June 1937. When he found no scientific prospects in Hungary, he decided to try his luck in Paris. While in Budapest, he had reconnected with Veronoka (Vera) Benedek, a young medical student with whom he had been romantically involved before he left for Kharkiv. The two decided to marry once Tisza established himself in France. The couple realized they needed a backup plan in case Tisza could not find employment in Paris, so they both applied for visas to immigrate to the United States. In Paris, Tisza, a highly regarded young theoretical physicist who had studied with the great Landau, had little difficulty in getting a reception committee fellowship, which he held at the Collège de France. Vera joined Tisza there by the end of the year, and in 1938 they were married. It was at the Collège de France in 1938 that Tisza made one of his most important contributions to physics. He worked with the German refugee physicist Fritz London to develop a theoretical understanding of the behaviour of liquid helium and superconductivity. He created the two-fluid theory of liquid helium "that explained the unusual behavior of liquid helium, which results when helium gas is cooled to within a few degrees of absolute zero."[19] After his fellowship, Tisza was employed at the CNRS in Paris. At CNRS, Tisza had an opportunity to meet many visiting physicists; among the visitors was Charles Squire, from the Massachusetts Institute of Technology. Tisza assisted Squire with his research, and the men became close friends.

With the German army approaching the French capital in June 1940, Tisza's laboratory was ordered to evacuate to Toulouse. Although

Tisza's lab was in Vichy France, he was not molested in any way, despite being Jewish, likely because of his Hungarian nationality. Tisza had no illusions about his long-term safety and made plans to escape to the United States. He was assisted by his close connections to the diaspora of Hungarian physicists, in particular his close friend Edward Teller. Tisza's survival, however, was also the result of a certain amount of serendipity. Charles Squire wrote from MIT, promising to write an affidavit in support of his move to the United States and offering a loan to pay for his voyage across the Atlantic. In October, Tisza received a telegram from the US consulate in Marseille that his and Vera's visas had been approved. It took some time to secure the necessary travel documents, but in early February 1941 he and Vera left France by train, headed for Lisbon. By March they had reached New York, where they took one final train journey to Cambridge, Massachusetts, to stay with relatives and to reconnect with his close friend in the physics department at MIT.[20]

After fleeing the Soviet Union, the mathematician Stefan Bergman, an expert in kernel functions, arrived in France in late 1937. Bergman soon found a position at Institut Henri Poincaré, a leading mathematics research institute at the University of Paris, "where he wrote an important two-volume monograph on complex analysis." No details are available about Bergman's escape from France, but it must have been a very close call; he only arrived in Boston on 12 June 1940, after landing in Yarmouth, Nova Scotia. His biographer states that Bergman was sponsored to come to the United States by Professor Richard von Mises, his doctoral supervisor. Von Mises, himself a scientific refugee, had been one of a large group of German scholars hired by the University of Istanbul in 1933, before accepting a position at Harvard University in 1939. He was one of the many refugee scholars who helped his friends and colleagues to come to the United States.[21]

Bergman and Tisza both found permanent and secure places of refuge. Guido Beck would also escape the Soviet Union and survive, but among the wandering scholars, his quest for a safe haven was among the longest and most convoluted. Beck began his post-Soviet adventures at the Niels Bohr Institute in Copenhagen. Realizing that Denmark's small academic community was already overwhelmed with refugees, he departed for Paris in January 1938, hoping that he would be offered another position in Kansas. In January 1938, the university did apply for a grant from the AEC, but withdrew its application when it learned that it had committed to making Beck a permanent faculty member after the committee's funding expired. Beck, however, remained unaware of these events for the next six months.[22]

In Paris, Beck survived with a small grant from the reception committee. His monthly stipend was so small that he was forced to leave his wife in Copenhagen and his mother in Prague. In 1932, Beck had been a high flyer, but six years later in Paris, according to Louis Rapkine, the head of the French committee, "prospects are rather dark for him, for we are invaded with exiled theoretical physicists, who ... according to our advisors, are superior to Guido Beck."[23] Although born in what in 1938 was Czechoslovakia, Beck held Austrian citizenship. The German annexation of Austria in March rendered his passport void, and as a Jew he was unlikely to be given a German replacement. That left him effectively stateless. In January 1939, Rapkine's committee found Beck a position as the theoretical physicist at the National Centre for Scientific Research, an academic backwater in Lyon.

France's declaration of war in September led to Beck's internment. His imprisonment was brief, however, and he was allowed to return to Lyon just before the German invasion of France. Lyon was part of Vichy France and was soon flooded with scientists, both French and refugees, both Jews and gentiles, fleeing Paris. Beck used his connections in Lyon to help many of the displaced scholars escape Vichy by way of French North Africa. Finally, in December 1941, he managed to finagle a visa to Portugal, simultaneously arranging a temporary teaching appointment at the Universities of Coimbra and Porto.[24] While in France and Portugal, he tried desperately to arrange visas for his mother and the rest of his extended family to leave Czechoslovakia. His efforts failed. All were sent to Theresienstadt concentration camp and perished in the Shoah.[25]

After two years in Portugal, the authoritarian government of António de Oliveira Salazar demanded that all Jewish refugees leave the country. Beck was forced to flee again. He finally found what he hoped was a permanent research position, at the Observatorio Astronómico in Córdoba, Argentina. Beck was forced to abandon theoretical physics and became an astrophysicist. In 1946, however, he became an adviser to Argentina's ill-fated atomic energy and bomb program. Beck arranged for Heisenberg, his former mentor, to visit Argentina to advise on the project. In February 1947, an article appeared in the American magazine *New Republic* outlining the Argentinian atomic bomb program, Beck's role in it, and Heisenberg's impending visit. Heisenberg's visit was immediately cancelled. Beck wrote a letter to the *New Republic* denying that he was involved in state-supported atomic research and stating that Heisenberg was coming to Argentina to merely advise on the training of young physicists. Henry Wallace, the politician who had recently assumed the editorship of the *New Republic*, disputed Beck's claims. Wallace concluded his letter to Beck: "It distresses me deeply, Dr. Beck, that scientists like yourself, perhaps innocently, are being edged or forced into actions which may result in a world catastrophe at some future time."[26]

Beck stayed in Córdoba for eight years before again being forced to flee what he described as the persecution of professors by the Perón regime. Finally, Beck found permanent refuge as Titular Professor at the Centro Brasileiro de Pesquisas Físicas in Rio de Janeiro.[27] Since leaving the Soviet Union, Beck had fled four different countries and worked in seven different universities and research institutes. He was never able to regain his past promise as a theoretical physicist; instead, in South America as in the Soviet Union, he devoted his time to educating a new generation of physicists.[28] While Beck was forced to endure a multicountry and multiyear path to finding a secure haven, he was much luckier than other refugee academics trapped in France after the German conquest.

Edgar Lederer and his family arrived back in France in December 1937. Lederer also was able to find work quickly, this time at the Institute of Physical and Chemical Biology in Paris. In March 1938 he was allowed to submit a thesis to the University of Paris, even though it had been written in Russian while he was in Leningrad. He received his second doctorate (Doctorat en lettres), the equivalent of the German habilitation. Lederer was bemused that he had received the degree without any formal time in a French doctoral program. This qualification ensured Lederer's acceptance as part of the French scientific community. In December, Lederer became a naturalized French citizen.

Nine months later, when war was declared against Germany, Lederer was conscripted into the French army. His service in the army was brief and unpleasant. His unit was made up of naturalized French citizens, and the non-commissioned officers treated their "foreign" soldiers quite harshly. The birth of Lederer's fourth child allowed him to apply for demobilization, and before the German invasion, he was back at the Institute of Physical and Chemical Biology; since the start of the war, this institute had become part of the CNRS in Paris.

In June 1940, just before Paris fell to the Germans, Lederer and his family fled in an old car with the family's mattresses tied to the roof. The Lederers arrived at the Pyrenees and were seeking a way to cross them into Spain when the armistice between France and Germany was signed. When it became clear that the Germans would not immediately be occupying southern France, Lederer wrote to Claude Fromageot, Professor of Biological Chemistry at the Faculty of Sciences at Lyon, for a research position. Fromageot hired Lederer and placed him in a senior research position. Lederer was able to work at Lyon until 1941, when the Vichy French regime passed anti-Jewish legislation that forced the dismissal of all Jews from government positions. Fortunately for Lederer, for some time he had been working for the perfume industry, studying natural animal perfumes. His work was so promising that he received a larger contract to support himself and two researchers; he was also provided with a kilo of ambergris, the intestinal excretion of the sperm whale, valued at more than $50,000. Using these private funds, Lederer continued to work in Fromageot's laboratory.

As a Jew in Vichy France, life remained uncertain for him and his family. Even so, while many others were fleeing France, Lederer passed on an opportunity to go to the United States in 1941. Louis Rapkine, the Parisian-based Jewish Canadian chemist with close connections to the Rockefeller Foundation, arranged for a US visa and a scholarship for Lederer. Lederer would later admit that in Lyon he had been "stupidly optimistic," as well as exhausted from all the moves he had endured since 1933. Of course, in 1941 he had no idea of the fate that awaited most of Europe's Jews. Ultimately, he decided to stay in France because he had been unable to secure permission for his parents, who had migrated to France from Austria after the Anchluss in 1938, to accompany his family to the United States.

As the situation became more dangerous for Jews in Vichy, Lederer took desperate measures. Max Roger, one of the perfume executives who supported Lederer's research, became mayor of Neuilly and was able to arrange fake ID papers for the Lederers. Lederer became Edouard Lefèvre, born in Abbeville, a town whose town hall had been destroyed in 1940 along all of the civil registers. This kept the family safe, although Lederer was once almost arrested by the Milice française, the paramilitary police of the Vichy regime. In May 1944 he had another close call when an Allied bomb exploded on Claude Fromageot's desk in an office adjacent to Lederer's laboratory. Four other employees were not so fortunate and were killed in the explosion. Lederer was at the time in central France, seeking safe hiding places for his children. Miraculously, the entire Lederer family survived the war.[29]

In October 1937, Helmuth Simons, the German biochemist, arrived in Paris from the Soviet Union. Almost nothing is known about Simons's time in the Soviet Union. While there, he had broken off all contact with those in Britain who had helped him get there; this ingratitude so infuriated A.J. Makower, head of the Professional Committee for German Jewish Refugees, that when he learned of Simons's arrival in Paris he refused "to take any interest in his case." Makower advised the SPSL to refuse to help Simons further, advice they appeared to have heeded.[30] The story of Simons's survival in wartime France is perhaps the most incredible of any scientific refugee. Most of that story, particularly after his arrival in Marseille, is based on Simons's own writings. His propensity to lie means that his accounts must be treated with a fair degree of caution. Still, there is enough external verification of elements of Simons's story, such as his son joining the French Foreign Legion and the need to pay bribes to book passage on ships leaving Marseille, that it cannot be dismissed out of hand.[31]

Having been cut off from British assistance, Simons sought it from the AEC. In a number of documents and letters sent to the AEC and potential employers in the United States, Simons neglected to mention his time in the Soviet Union, suggesting that he had moved directly to Paris from London in 1935. Why he once again engaged in deceit in his quest for employment is unclear, but he may have understood that many Americans did not want entanglements with Communists. It is also possible that the nature of Simons's work in the Soviet Union required him to hide it from potential academic employers in the West. We know that Simons had been employed at the Military Medical Academy in Leningrad, and, given his specific expertise, it is possible his research may have been in bacteriological warfare. As we shall see, he would retain an interest in this type of military research.

Simons's application to the AEC in 1938 was rejected, despite his academic qualifications and record of scholarly publication. Overwhelmed by the large number of applications from equally desperate refugee scholars, the AEC stuck to the letter of its mandate, offering grants only to people who had been dismissed from German university positions. A little over a month later, the AEC expanded that mandate to include all university faculty dismissed as a result of Nazi persecution, but even then, Simons did not qualify. The American organization refused to find a position for Simons in the United States, but when they learned later that he could speak Spanish, the AEC put his name forward for several positions in Latin America; Simons, however, was not selected for any of those positions.

In Paris, Simons found part-time employment at the Pasteur Institute, conducting research in the parasitological and haematological laboratories.[32] He was able to develop an original technique for detecting the blood parasites that caused trypanosomiasis (African sleeping sickness).[33] Despite his success, the work did not pay very well, and, as he had done in the past, he eked out a living taking every quasi-scientific job he could find. In 1939, he took a piecework job writing abstracts in English of articles in the *Comptes Rendus Societé de Biologie* for the American publication *Biological Abstracts*.

At the start of the Second World War, the French government interned Simons because of his German citizenship, as it had with Beck. The sanitary conditions in the camp were "quite indescribable."

On 1 December 1939, Simons was released from the camp because his son had volunteered for the French Foreign Legion. Simons returned to Paris. On 14 May 1940, with the German invasion of France, Simons was interned once again, the result "of the new psychosis-wave" of the so-called "5th column" and the "parachutists" that swept over the country. As a father of a "war hero," Simons was supposed to be released from the Paris cycling track where he was first interned. Just moments before his name was to be read out announcing his release, an air raid caused such confusion that all remaining prisoners were sent to a camp at La Braconne in southwest France. There Simons languished with 1,200 other men who had been rounded up by the French authorities. On 23 June at 11 p.m., with the German army just 20 kilometres away, the guards opened the gates of the camp. In the pouring rain, the men, all refugees from the Nazis, began fleeing the approaching German army. The following day, the group was machine-gunned by a German army unit and twenty-four men were killed. The group broke up, and eventually Simons managed to find work on a farm, which saved him from starvation. After only eight days of refuge, however, German soldiers began to occupy the area, and once again Simons was forced to flee.

With trains and buses not running, Simons decided he had no choice but to walk to safety. On 25 June, carrying 20 kilograms of luggage on his back, Simons headed for the Mediterranean seaport of Marseille, 887 kilometres away. He walked mainly at night to avoid Marshall Pétain's "gangster-gendarmes." It took him thirty days in blistering heat to reach the city. Simons had chosen Marseille because of its warm climate. He anticipated that Paris would soon face a terrible winter because of a looming shortage of coal and food. Marseille had the added advantages of being a university town as well as a point of embarkation for ships to the United States, if he could secure a visa. Simons explained to John Flynn, the editor of *Biological Abstracts*, that "for many hundred thousands of German, Austrian, and Czechoslovakian refugees (Jews as well as "Aryans") there is no more any possibility of working and breathing." This was, he explained, "only the very first consequence of Hitler's conquest of France."

Simons was entitled to lodgings at Fort Saint-Jean, where the Foreign Legion operated a training camp. He was allowed to move freely about the city as long as he returned to the fort by 9 p.m. Police sweeps through the city looking for Jews were frequent, and Simons claimed that he was arrested no less than thirty-three times. He was usually released in a few hours, after the police could confirm his status as the father of a soldier in the Foreign Legion. Sometimes he was held for several days or even weeks; each time, the police released him with an apology. They claimed they had simply forgotten he was being detained.

Simons did everything he could think of to escape France. He begged John Flynn to write to the US consulate in Marseille in support of his application for a US visa. He asked Flynn specifically to inform the consul that his employment with *Biological Abstracts* would continue in the United States and that his work has "special value for the [*sic*] Franco-American cultural relations between the scholars of both countries."[34] *Biological Abstracts* could not offer Simons employment, especially given that most French scientific journals had ceased to publish in 1940. Flynn, however, did what he could for Simons. When he learned that the AEC only offered grants to universities seeking to hire specific refugee scientists, he wrote to a number of American universities that he thought could benefit from Simons' expertise. Unfortunately, Flynn's efforts came to nought, and Simons remained trapped in Vichy France. Simons claimed that he did manage to arrange for a free berth on a ship to New York via Lisbon, but that corruption on an unidentified refugee committee delayed the day of his scheduled departure until April 1943. Top bureaucrats on the refugee committee – he does not identify which one – accepted bribes from wealthy refugees. The bureaucrats then reallocated all of the early departure dates, including his, giving them to those who had paid them. It is impossible to identify the specific organization Simons mentions, since a number refugee organizations operated in Marseille. These groups included "the American Friends Service Committee (AFSC), Jewish organizations HICEM, and Jewish Labour Committee (JLC), and the leftist Centre Américain de Secours (CAS). The second group was composed of revolutionary Communists and anarchist militants."[35]

On 26 August 1942, Simons was arrested by French detectives and this time was immediately sent to the Les Milles detention camp to await transportation to his almost certain death in Poland. From the camp, Simons witnessed horrific scenes of the first transports. Families were forcefully separated before being brutally stuffed into trucks for the short drive to where train cars waited to carry people to their doom. "Whenever such transports left," Simons wrote, "we were able to hear the near-bestial cries of despair of the inmates." The camp was a reflection of the strange realities of Vichy France. The United States was not yet at war with Vichy, and the American Quaker Committee managed to set up a telegraph office inside the camp. Simons was able to cable his son in Morocco. Soon "a passionate appeal of protest" arrived from the Foreign Legion, demanding that all relatives of legionnaires be released. Simons was one of a handful removed from the camp, all of whom had been provided with a "pledge of honour" that they would not be handed over to the Germans. He and the others were then transferred to Rivesaltes, an isolated camp near the Spanish border, where they were promised there would be no transports out of France. This promise was soon broken, and Jews were shipped to Poland. The commander at Rivesaltes, who was a secret member of the Resistance, finally persuaded Vichy officials to release the relatives of French soldiers, just days before Simons was scheduled to be deported. American Quakers working in the camp gave Simons new clothing to replace the rags he now wore, and he was able to return to Marseille. While there he was repeatedly arrested, only to be released after the Foreign Legion intervened. By this point, Simons was secretly working for the Resistance. He employed his "skill and knowledge of chemistry to forge innumerable quantities of passports and to supply endangered politicians and Jews with forged certificates of birth."

Simons's final arrest occurred in late February of 1943, when he was rounded up as one of twenty intellectuals selected by the Germans to be executed in revenge for the murder of a German soldier. In fact, the German had been shot by an Italian soldier in a fight over meat supplied for rations. Unwilling to allow knowledge of the actual circumstances of the murder to leak out, the Germans claimed that the soldier had been killed by the Resistance. Simons and the other prisoners were subjected to brutal beatings by French police acting under the orders of the Germans. They were then taken to the basement cells in police headquarters, where they were placed in large upright wooden boxes – what the prisoners called wooden coffins. They were to be confined that way until their execution. On 4 March 1943, two nights before his scheduled execution, Simons noticed that his cell door was open. He emerged to find that three others, two Americans and an Englishman, had also been released. The men were certain they were going to be shot "trying to escape," but instead a guard approached and told them to follow him.

The guard led them through a labyrinth of underground passages to a small shed near the harbour. He explained that the Resistance had paid 40,000 francs for each man to allow them to escape. The guard then gave each man an envelope containing 10,000 francs, "ingeniously forged" identification papers, and specific instructions for meeting a contact who would help them escape from France. Simons was sent to the main railway station, where he was to meet his contact. A young woman, cleverly disguised as an elderly nun, met Simons and handed him another envelope with a train ticket for the Swiss border, along with instructions on how to cross over into Switzerland.[36]

Image 14. Helmuth Simons with the Swiss-Jewish-Communist artist Alis Guggenheim in 1943, after his release from a Swiss labour camp. Simons's gaunt appearance is likely the result of his harsh treatment during his time in France. (Schweizerisches Sozialarchiv, Alis Guggenheim Papers, F 5090-Fa-101.)

Simons's ordeal, however, was not yet over. Because he had illegally entered their country, the Swiss authorities placed Simons in a labour camp with other Jewish refugees. Simons soon received an offer of employment from Zurich Polytechnic Institute, but only if someone in the Jewish community was willing to post a surety bond. Simons's cousin, who lived in Zurich, knew the wealthy German art dealer Hans Wendland and asked him to provide the Jewish community with the money for Simons's release. Wendland was heavily involved in the Nazi looting of art, yet he willingly provided the necessary funds.[37] Simons was released, and was safe for the first time since the Germans invaded France three years before. He would later act as a clandestine source of information on German biological and chemical warfare for Allen Dulles, the agent in Switzerland for the US Office of Strategic Services. On 8 December 1943, Dulles telegrammed Washington that Simons had reliable information that the Germans were poised to use bacteriological weapons, likely in the form of the poison bacillus botulinus. Professor Simons, Dulles informed Washington, "appears to have reached his conclusions through deductive evidence rather than exact new evidence from Germany."[38] While Simons lacked direct evidence, Dulles believed him because he claimed that before he left Germany he had worked for I.G. Farben, the huge German chemical and armament manufacturer. If Simons told Dulles this, he lied. Simons's CV makes no mention of him ever having worked for the company.[39]

This information caused a near panic among the Allies, who quickly devoted huge amounts of scientific and medical resources to counter a bacteriological attack. The United States and Canada both created programs to develop and manufacture hundreds of thousands of doses of vaccines against the infection. This work continued right up to the invasion of Normandy in June 1944, even though Ultra decryptions of German secret military communication revealed no sign of any impending plan to use bacteriological weapons. Simons had proven himself not just unreliable, but a liability in the fight against Hitler.[40]

Simons had suffered a harrowing ordeal, but at least he survived. Two physicists who found refuge in France after their expulsion from the Soviet Union – Emanuel Wasser and Fritz Duschinsky – did not. We left Wasser trapped in Poland in 1938. Almost nothing more is known about him and his family except that at some point, likely before the German attack on Poland, he was able to migrate to France. In the late spring of 1940, he fled south to Vichy, where the authorities interned him, first at Saint-Cyprien and later at Les Milles. His wife Sara and his daughter Elizabeth lived nearby in the southern French town of Haute-Garonne. While Emanuel was interned, a family friend or relative in Pittsburgh, Pennsylvania, deposited $1,850 with the Jewish Transmigration Bureau in New York in a desperate attempt to facilitate the family's migration to the United States. These efforts apparently did not succeed.[41] On 20 January 1944, Wasser was part of Transport 66 from Drancy camp in France to the Auschwitz-Birkenau extermination camp in Poland. He did not survive. The fate of his family remains unknown, but there is no evidence they succeeded in leaving Europe. It is more than likely that they too perished in the Shoah.[42]

Even less is known about Fritz Duschinsky after he was forced to leave Leningrad in October 1937. He returned to his native town of Gablonz, Czechoslovakia, but after that, his movements were unknown until after the Second World War. After the conflict, the SPSL tried to trace all the scholars in its files whose fate was unknown. In May 1947, Ilse Ursell, the postwar secretary of the SPSL, began work to ascertain the fate of Duschinsky. She used the lists kept by various refugee organizations and discovered an Erich Duschinsky, native of Gablonz, living in Britain. Ursell wrote to Erich asking him if he was related to the physicist and if he had any news about his fate. She concluded her letter: "I sincerely hope he was able to get away from the continent before the outbreak of war."[43]

In January 1948, six months after Ursell's letter, Erich Duschinsky finally replied: "Dr. F. Duschinsky was my brother. Unfortunately, I can only report that he has not returned from Nazi-occupied France, and vague information is available that he was taken to one of the extermination camps on account of a denunciation, as late as 1944, in Paris, where he was hiding until then." Erich explained that he had not been able to obtain any other particulars of his brother's fate, nor had he been able "to trace any remnants of his scientific work."[44]

Ursell replied to Erich, expressing her condolences over his brother's death after having survived the Nazis until 1944. She could have added that Duschinsky had survived Stalin as well. All Ursell could offer as a means of comfort to Erich was copies of a few scholarly articles written by Fritz contained in his SPSL file. Fritz had sent offprints of his articles to the society during his unsuccessful efforts to apply for assistance prior to his departure for the Soviet Union. Erich was indeed grateful for these reminders of his brother.[45] It is doubtful that Erich ever knew the actual fate of his brother. Fritz was murdered in Auschwitz on 12 January 1942.

Seeking Safe Havens Elsewhere in Europe

France, of course, was not the only country occupied by the Germans. Two other refugee scholars found themselves having to flee once again from the Nazis. The physicist Warner Romberg used his now fine-tuned survival skills to keep himself alive and in the academy after his contract in Dnepropetrovsk in Ukraine was terminated in 1937. Romberg first arrived in Warsaw, then quickly made his way to Prague. From Prague, he applied to the SPSL for a grant and a new position. Since he had left for the Soviet Union shortly after completing his doctorate at Munich, the aid organization had no basis on which to judge his scientific abilities. The SPSL asked several refugee scientists familiar with Romberg's graduate work, including the physicists Hans Bethe and Rudolf Peierls, to comment on Romberg's abilities. Bethe and Peierls judged Romberg "rather average" and "not at all brilliant." He was highly competent in the mechanics of quantum calculations; even so, he was unlikely to find a university position in either Britain or the United States.[46] On the basis of these lacklustre evaluations, the SPSL rejected Romberg's application.

While in Prague, Romberg developed contacts with the Czech scientific community and was able to find some part-time teaching and research work. But he was well aware that the country was under threat from Hitler and that he would end up in a concentration camp if he did not find another country. In September 1938 the infamous Munich Accord deprived Czechoslovakia of its border defences, and few doubted that the country's days as an independent state were numbered. Romberg, desperate to escape anywhere, used his scientific contacts in Prague to connect with Egil Andersen Hylleraas, professor of theoretical physics at the University of Oslo. He offered to work as Hylleraas's assistant. Hylleraas, unwilling to pass up a German-trained physicist who was willing to go to the back and beyond of Oslo, found funding to employ Romberg as a research assistant.

Romberg arrived in Oslo on 20 November 1938, having somehow raised enough money for a plane ticket so that he could avoid being intercepted by the Gestapo. He immediately began collaborating with Hylleraas investigating "the hydrodynamical equations for an ideal incompressible fluid on a rotating sphere which is subjected to the influence of tidal forces." In 1941, the results of the project were published in a collection of research papers by Hylleraas, but by then Romberg had once again been forced to flee the Nazis. In the spring of 1940, he bolted from Oslo one step ahead of the German army, crossing the border into Sweden. From 1940 to 1944, he lived in the Swedish university town of Uppsala. His escape was noted by the German authorities; unable to arrest him, they revoked his German citizenship in 1941. Two years later, they revoked his doctorate as well. Romberg was unfazed by these measures, and as soon as Oslo was liberated late in 1944, he returned to Norway to resume his duties at the university.[47]

Heinrich Luftschitz's tale of survival is remarkable, but it too is tinged with tragedy. The chemical engineer, one of Europe's leading experts on the chemistry of concrete, was expelled from his teaching position at Sverdlovsk in the Urals in 1937. Unable to return to Germany, he once again settled in Yugoslavia. He had been born in Karlstadt (now Karlovac, Croatia), then part of the Austro-Hungarian Empire. In 1937, Karlovak was part of Yugoslavia, and Luftschitz was entitled to citizenship. He worked in the cement industry until Italy invaded Yugoslavia in 1941. The Italians decided Luftschitz was an Italian citizen, and they interned him for racial reasons. He was held at a number of internment camps in Italy. In February 1944 he was transferred to a new camp at Castello Guglielmi on the Isola Maggiore in Lake Trasimeno. The camp was run by the Italian Fascist Public Security Police, and most of the prisoners were Jews from the region, transferred from a camp in Perugia. As concentration camps went, it was a pleasant place. The commandant ensured that the inmates received the same rations as other Italian civilians and allowed his charges to wander the island at will during daylight hours. The guards were poorly motivated auxiliary agents and had little interest in supporting German forces in the region.

In June, after Rome had fallen to the Allied armies, the camp commandant received instructions to surrender his charges to the Germans before the approaching Allied armies could liberate them. But on the night of 11–12 June, some of the guards decided to abandon their posts and return home to their families on the mainland. They offered to take with them any willing Jewish prisoners and to get them as close to the Allies' lines as possible. Luftschitz was one of five Jews to accept the guards' offer. One week later, while crossing the battle line between the German and Allied armies, likely the "Trasimene Line," Luftschitz was severely wounded in the right hand and arm by a German hand grenade. He spent three months recuperating at a convent hospital in the nearby city of Orvieto.[48] Luftschitz had survived, but he would never return to an academic career. When the war ended he felt he had no choice but to abandon any hope of returning to his cherished research and teaching position in Dresden, which had been destroyed by Allied bombing. In 1945, now sixty-one years old, he returned to Yugoslavia, where he went back to work in the cement industry.

In 1946, Luftschitz's whereabouts were traced by Ilse Ursell at the SPSL. Working through a refugee search organization, she found an address for Luftschitz in Trieste. Ursell asked the Red Cross to contact him and make it possible for him to inform the society of his situation.[49] Luftschitz was pleased to hear from the SPSL. He informed Ursell that the war had shattered his family; two of his brothers had been killed in Germany. He was still recovering from wounds suffered during his escape. He hoped he could receive financial and moral help. In a follow-up letter, he wrote to the SPSL asking whether there was any possibility of his being compensated for his illegal dismissal from his research and teaching positions in Germany. Specifically, he hoped he could claim the pension that had been stolen from him by the Nazi government.[50]

Ursell did considerable research on Lutschitz's behalf, but to no avail. She informed him that, as yet, no provision had been made for paying pensions to people living outside of Germany.[51] Luftschitz's surviving brother, Giuseppe Ludovisi-Luftchitz, who also lived in Trieste and had a better command of English, wrote several further appeals to the SPSL on behalf of Heinrich. Giuseppe pleaded for the society to help his brother find any position anywhere outside of the Communist world. On 20 March 1949, Luftschitz's brother made one final appeal to the society. He wrote: "It is a great *pitty* [*sic*], that a man, capable in every respect, is obliged to work in a cement factory, in Yougoslavia [*sic*] not affording him whether moral, nor financial satisfaction."[52] Reluctant to

give the Luftchitzes any false hope, Ursell replied with unusual can-
dour, telling them about the extreme difficulty of finding any position
for him "on account of his age," since most companies had retirement schemes that required their employees to retire at age sixty or
sixty-five and were unlikely to want to employ a sixty-three-year-old
engineer. Ursell concluded: "I am very sorry not to be able to hold
out more hope but under the circumstances it would not be wise or
kind to mislead either yourself or your brother."[53] Nothing more is
known about Luftschitz. Once one of the world's leading experts on
the chemistry of cement, his name disappeared from the historical
record at this point. He was defeated not by Hitler or Stalin, but by
time. Luftschitz's is certainly among the most tragic stories of those
who survived the Second World War.

Conclusion

Like so many other refugee Jews in Europe, these ten scholars who
fled the Soviet Union had to endure the Nazis for a second time.
For those who fled to Germany, persecution and violence had made
the country too dangerous to stay, but prior to the start of the war,
killings of German Jews remained the exception, not yet the norm.
Regarding the scholars who tried to return to Germany, both were
able to reach out to family or friends to escape Germany. Had they
failed to find another sanctuary, they likely would have died in the
Holocaust.

Six of the eight scholars who found themselves in Nazi-occupied
Europe also lived – a much higher survival rate than for the general
population of displaced Jews in occupied Europe. In large measure
their survival was simply a function of the country in which they sought
refuge. Romberg escaped to neutral Sweden just ahead of the German
army. Luftschiftz was never under the control of the Germans and managed to escape to the Allied lines just before his Italian-run camp was
turned over to them. France had one of the highest survival rates of Jews
in occupied Europe; 90 per cent of French and 60 per cent of foreign
Jews survived. Whether or not the scientists survived depended on the
same factors that determined the fate of other Jews who had recently
arrived in France. Lederer and Simons – the former through citizenship,
the latter through his son's military service – became de facto French
Jews. Bergman, Tisza, and Beck were able to use their connections with
the international academic community to escape from France. Duschinsky, Wasser, and likely the latter's family perished in the Holocaust. They,
like so many other murdered refugee Jews, likely "spoke little, if any,

French," were unable to fit into French social and academic networks, and had little money. "These vulnerable foreign Jews were the first to be targeted by both Vichy and the Germans."[54] Add to this horrible toll the entire Gilde Family – Siegfried, Ruth, and Marina. Most of the refugee scholars fleeing the Soviet Union, however, managed to survive the Second World War and re-establish themselves elsewhere. It is to their stories that we now turn.

11 Survival and Triumph

For the scholars who managed to escape both Stalin and Hitler, the quest for a secure refuge where, if they were lucky, they could continue their academic careers was not an easy one. The Soviet Union had offered the possibility of a new beginning, but the promise had been a chimera that proved fatal to many. By 1938, competition among refugee scholars for academic jobs was fiercer than ever as the numbers of those seeking employment continued to grow and the number of positions began to shrink. Moreover, some places of refuge, even if they were beyond the reach of Nazi Germany after its conquests in the first two years of the war, offered at best a precarious existence.

Sweden and the Ensnared

Sweden was the first place of safety for four of the academic refugees fleeing the Soviet Union, most of them from Leningrad. Most of them left the Soviet Union by crossing into the Baltic States, but these small republics offered little opportunity for them. Ernst Emsheimer and Wolfgang Steinitz and their families went from Leningrad to Stockholm in 1937. The Steinitzes were joined there by Wolfgang's brother-in-law, the physicist Hans Jürgen Cohn-Peters. Cohn-Peters had worked with his former doctoral supervisor, Fritz Lange, in Kharkiv, but had fled or been expelled to Sweden by the beginning of 1938.

All of these scholars settled in Stockholm and soon immersed themselves in the small but active community of Communist and socialist German expatriates living in the Swedish capital. Sweden offered little opportunity for academic jobs for these refugee scholars, since, according to Swedish musicologist Jan Ling, "Sweden in the late 1930s was strongly pro-German and in the leading circles many people – even intellectuals – sympathized with Nazi ideology."[1] Being both Communists and

Jewish made it impossible for any of this group to find permanent academic employment before the end of the Second World War. Moreover, the Swedish security services carefully watched all of them, fearing, not without reason, that some were active Soviet agents.[2]

Given this less than warm welcome, only Emsheimer was able and willing to make the sacrifices necessary to find a permanent place in the Swedish academic community. For twelve years after arriving in Stockholm, he was employed as an underpaid researcher at the National Ethnographic Museum, in conditions so harsh that his future graduate student, Jan Ling, compared his position to that of a "galley slave."[3] At the museum, Emsheimer discovered a wealth of anthropological musical material that Swedish researcher Sven Hedin had gathered in Mongolia before the war. In 1943, working with Swedish scholar Haslund-Christensen, Emsheimer published an analysis of Mongolian music and instruments that, according to Ling, "set a methodological pattern for ethnomusicological research the world over."[4] Despite this groundbreaking work, Emsheimer would have to wait until 1949 before receiving a permanent academic position as head of the Museum of Ethnic History. By this time, Steinitz and Cohn-Peters had left Sweden, both migrating to East Germany.

Outside of this group of politically active refugee scholars in Stockholm was the mathematician Herman Muntz. In October 1937, after being expelled from the Soviet Union, Muntz first went first to Tallinn in Estonia, where he was offered a visiting professorship at the Technical University. When he learned that he would have to teach his courses in Estonian, he had no choice but to turn down the offer. In February 1938, he and his wife moved to Sweden, where they requested political asylum. Even before his arrival in Sweden, Muntz had applied to the SPSL for assistance in finding a permanent academic position elsewhere. He had a strong case, based on his remarkable career as a mathematician, and he provided an impressive list of references, among them Albert Einstein and some of the leading mathematicians in Europe.[5]

Muntz, however, did not receive a single job offer. When he left the Soviet Union, he was already fifty-three years old, and just as with Heinrich Luftschitz, his age worked strongly against him. As well, like Emanuel Wasser, Muntz was not a refugee from a German university; he had never held a permanent academic appointment before the Soviets offered him a professorship in Leningrad in 1929. Although Muntz was now a refugee from the Soviet Union, he was not in immediate danger because he had been accepted as a refugee in Sweden.[6] Another issue that came to haunt him was that Einstein, his former collaborator, expressed some serious concerns about Muntz's inability to submit his scholarship to "a proper level of critical analysis," and about his mental health issues from almost twenty years earlier. In previous years, Einstein had written hundreds of uncritical letters of reference for displaced scholars, many of whom he barely knew. By 1938, Einstein was well aware of the highly competitive nature of job competitions open to former German academics, and of the need to be brutally honest in his reference letters if they were to be at all useful in swaying hiring committees to select truly outstanding candidates.[7]

As a result, Muntz was forced to remain in Sweden, and while he kept his interest in mathematics, his active research career came to an end. He survived by cobbling together small grants from the Swedish academic refugee organization and landing a few research contracts, and by giving private lessons (as he had in the 1920s in Germany). His income was just enough to provide him and his wife Magdalena with a small apartment in a good district in the city. Magdalena died in 1949, and after her death Muntz became a recluse, suffering from failing eyesight and increasing ill health. He became a Swedish citizen in 1951. In his last years, he was legally blind and was supported by a small pension from a Swedish Jewish charity. In 1956, he died at age seventy-one, virtually unknown in Sweden and all but forgotten by the international community of mathematicians. Only one obituary, in a leading Swedish newspaper, marked Muntz's death. It was written by Professor Folke Odqvist, a mechanics specialist and one of the few Swedish academics to have befriended Muntz. Odqvist neatly captured the tragedy of Muntz's life in Sweden:

> Herman Muntz is dead. In spite of the fact that he lived in Sweden for 18 years, the last five years as a Swedish citizen, there are probably not many Swedes outside his nearest circle of acquaintances, that knew that we had among us a mathematician of international fame who was thrown up on our calm shore by the storms of the times, his life saved but with his scientific activities broken.[8]

Odqvist concluded with a tribute to his friend's devotion to his beloved wife:

> Herman Muntz lived in an exceptionally harmonious marriage and his wife Magda meant much to him, not in the least in order to keep his floating spirit down to earth. After her death in 1949 he only seldom saw his friends and he went every day to her grave in the Jewish cemetery with fresh flowers as long as he could. Now he is gone. Let this be a modest flower of memory from his Swedish friends. Let his memory be blessed.[9]

Britain and the Second Wave of Academic Migration

Muntz was not alone in finding it impossible to land an academic position after being expelled from the Soviet Union. A major reason for this difficulty was that by 1938, Great Britain, which had in the early years of the crisis found more research and teaching positions for academic refugees than any other country, was reaching a saturation point and was less and less able to absorb any more. Among those scholars who fled the Soviet Union in 1937–38, only Kurt Zinnemann found both refuge and a university faculty position in Britain. The Soviets released Kurt and Ena Zinnemann from prison in May 1938. The NKVD provided Ena with her underwear, fur coat, hat, and rings. Kurt was given back the suit he had been wearing when he was arrested. Both showed signs of the near starvation diet they had endured in prison; Ena was swollen from malnutrition, while Kurt was terribly thin. In Kiev they were placed on a train, which took them into Poland, where the German Red Cross provided the couple with a train ticket to Berlin and enough money for one day's worth of food.

The Zinnemanns had no illusions about what would happen to Kurt if they returned to Germany, and they were determined to avoid going there whatever the cost. Ena was able to sell her gold rings to some Red Cross workers and so was able to purchase tickets to Warsaw. In Warsaw, Kurt approached the German consul, asking for an emergency loan, which he promised that his father in Frankfurt would repay. The consul shook Kurt's hand, even though he knew he was a Jew, and gave him enough money for them to continue their journey. The consul's parting words were: "I wish you luck." The couple finally found refuge in Kraków with Kurt's uncle, Dr Marek Margulies. Kurt was able to contact his best friend Franz Gugenheim, who by 1938 had established a medical practice in London. Three years earlier, Gugenheim had been one of fifty German general practitioners given special permission by Scottish medical authorities to emigrate from Germany and continue to practise medicine. Gugenheim's parents were able to purchase tickets for Kurt and Ena to sail from Danzig to Britain, their entry into the country sponsored by Franz. Their voyage was somewhat perilous, for the ship travelled through Germany's Kiel Canal, which links the Baltic to the North Sea. Fortunately, the German government still respected international law enough not to inspect the ship. There is little doubt that had they done so, Kurt would have been arrested.

The couple intended to stay in Britain only briefly before trying to migrate to the United States. Even so, Kurt applied to the SPSL for assistance in finding a medical research position. The SPSL contacted

J.W. McLeod, professor of bacteriology at Leeds University Medical
School. While in Kharkiv, Kurt had corresponded with McLeod about
his research into a major diphtheria epidemic that had swept through
the city soon after he and Ena arrived in the Soviet Union. McLeod, one
of the world's leading experts on the disease, had been impressed by
the couple's findings and quickly offered Kurt a research assistantship,
at £250 per annum. McLeod explained to the British government that
no British subject with suitable qualifications had applied for the job
after extensive advertisements. By early July, Leeds had offered Kurt the
assistantship, on the condition he receive British government approval.
The SPSL applied to the British government for a work permit on Kurt's
behalf; outlining the steps taken by Leeds to find a British candidate,
as well as Kurt's desperate situation. Walter Adams explained: "As a Jew
and because he has been in Russia it would be particularly dangerous for
him to return to Germany. It is therefore a matter of saving his liberty if
not his life that he should be re-established outside of Germany."[10] The
hiring process occurred so quickly that it is doubtful that an extensive
job search for a British candidate had actually taken place, but stating
that a search had occurred was required by British law. As long as no one
complained to the Home Office, the authorities were willing to look the
other way to assist a qualified and desperate refugee. By August, Kurt,
assisted by Ena, had started work at Leeds. Their first research paper,
published in January 1939, examined the virulent diphtheria epidemic
that had swept through Kharkiv shortly after their arrival. Since all
of Kurt's scientific notes were destroyed by the NKVD, it can only be
assumed that the article was based on memory and on information he
had sent to McLeod.[11]

For the Zinnemanns, the position at Leeds not only rescued them
from a desperate situation but also put them in a position to save Kurt's
parents, Leah and Lazar. Kurt's parents owned an Etam lingerie fran-
chise shop on Kaiserstrasse, the high-end shopping street in Frankfurt
am Main. Their shop had been attacked on Kristallnacht, its windows
smashed and the fine panties and stockings stolen as gifts for the rioters'
girlfriends and wives. Leah and Lazar were terrified to leave their home,
and a family friend had to bring them food. Kurt was not yet in a position to sponsor his parents, but a colleague, Dr Maurice Gordon, having
learned of the situation, agreed to stand as surety, allowing Leah and
Lazar to migrate to Britain in January 1939. Leah packed in such haste
that she forgot all the family photographs; however, she hid some silver-
ware, including the family' s Pesach cup, and a few pieces of jewellery at
the bottom of a small bag, then filled rest of the bag to the brim with pins
and needles, to discourage any search by the Nazis.[12]

Having finally reached safety, the Zinnemanns faced one final indignity. In May 1940, like Herbert Fröhlich, Kurt, Ena, and Lazar were all interned on the Isle of Man. While the internment camp was crowded, it was far better in every way than what the couple had experienced in the Soviet Union. Still, Kurt and Ena were again separated in the men's and women's camps, only occasionally being able to see each other. They spent the next ten months in the camps, although Lazar was allowed to return to Leeds on his sixty-fifth birthday in November.

The Zinnemanns were extremely fortunate to have found refuge for themselves and his parents and an academic position for Kurt. Other German academic refugees from Stalin who managed to reach Britain had far less success. In early 1938, Walter Zehden was forced to exit the Soviet Union when his visa expired, leaving behind his Russian wife Zlata Aleksandrova and their young child. Without a visa to travel anywhere else, Zehden went to Berlin; desperate to leave Germany, he turned to J.G. Crowther, the British journalist who had written about Zehden in his book *Soviet Science*. In May, Crowther offered Zehden a place to stay in his flat in London, for several months if necessary, while he searched for a position in Britain.

When Zehden arrived in Britain in May, Crowther introduced him to Mark Oliphant, an Australian-born professor of physics at the University of Birmingham. That university had already hired several prominent refugee scientists, including the nuclear physicist Rudolf Peierls. Oliphant was greatly impressed by Zehden and immediately offered him a position. Much to Oliphant's surprise, however, when he approached university officials for final permission to use funds from an endowment to hire Zehden, they turned down the request, telling Oliphant they had "already done as much as they could for refugees" and that in the future they would be using the endowment to fund "British subjects only."[13] Britain's universities were reaching a saturation point in terms of academic refugees. Fortunately for Zehden, however, his skills at building, modifying, and maintaining scientific instruments were applicable to the private sector. On Zehden's behalf, Crowther and Oliphant approached Adam Hilger, a prominent manufacturer of scientific equipment. Hilger promptly hired Zehden, who would spend the rest of his career working in industry. There were major advantages to being in the private sector. In the spring of 1940, when German refugee scholars were being rounded up for internment, the management at Hilger's firm successfully appealed to authorities that Zehden was indispensable for the war effort. So Zehden avoided internment and found employment in Britain; sadly, though, he remained cut off from his family in the Soviet Union.[14]

To the Promised Land: The Ensnared in the United States

The United States remained the Promised Land for most of the academic refugees. Immigration to that land, however, was severely constrained by the limited quotas of visas given to people from each country. There were ways to jump the visa queues, but only if one had a job offer, money, or a willing sponsor of sufficient means. Michael Sadowsky was the first of those expelled from the Soviet Union to succeed in finding refuge and an academic position in the United States. His advantage was that he had once been employed by the University of Minnesota. After being laid off in 1933, he had maintained regular correspondence with his former colleagues there. Having fled the Gestapo in 1937, the Russian-German mathematician found himself trapped in Palestine. Unable to find employment there, he and his family lived with his wife Hilda's parents and brothers in the German settlement of Kfar Sirkin. Soon after, to be closer to a post office, the family moved to a one-room house in East Jerusalem. Michael now began a letter-writing campaign to solicit assistance from his former colleagues in Minnesota and from a growing list of German refugee mathematicians in the United States. In early September 1937 he once again begged the SPSL and the AEC for "urgent" assistance before he became completely destitute.[15] Finally, after months of effort, Sadowsky received a job offer from the Illinois Institute of Technology. It was, Hilda Sadowsky later recalled, "not a very significant job, considering Michael's abilities, but he grasped it."[16]

Job offer in hand, Sadowsky applied to the American consulate for a visa. Authorities at the consulate raised questions about Michael's Russian birth and his time in the Soviet Union, demanding proof that he was not a Communist. He again reached out to his former American colleagues, who sent letters of support to the consulate asserting that he had never expressed any interest in Communism and that he had gone to the Soviet Union with great trepidation and only because of his desperate circumstances. Among the letter writers were Guy Stanton Ford, acting president of the University of Minnesota, and Oswald Veblen of the Institute for Advanced Study at Princeton. Albert Einstein also wrote, attesting to Sadowsky's character. With such powerful support, Michael's visa was granted, and the family arrived back in the United States on 28 January 1938.[17] Sadowsky was the only one of the Ensnared who was questioned about his possible Communist affiliations by American diplomats. Fortunately for him, influential Americans were willing to attest in his favour. Also, he had never been a Communist. Many of the other migrants to the United States carefully hid their past affiliations or sympathies and were lucky not to have been questioned about their time in the Soviet Union.

Surprisingly, the AEC played a smaller role than the SPSL in facilitating the migration of the Ensnared to the United States. Around the time that those fleeing the Soviet Union began arriving in Western Europe looking for another safe haven, the British academic assistance organization began a new "unofficial" program to aid migration to the United States. The SPSL was becoming increasingly frustrated with the AEC's perceived failure to place sufficient numbers of academics in the United States. To circumvent the AEC, the SPSL began providing some scholars with small, one-time grants that included a return ticket to the United States and subsistence for a number of months. Supposedly the scholar was travelling as a tourist, but in fact he was seeking a position in industry or academia. According to one SPSL executive member, the scheme worked, and most of these academic "tourists" found positions in America.[18] Marcel Schein, the expert on cosmic rays, and Ernst Simonson, the industrial physiologist, both benefited from this scheme.

Schein and Simonson followed remarkably similar paths from the Soviet Union to the United States. Both found themselves in Prague after leaving the USSR, then were forced to leave Czechoslovakia, then applied to the SPSL in Britain for assistance. In Britain, the scientists were ranked highly by the SPSL and offered funding, but neither man could find employment there. So they were provided with money to travel to the United States and seek academic positions there. After lengthy searches, both men found temporary scientific work that eventually turned into permanent employment.

In July 1937, while still in Soviet Union, Schein applied to the SPSL for support. Walter Adams at the SPSL knew nothing about Schein, so, for information, he reached out to the physicist community in Britain, which by then included two now well-established refugees, Franz Simon at Oxford and Rudolf Peierls at Birmingham. The general consensus was that Schein was an excellent experimental scientist, but that there were no prospects for him in Britain because his research focused on cosmic rays. In October, however, Arthur Holly Compton of the University of Chicago wrote to the SPSL stating that he was interested in again working with the Czech physicist. Schein was offered £150 to travel to the United States – a considerable sum, given that the annual SPSL fellowship grants to unmarried scholars was just £100 more. Schein sent his family to live with relatives in Switzerland while he proceeded to the United States. He arrived in New York in late January 1938, armed with a letter of introduction from the SPSL and enough money for him to live for three to four months. Schein began his trip meeting with people at the AEC, where he gave the first eyewitness account of the arrests and expulsions of the refugee scholars from the Soviet Union received by the American organization.[19]

Schein then continued on to Chicago, where he learned that Compton had left on an extended cruise in southern waters and was not expected back until April 1. Compton had left no word about hiring Schein and may not have been aware of how quickly he would be able to arrange his crossing to America. Schein spent a desperate three months seeking an academic appointment, including a trip to California where he visited Felix Bloch, the Stanford-based refugee physicist. Bloch drove Schein to universities throughout the region in an equally unsuccessful quest for employment. Having failed at every turn, and almost out of money, Schein returned to Chicago hoping that his funds would hold out until Compton returned. When Compton arrived in Chicago he found Schein in a desperate state. Compton's wife Betty recalled: "I'll never forget when he came after his harrowing experience. He looked just haunted because wherever they'd been they couldn't go down the street and look in the window without being shadowed. It was horrible. He was always so thankful to have come [to Chicago]."[20] Compton took pity on Schein and managed to scrape together enough funds for a one-year research appointment at Compton's Cosmic Ray laboratory. Within a year the temporary appointment was converted to a permanent position.[21]

Ernst Simonson arrived in Prague in the summer of 1937. His impressive academic credentials, which by this time included eighty published articles, led to him being hired to organize a physiological laboratory at the Central Psychotechnical Institution of Czechoslovakia. It took Simonson a year to get the new laboratory organized. By the late autumn of 1938 the Munich Crisis and the subsequent dismemberment of Czechoslovakia had rendered Simonson's position there untenable. In the immediate aftermath of the notorious Munich Agreement, government funding for universities was dramatically reduced. At the institute, three fifths of the academic staff were dismissed.[22]

In late 1938, Simonson wrote to a number of British physiologists about his increasingly precarious situation. Among the scientists Simonson contacted was A.V. Hill, Nobel laureate, secretary of the Royal Society, and a founding member of the SPSL. Hill and other British scientists assessed Simonson's work as having immense value in the field of industrial physiology, and in November 1938 the SPSL offered him a one-year research fellowship. Despite the best efforts of Hill and other British physiologists, however, no research facility in Britain had space to accommodate Simonson. Hill concluded that the only hope for Simonson lay in the United States. The awarding of a fellowship should have guaranteed Simonson entry into Britain, but without the name of an institution willing to host him, a bureaucratic muddle was created, with the Home Office unwilling to approve a visa even though the British diplomats in Prague supported Simonson's application. When Simonson was able to provide the name of an uncle who had already immigrated to London and who was willing to sponsor his migration, the consulate approved a visa for Simonson, but not his wife and child. The reason for this exclusion is unclear, but it may be that Simonson's uncle did not have sufficient funds to post a guarantee for the entire family. Whatever the cause, Simonson refused to leave his family behind. In January, the Czech government ordered the expulsion of all German citizens, and being a German Jew, Simonson could not return to Germany. Simonson's dangerous situation was emphasized in a letter that Kurt Zinnemann, Simonson's friend and colleague in Kharkiv, wrote to the SPSL pleading that they intervene on Simonson's behalf.[23] By mid-January, Simonson and every other German and Austrian in Czechoslovakia faced imminent deportation either to Germany or in a mass transport to South America. As of mid-February, Simonson was still stuck in Prague, uncertain of his status.

Simonson's situation became even more confusing when he obtained an immigration visa to the United States. He had applied for the visa seven months earlier but had never thought he would get one so quickly. The US visa should have smoothed things over for him, but the British consul was unwilling to provide a transit visa while Simonson's earlier applications were still pending. Simonson found himself in a bureaucratic Catch-22. He needed to get to Britain to receive the SPSL funds, which he hoped he could use to go the United States, but he could not get a transit visa to Britain since he had initially applied for a work visa. In London, the SPSL was unable to solve the conundrums concerning Simonson's status. So instead, the society decided to convert three months of Simonson's fellowship into a travel grant for him to establish himself in the United States.

Fortunately, Simonson had learned by now to explore every possible avenue of escape. For several months he had been in correspondence with Helen Lawson, editor of the *Journal of Industrial Hygiene and Toxicology* at the Harvard School of Public Health, about possible employment in the United States. Lawson managed to connect Simonson with Dr Norbert Enzer of Mount Sinai Hospital in Milwaukee, "who had recently made a study on the respiratory and cardiac behavior of silicotics, work quite closely allied with Dr. Simonson's interests." Enzer was willing to offer Simonson an unpaid research position in his laboratory, if he could get to Wisconsin. Lawson believed this was the opening Simonson needed to find employment in the United States, since once he had made a few friends in the academic community a job for the acclaimed scientist would surely follow.[24] With British funds and a US visa, Simonson finally had the means and an escape route, but time was quickly running out. He was able to book passage to New York via London, leaving Prague on 22 March, but seven days before his departure, the Germany army occupied Prague. It was suddenly "very difficult" to leave Prague, but Simonson and his family somehow managed to escape, arriving in the United States on 2 April.[25]

Simonson's ordeal, however, was not quite over. Dr Enzer had no funds to pay Simonson. When the three-month grant from the SPSL expired, Simonson and his family would be left destitute. The SPSL agreed to extend the grant for one more month but warned Simonson that no further extension would be possible. Simonson, however, found that his quest for paid employment was more difficult than he had anticipated. He discovered that industrial physiology did not exist in the United States, and he reasoned that he would need to persuade Americans of the great value of his research. To do so he began writing a two-volume textbook on the subject, which Enzer supported by correcting the German scientist's English. Somehow Simonson managed to survive until the end of 1939. In early 1940, he received a one-year grant from the Emergency Committee in Aid of Displaced Foreign Physicians, a small New York–based relief organization modelled after the AEC. The grant gave Simonson the time he needed to persuade Mount Sinai Hospital and the nearby Marquette Medical School to hire him into a paid position. It had taken Simonson three years after leaving the Soviet Union, but he had finally re-established his scientific career.[26]

Charlotte Schlesinger also found herself in Prague after her escape from the Soviet Union in mid-December 1937. Before fleeing, she had helped Charlotte Houtermans make her desperate escape from Moscow after Fritz's arrest. In Prague, Schlesinger stayed with Rudolf Schwarzkopf, her uncle. Schwarzkopf had been a wealthy, energetic movie producer in Berlin; now Schlesinger found him old, "but his eyes still sparkled." Schlesinger was startled by the sharp contrast between the gloom of Moscow and the radiance of Prague, which at that moment was filled with Christmas lights, markets, and music, and with frivolous people preparing for the holidays. The city, however, was also filled with desperate German refugees selling matches, pins, and shoe polish to survive. Most had no documents and would be sent back to Germany if they were arrested by the police. Schlesinger bought so many matchsticks that she soon had a two-year supply.

Schlesinger and Schwarzkopf realized that they had to plan a way for her to escape before her German passport expired. Charlotte's brother had already succeeded in migrating to Britain, but a British visa would take too long. US visas required an even longer wait unless an American sponsor could be found. Schwarzkopf had friends in the German film industry who had migrated to Hollywood. He wrote to them, and on 26 July 1938, Schlesinger had her American visa. Her brother sent her money to pay her passage to the United States via Britain. She said a tearful farewell to her uncle, knowing that it was likely the last time she would ever see him. Schwarzkopf had raised Schlesinger, given her a musical education, and was now using the last of his resources to save her life. When Schlesinger arrived in New York in late August, waiting for her at the dock was Elsa Houtermans, Fritz's mother. Elsa had escaped Vienna several years earlier by taking a low-paying position teaching Latin, French, and German at Foxhollow, an exclusive private girls' school in Massachusetts. Charlotte Houtermans had written to her mother-in-law asking her to look after Schlesinger. Elsa arrived in the school owner's chauffeur-driven limousine and with an offer to teach music at Foxhollow. Although teaching music at a school was a comedown for Schlesinger, she was safe and, unlike so many of her contemporaries, alive.[27]

By early 1939, after her futile efforts in London to free her husband from a Soviet prison, Charlotte Houtermans was destitute. She had survived on a small SPSL grant of just £10 per month, but with no sign of Fritz and no chance of employment in Britain, the only hope for her and her two children was in the United States. The SPSL offered to pay for her and the children's passage to New York, but the American embassy, overwhelmed by the flood of displaced Europeans seeking refuge, would not even accept Charlotte's immigration application. American diplomats informed her that her only hope was to have a relative in the United States sponsor the family and post an Affidavit of Support; if this could be done, then given her "extraordinary circumstances" they would most likely grant her a visa. Finding a close relative was not a problem: Elsa Houtermans was already there. However, her salary was too low to enable her to post a bond for her daughter-in-law and grandchildren. Charlotte wrote to Robert Oppenheimer, her would-be suitor in graduate school at Gottingen, asking him to co-sign the bond with Elsa. Oppenheimer immediately agreed to help, and by the end of March, Charlotte had the much coveted visa and was on her way across the Atlantic.

Charlotte Houtermans's stay with Elsa was brief. Her contacts in American women's colleges from her two-year stay more than a decade earlier soon brought an offer of a one-year teaching position at Vassar. After the grant from Vassar expired, Charlotte taught at a variety of prestigious women's colleges on a series of temporary teaching contracts until, finally, in 1946 she was offered a full-time tenured position at Sarah Lawrence

College in Bronxville, New York. She would teach and conduct research at Sarah Lawrence until she retired in 1968. Charlotte took on the role not only of a college professor but also of a successful single mother at a time when society frowned on such things.[28]

A few other scientists managed to make their way to the United States from the Soviet Union, but their routes to the country and their subsequent lives remain a mystery. Hans Baerwald, the electrical engineer, was turned down for aid by the SPSL and the AEC committee because except during his time in the Soviet Union, he was an industrial scientist and therefore did not qualify under their mandate. Moreover, Baerwald had wealthy relatives in the United States to sponsor his migration. Baerwald's relatives apparently did sponsor it, and by 1939 he was working for an electronics company based in Cleveland, Ohio.[29]

Saul Levy, the experimental physicist who had worked with Walter Zehden first in Berlin and then in Moscow, fled to Lithuania, the country of his birth, after his expulsion from the Soviet Union. In late September 1937, shortly after arriving in Lithuania, uncertain of how to proceed, he wrote to Professor Michael Polanyi, by then at Manchester University, asking for help. Polanyi had known Levy slightly in Berlin before they both fled Germany, and Levy reached out to him because he was the only former colleague whose address he knew who had managed to settle in Britain. Polanyi forwarded Levy's plea for assistance to the SPSL, recalling Levy to be "a very able man" who was "the main support for Ladenburg while he was his assistant."[30] Despite Polanyi's recommendation, the SPSL quickly concluded that there was nothing they could do for Levy. On learning of the SPSL's decision, Levy asked for help reaching the United States, but in his case all the SPSL would offer was advice that he needed to find an American citizen to sponsor his visa. Nothing more is known about Levy, except that by January 1940 he was in New Jersey. It seems likely that Rudolf Ladenburg, Levy's mentor in Berlin and by 1937 a US citizen and professor at Princeton, helped Levy enter the United States. After publishing one article in *Journal of Applied Physics* in early 1940, Levy disappeared from the historical record. It is unknown whether he was able to continue his career as a scientist.[31]

Academic Careers after the Second World War

If Levy's scientific career came to an end in 1940, his fate was typical of the majority of the twenty-seven scholars that had sought refuge in the Soviet Union in the 1930s and lived through the Second World War. Of the survivors, only a dozen would be able resume working at a university or a research centre, and of this group, only a very few would have careers that came close to the potential they had shown prior to their forced migration. It is not possible to measure lost potential; no one can know for certain what promising young scholars like Hans Hellmann, Siegfried Gilde, and Fritz Duschinsky would have achieved had they not been murdered. Similarly, no one can ever know if Guido Beck could have made an important breakthrough in quantum mechanics had he not spent almost twenty years seeking a safe and stable refuge.

Hans Baerwald, Martin Ruhemann, and Walter Zehden never returned to the academy, but all became successful industrial scientists or engineers. Helmuth Simons, the long-suffering biochemist, appears to have failed to find a permanent scientific position either at a university or in private industry. In 1947, he spent a year teaching at the Philadelphia College of Pharmacy and Science in the United States. Despite his impressive scientific accomplishments, however, his stay in Philadelphia was brief.[32] By the early 1950s, he was forced to settle in Bern, Switzerland, where he conducted research for at least one pharmaceutical company and received research funding from various private foundations. He published a handful of scientific papers, including two that focused on blood parasites as a potential cause of multiple sclerosis. Simons stopped actively publishing around 1960. He died in 1969, age seventy-five.[33]

Even those who were able to resume careers at a research-oriented university or laboratory never quite achieved their full potential. In 1951, J.W. McLeod wrote a brief letter of reference for Kurt Zinnemann, describing him as "a worker of marked ability with a thorough background in bacteriological training." McLeod thought it necessary to explain why Zinnemann still only held a junior academic position. In discussing Zinnemann's tumultuous life, McLeod showed himself to be a master of British understatement: "There is no doubt that a man of similar competence who had worked as a bacteriologist without having his career disorganized on account of obstruction based on political and racial considerations would at the present time have reached a position senior to that which he now occupies."[34]

Depending on how one measures the interruptions in Zinnemann's career, he lost anywhere between seven and ten years of productive work. In 1933, his research career had barely started when he was dismissed by the University of Frankfurt. It was not until 1935 that the Zinnemanns escaped to Kharkiv, and their research program in the Soviet Union was just being established when they were arrested in the spring of 1937. They arrived in Leeds in August 1938 but were interned in May 1940, and were able to return to the university only in February 1941.

Their release from the camp on the Isle of Man did not result in an immediate return to active research. In the eighteen months prior to his internment, Zinnemann produced three articles, but he would not resume active publishing until 1943. Despite this great handicap, Zinnemann went on to have a distinguished career as a medical scientist. He became an internationally recognized expert on meningitis and tuberculosis and was the author or co-author of at least fifty scientific papers, all but one after 1938. In 1970 he was elected by the University of Leeds to be professor in clinical bacteriology. Zinnemann retired from the university in 1973, although he continued research and worked as a consultant microbiologist for the National Health Service. He died in 1988, age eighty.

Ernst Simonson achieved more than most of the refugee scholars. In large measure this was because his work had been less disrupted by the political turmoil of the 1930s. He was already well-established in the Soviet Union when he was dismissed by the University of Frankfurt for racial reasons. The only significant interruption in his scientific research was the two years between his flight from the Soviet Union in the summer of 1937 and his arrival in Milwaukee. After he finally established himself in a safe haven, Ernst Simonson managed to have a notable career in medical research. He spent five years as a researcher at Mount Sinai Hospital in Milwaukee. In October 1944, he was appointed associate professor of physiology and hygiene at the University of Minnesota. To the end of his life, he maintained a keen interest in applied physiological research. His detailed monograph *Physiology of Work Capacity and Fatigue* was completed and edited at the age of seventy-three. The manuscript for a follow-on volume, *Psychological Aspects and Physiological Correlates of Work and Fatigue*, was sent to the publishers only weeks before his death in 1974.

While still in Milwaukee, in addition to his ongoing work in physiology, Simonson became interested in electrocardiographic research. He expanded his investigation into the use of the ECG after he arrived in Minnesota, and he was eventually cross-appointed with the Department of Biophysics. After Simonson's death a group of leading cardiologists stated that "he did more perhaps than any other electrocardiographer to promote measurement in ECG research and clinical application." Simonson's 1961 monograph, *Differentiation between Normal and Abnormal in Electrocardiography*, was "a milestone in ECG literature."

Perhaps even more remarkable, after having been forced to flee the Soviet Union in fear for his own life and that of his family, in the late 1950s he began a form of "intellectual détente and scientific collaboration"

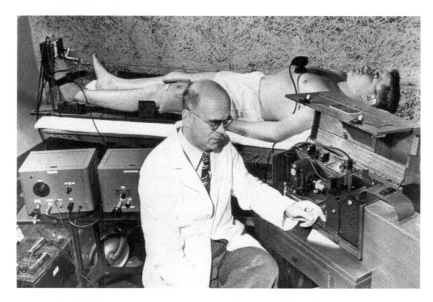

Image 15. Ernst Simonson recording a ballistocardiogram, Laboratory of Physiological Hygiene, Minneapolis, Minnesota, 1948 (image courtesy of University of Minnesota Archives, University of Minnesota – Twin Cities).

with Soviet scientists. He received a grant from the National Library of Medicine to translate Russian cardiovascular literature and published a series of articles in English that summarized it. Simonson won a host of awards and honours; possibly the most meaningful, for him, was that the year before his death he was reinstated as a full professor by the University of Frankfurt, the very institution that had dismissed him forty years earlier.[35]

A similar pattern of long-term career achievement being linked to the amount of time a scholar was unable to work can be seen among scholars in other fields. Perhaps the most accomplished of the fourteen surviving physicists was Herbert Fröhlich, whose charmed life resulted in minimal career disruption. He spent less than one year in Munich after his dismissal in 1933 before he found employment in Leningrad. He was expelled from the Soviet Union in 1935, but the research he had conducted there led to an almost immediate one-year fellowship for him, which was followed by his appointment to Bristol. Like Zinnemann's, Fröhlich's research career was briefly interrupted in the summer of 1940, when he was interned. Fröhlich was joined on the Isle of Man by all of his German colleges at the University of Bristol. His university and the SPSL worked tirelessly to have him and the other German physicists freed. He had resumed his work at the university by the start of 1941.[36]

Having triumphed over Hitler and Stalin, Fröhlich went on to a long and distinguished scientific career in which he made major contributions in an exceptionally wide range of topics, including nuclear and particle theory; he was also the "acknowledged master world wide on the (non-electronic) theory of dielectrics." After the war, Fröhlich moved to the University of Liverpool to establish a new research department in theoretical physics. There, he made important contributions to the understanding of polarons (a quasiparticle that allows for an understanding of the interactions between electrons and ions in a solid material), the theory of superconductivity, and, perhaps his greatest achievement, the application of quantum mechanics to biology. Among many other honours, in 1951 he was elected as a Fellow of the Royal Society. He died in 1991, at the age of eighty-five.

Marcel Schein also had a successful scientific career after he left the Soviet Union. Almost immediately after arriving in Chicago in 1938, he continued the research program he had undertaken in Odessa, examining the nature of the still very mysterious cosmic rays. For the next three years, using specially modified weather balloons, he conducted a series of experiments in the highest layers of the atmosphere to elucidate the physical properties cosmic rays. The prevailing view at the time was that cosmic rays consisted of both protons and electrons. Schein's experiments proved, instead, "that electrons did not appear in any appreciable amount among primary cosmic rays, and that the incoming radiation consisted most likely of protons." According to physicist Bruno Rossi, the results of these experiments were among "the milestones of cosmic-ray research."

After the Second World War, Schein conducted a series of increasingly large, elaborate, and expensive research programs to investigate the "remarkabl[e] and unbelievabl[y] intricate process that occur[s]" when cosmic rays strike atomic nuclei.[37] In 1959 he launched his largest and what proved to be his last scientific project, dubbed Operation Skyhook, or "the International Cooperative Emulsion Flights." Skyhook, funded by the Office of US Navy Research and the American National Science Foundation, involved mobilizing a naval force including an aircraft carrier, six destroyers, and four radar-equipped weather planes. The aircraft carrier was used to launch three enormous balloons, five hundred feet in height and three hundred in diameter, carrying nuclear emulsion plates, which would be struck by cosmic rays. One of these balloons rose to over 116,000 feet. The weather planes tracked the balloons, and the destroyers were used to recover them. Tragically, shortly after his return to Chicago, before any of the plates could be developed and analysed, Schein suffered a massive heart attack. He died three days later, on 21 February 1960. He was just fifty-seven years old.[38] Though his career was cut short at its height, he is remembered to this day as a pioneer of research into cosmic rays, high-energy physics, and elementary nuclear particles.[39]

Laszlo Tisza's most productive period as a theoretical physicist ended when he left France for the United States in 1941. His two most influential scientific papers on superconductivity were published in 1940. On his arrival in Boston, despite considerable support from other physicists, including fellow Hungarian and close friend Edward Teller, George Gamow, the Soviet refugee, and Charles Squire, Tisza's friend on the staff of MIT, Tisza's initial search for any sort of academic position proved futile. John Clark Slater, head of the physics department at MIT, explained to the young Hungarian that the major stumbling block was Tisza's lack of teaching experience. As Second World War scientific research programs such as into radar began to recruit American physicists away from universities, however, Tisza suddenly found himself in demand as a substitute. He was hired by Slater on a series of one-year contracts until the war ended in 1945, at which time Tisza reached MIT's time limit on temporary lecturers. Fortunately for him, the growing prestige of physics resulted in MIT expanding its physics department, and he was promoted to assistant professor. Three years later he was promoted to associate professor, with tenure.

At MIT, Tisza became known mainly as a teacher and a graduate supervisor. Although he continued to work in various aspects of theoretical physics, he made no significant research contributions after 1940. It is difficult to know how much the various disruptions in Tisza's career affected his scientific production, since it is not uncommon for theoretical physicists to do their most important work earlier rather than later in their careers. Tisza remained at MIT until he retired in 1973, and continued to be active in physics right up to just before his death in 2009, at the age of 101.[40]

Fritz Houtermans was never able to relive the pre-1933 glories of his research career. Having survived the war in Germany and avoided arrest by the Soviets in its immediate aftermath, Houtermans languished in Gottingen, where the physics department was a poor shadow of its former self. At the time, physics research in Germany was severely hindered by restrictions imposed by the Allied Control Commission. Moreover, there was almost no money for research, scholarships, or even basic upkeep of the university infrastructure. In 1952, to get away from postwar Germany, Houtermans accepted a professorship in experimental physics at the University of Bern. Bern was little better equipped than Gottingen, he wrote to a colleague: "If you want to see an authentic early twentieth century laboratory visit me."[41] So he rebuilt the applied physics laboratory at Bern from the ground up, turning it into a major European centre. He did groundbreaking research in applying nuclear physics to geology and archaeology; he also imported work from other areas, such as cosmic rays. While at Bern, he restored some of the lustre to his scientific career, but his personal life remained at best chaotic.

Fritz's divorce from Charlotte during the war had forced Charlotte to resume her own career as a scientist. In the United States she remained single, raising their children, teaching, and, when possible, conducting research. In the early 1950s she returned to Europe three times, meeting with Fritz during each visit. For the children it was the first time they had seen their father since before the war, as well as their first opportunity to meet their three half-siblings. By this point, Fritz's second marriage had broken down and he was once again divorced. Charlotte and Fritz's first meeting since they were separated in Moscow was stormy. Gradually, despite Fritz's betrayal and sixteen years apart, Charlotte allowed herself to be charmed by him once again, and in August 1953 they remarried. Fortunately, Charlotte had not resigned her position in the United States, instead she had taken a one-year leave of absence. She soon learned either that they had grown apart over the years or that she was less willing to tolerate the more negative aspects of his personality. Fritz had become a heavy drinker, and when he proved unable or unwilling to stop, Charlotte realized she had made a terrible mistake. Six months after her remarriage, she divorced him and returned to Sarah Lawrence College.

Fritz would marry one more time – this time it was the sister of his sister-in-law. They would have one son together, and Fritz would adopt his new wife's daughter. Houtermans's heavy smoking and his growing problem with alcohol resulted in increasing health problems when he was in his late fifties. He survived lung cancer, but in 1966 he suffered a pulmonary embolism and died, aged just sixty-three.[42]

Werner Romberg also had a successful academic career after the war. Turning away from physics, he became a noted expert in the applications of mathematics to computing. After his wartime exile in Sweden, he returned to Oslo in 1944 and resumed his position as assistant to Harald Wergeland. In 1948, he became a *dosent* or assistant lecturer in physics at the Norwegian Institute of Technology in Trondheim. His academic work focused on applied mathematics in the field of numerical analysis, particularly with regard to computers. In 1960 he was named the chair in applied mathematics at the institute, with a specific focus to computing.

Romberg offered the first university courses on digital computing taught in Norway.

In 1968, Romberg was appointed professor of mathematical methods in science and numerical analysis at the University of Heidelberg, which made him the only refugee academic who had fled to the Soviet Union to return to an academic position in West Germany. Romberg had started his undergraduate degree at Heidelberg in the late 1920s. During the 1930s, the university had been at the centre of German academic support for the Nazis. Romberg's return brought him full circle: he had fled Germany for Ukraine in 1933, then he had been forced to leave the Soviet Union in 1937. He then eked out in existence in Prague, before his appointment to Oslo, which was followed by his wartime exile in Uppsala, his return to Oslo in 1944, and then his appointments in Trondheim. By then, he had been in exile from Germany for thirty-five years. In 1969 he also became the scientific head of Heidelberg's Computer Centre. He continued to teach and do research there until his retirement in 1978, aged sixty-nine. He too enjoyed a long retirement, dying in 2003, just short of his ninety-fourth birthday.[43]

Michael Sadowsky was able to resume his career in mathematics after his flight from the Soviet Union. After fleeing first to Germany and then to Palestine, on 26 January 1938 he returned to the United States, joining the faculty at the Illinois Institute of Technology. In 1953 he become a full professor of mathematics at Rensselaer Polytechnic Institute. He continued to do valuable work on the mathematical theory of elasticity, and his studies of three-dimensional problems brought him international fame. In 1960 he began pioneering work in micromechanics, during which he studied mathematically the force transfer mechanisms in composite materials, the classification of microfibres, and the stress behaviour of both types of materials. He remained at Rensselaer until his retirement in 1967, but died a few months later, just sixty-five years old.[44]

Despite his towering international reputation as a mathematician, it took Stefan Bergman more than a decade before he found a permanent university position in the United States. He taught first at Yeshiva College before moving to Brown University in 1941. He contributed to the war effort, writing a series of mathematical studies for the National Advisory Committee for Aeronautics related to supersonic and near-supersonic flows. After the war, Bergman lost his position at Brown because he was unable to communicate mathematical ideas to undergraduates. One of Bergman's colleagues wrote to the president of the university that while Bergman "indeed has a touch of genius ... With all his desire to accommodate himself to his environment and his colleagues, he is alien and will never achieve any popular favor." Like a number of refugee scholars, Bergman found it impossible to make the cultural shift to American universities, which required most faculty to be adept at teaching undergraduates.[45] Bergman then joined von Mises, his doctoral supervisor, at Harvard, as a visiting professor. He stayed at Harvard for seven years while he sought a university research position with little or no teaching. In 1952, Stanford University offered him such a situation. Bergman was then fifty-seven years old and incredibly fortunate to have found employment. He would stay at Stanford for the rest of his career. Throughout his years in the United States, he continued to develop his concept of kernel function, as well as the theory of linear partial differential equations with complex variables. He died in 1977, aged eighty-two. After his death, Bergman's wife established an important mathematical prize in his name.[46]

Edgar Lederer's decision to stay in France during the Second World War may have been foolhardy, but it meant that after surviving the Vichy government and the German occupation he did not need to re-establish himself in yet another country. Already well integrated into the French scientific establishment, he was able to parlay his fruitful wartime research on the chemistry of perfumes into a long and successful research career. In 1947 he returned to Paris, where at his laboratory he worked with a number of collaborators on important practical problems in biochemistry. For example, one project demonstrated the structure of the pheromone of queen bees, which then was synthesized. He also began work on the chemistry of bacteria. His work focused mainly on the tubercle bacillus; he isolated various molecules in the bacterial walls, some of which played an important role in immune defence. This work came to be central to his later scientific endeavours.

In 1954, Lederer was appointed a lecturer at the science faculty at the University of Paris. Four years later he was promoted to professor and named chair of biochemistry. That same year, the CNRS appointed him to direct the new Institute of Chemistry of Natural Substances, at Gif-sur-Yvette. In 1960 the dean of science at Paris-Sud University asked Lederer to establish a new Institute of Biochemistry at Orsay. He moved his professorship to Orsay and divided his time between Orsay and Gif. Both institutes became major international centres in biochemical research, fuelled by Lederer's ability to generate funds to purchase the most sophisticated equipment, including for mass spectroscopy and nuclear magnetic resonance imaging. It is not possible to outline all the research breakthroughs achieved at these centres. Despite his administrative and teaching responsibilities, Lederer continued his own successful research. He conducted groundbreaking research on muramylpeptides, compounds with a wide range of therapeutic properties, used not only as vaccine adjuvants but also as antibacterials.

Lederer won numerous awards, including, in 1964, the August Wilhelm von Hofmann Gold Medal of the German Chemical Society, and, in 1974, the CNRS Gold Medal. He was elected to a number of foreign scientific academies, but not until 1982 was he made a member of the Académie des Sciences in Paris. The delay in his election can only be explained by the fact that in some circles of French academe, Lederer was still not fully accepted as a French citizen.[47] Lederer retired in 1978 and died ten years later, age eighty.

Ernst Emsheimer's decision to remain in the country to which he had fled after leaving the Soviet Union – in this case Sweden – played an important part in his long-term scholarly success. As we have seen, Emsheimer persevered through years of low- paying research work before being named head of the Museum of Music History (now the Musikmuseet) in 1949. Even this appointment, however, did not ensure his long-term success. Until the 1960s, the museum was poorly funded and grants were "insufficient and accidental." Consequently, his salary remained low. However, his doggedness and hard work turned the museum's fortunes around, and it became world-leading centre in its field. He used whatever funds were available to make "prudent purchases of instruments," plan well-received exhibitions, and establish an international research program in the history of music.

Emsheimer did much to revive interest in medieval and Renaissance music. He organized concerts where prominent musicians played the same instruments that would have been used when the music was first performed. The concerts began with informative talks by leading musicologists, many of whom had been trained by Emsheimer. These concerts became famous in Sweden, so much so that Ingmar Bergman consulted Emsheimer about the music to be used in his film *The Seventh Seal*, which was set in the time of Black Death. Perhaps above all else, Emsheimer taught and influenced several generations of Swedish musicologists and music ethnologists. Jan Ling, one of Emsheimer's former students, reflected:

> We particularly remember Ernst Emsheimer as teacher and adviser, his profound knowledge, his bibliographical insight, and his accuracy in theory and method combined with a unique capacity to listen to and understand viewpoints which he himself did not hold. His modesty sometimes baffled those of us who regarded him as a great authority. But it taught us to respect and have faith in the capacity of other people.

Emsheimer died in 1989, at eight-five years of age. Perhaps it is the mark of his success that, unusually for a Swedish scholar, his passing was noted in four international commemorative publications, all of which expressed profound gratitude toward him, both as a scholar and as an individual.[48]

To these accounts of successful scholars must be added the remarkable story of Eva (Striker) Zeisel. As we have seen, in October 1938, Zeisel immigrated to the United States with her new husband. Despite initial poverty, which scuttled her desperate scheme to rescue Alexander Weissberg from Poland, Zeisel became one of the most influential ceramics designers in the United States. Her designs were both practical and beautiful and would eventually be found in many American homes. They were produced not only in the United States but also in Japan, Germany, Italy, and many other countries. Zeisel lived long enough to give a Ted Talk, in which she briefly mentioned her time in the Soviet Union. She died in 2011, aged 105.[49]

Conclusion

The scholars who survived the Second World War and enjoyed academic careers afterwards had some things in common. Most had good connections, be they personal or professional. Another factor was persistence, sometimes to the point of obstinacy. Edgar Lederer turned down an opportunity to migrate to the United States. Staying in France was dangerous, but by the time of liberation he had been accepted as part of the French scientific community. Ernst Emsheimer was willing to work for more than a decade in an underpaid and underappreciated post; it led to a successful career that reshaped Europeans' understanding of music history and music ethnography. The desperate trips to the United States made by Marcel Schein and Ernst Simonson in search of work were examples of persistence, and also of the good fortune a refugee scholar needed in order to survive and prosper. The one thing that links all of these ultimately successful academic refugees is that all of them were, if not brilliant, then highly capable scientists. Brilliance by itself did not guarantee success but certainly was a prerequisite for it.

The SPSL played an important role in supporting many of these scholars and, later on, in tracing at least some of those who disappeared during the war. In 1936, Herbert Fröhlich's SPSL fellowship became the foundation of his highly successful career in British physics. For the much larger second wave of those forced to flee the Soviet Union, only Simonson and Schein received SPSL fellowships, and neither of them was able to find place at a British university. These two cases, along with

Zehden's rejection for a post at Birmingham, indicate that by 1938, British post-secondary institutions had reached their saturation point and could take very few more academic refugees. The SPSL showed considerable flexibility in allowing Simonson and Schein to turn at least part of their fellowships into travel grants, which made possible their successful job searches in the United States. The AEC had almost no success in assisting these scholars, though it did its best to help a number of them.

Success in the academy or industry, however, may not have been the most important influence on this small group that survived both Hitler and Stalin. These scholars had experienced first-hand the Great Terror and the Holocaust, and the response of many of them to what they had witnessed would be remarkably influential in shaping the Cold War. The responses of the survivors to the struggle between West and East were remarkably diverse, reflecting and helping shape the great ideological battle that characterized this period.

12 The Ensnared in the Cold War

The scholars who survived the horrors perpetrated by Hitler and Stalin needed to decide how to make sense of the postwar world. Most of these scholars, especially those who had gone to the Soviet Union without any strong ideological or political beliefs, simply wished to be left alone to live their lives. Even before the Cold War, however, some of the formerly ensnared scholars found themselves embroiled in the great ideological battles between Communism and the Western liberal democracies. Those who had been politically engaged before their migration to the Soviet Union, especially Communists or fellow-travellers, found their faith in Marxism profoundly challenged by their encounters with the harsh realities of Stalin's regime. A small number somehow maintained their belief in Communism and would continue to fight for the cause. Another group would renounce the harsher tenets of Marxism-Leninism but continue to advocate for left-wing and socialist causes. Still others became among the fiercest critics of the Soviet Union and Communism. The ensnared scholars who remained politically engaged would influence the Cold War to a degree disproportionate to their small numbers.

Barbara and Martin Ruhemann: The Unrepentant Ideologues

Of the five scholars who continued to believe in Communism, the story of Barbara and Martin Ruhemann stands out as an extraordinary example of zealotry. In March 1939, Barbara Ruhemann, then living in London, wrote to the SPSL in London, demanding proof for the allegations made in December of the previous year in *Nature* concerning the "persecution" of refugee academics by the Soviet government. She also wrote to the journal's editor demanding to know the sources of this "grave accusation." The editor informed her that the story had been based on data supplied by the SPSL.

Since none of the society's publications provided such negative accounts of the treatment of foreign scholars by the Soviet Union, Ruhemann asked the SPSL to provide her with "any additional information which might throw light on the matter." She doubted that such proof existed, because to "anyone familiar with Soviet traditions the falsehood of this statement is obvious."[1] Esther Simpson, assistant secretary of the SPSL, replied to Ruhemann:

> Of these [refugee scientists in the Soviet Union] we know personally at least a dozen who were dismissed and compelled to leave the country, and we know of three cases of imprisonment, two of whom as far as we know being still in prison.
>
> There is no evidence whatever that these scientists ... have committed any acts against the laws of that country. They were not Soviet citizens, and that seems to have been adequate reason in most cases.[2]

Simpson was unable to convince Barbara Ruhemann, whom she realized was a dyed-in-the-wool Communist. Indeed, Ruhemann wrote her a follow-up letter in which she defended Stalin's government and argued that some innocents must suffer to further the cause of Communism.[3] Simpson, who herself had once been a party member, replied one more time, declaring that one did not serve the cause of Communism "by closing one's eyes to facts," and adding that "that facts are facts, whatever their motives or inspiration, and the result of these facts has been to swell the numbers of the unfortunate victims whose names are registered with us." Here the exchange ended; Simpson was far too busy to engage in an obviously futile effort to convince a Stalinist zealot of the suffering of refugee scholars forced to flee the Soviet Union. It is unclear whether Simpson was aware that Ruhemann was herself one of those refugee scientists forced to leave the Soviet Union.[4]

Barbara Ruhemann, of course, knew perfectly well what had happened to the refugee scientists in the Soviet Union. Alexander Weissberg, her close friend and principal instructor regarding the religion of dialectical materialism, was still languishing in a Soviet prison. Other foreigners she had known in Kharkiv, including Eva Zeisel, Konrad Weisselburg, and the Zinnemanns, had also been arrested. As well, she knew that many of her Soviet colleagues had disappeared or been forced to run for their lives. Barbara and her husband Martin had themselves been cast out of the Soviet Union. They had been allowed to leave, and perhaps had not been arrested because of their British nationality.

Barbara Ruhemann stands out as an extreme example of ideological obstinacy. Once in the West she joined a large group of Communist scientists and intellectuals, and other fellow-travellers, who refused to question the righteousness of the Communist cause. Nor would they accept any criticism of Stalin's regime, no matter how egregious his crimes. Left-leaning scientists in Western Europe knew about the widespread arrests of foreign and Soviet scientists. After the detention of Eva Zeisel in June 1936, Weissberg and Laura Striker had reached out to their broad network of academic friends and relatives pleading for letters of support. In the spring of 1937, however, Charlotte Houtermans, during her futile trip to the West seeking a way out of the Soviet Union for her family, had found that her leftist friends didn't believe they could be in trouble and showed no sympathy for their plight. Later, Charlotte and Arthur Koestler would encounter similar reticence among left-leaning scientists in France, Britain, and the Netherlands; many refused to take any action to pressure the Soviets to release Weissberg, Fritz Houtermans, and other imprisoned scientists. Some, like Niels and Harald Bohr, said they were reluctant to take any action that might place Soviet colleagues at greater risk. Many others were more ready to believe that those arrested must have been guilty than to accept that a massive violation of human rights was under way in the Soviet Union.

One such intellectual was the British physicist and Communist J.D. Bernal. As we have seen, in June 1939, Bernal met with the Soviet ambassador in London to make one final plea for Fritz Houtermans's release. Rather than pressing Houtermans's case, Bernal meekly accepted the ambassador's assertion that the charges against the German scientist were serious. In fact, Bernal was the worst possible person to advocate on Houtermans's behalf. He had more than the eyewitness account of Charlotte Houtermans to attest to recent events in the Soviet Union. Upon their arrival in Britain in late 1937, the Ruhemanns had quickly established connections with Bernal and other Communist scientists. J.G. Crowther, the science journalist, was Barbara's brother-in-law.[5] Crowther was a Communist, although his faith in the Soviet Union had been greatly shaken by the arrest or expulsion of so many of his friends. Crowther knew all of the leading British left-leaning scientists, including Bernal, J.B.S. Haldane, Joseph Needham, Lancelot Hogben, and Julian Huxley.[6]

Just prior to his meeting with the Soviet ambassador, Bernal published *The Social Function of Science*, a detailed sociopolitical analysis of how best to organize research to serve society without "killing its necessary freedom and flexibility." Bernal used the Soviet Union's organization of science as his model to be emulated, arguing that Soviet-style central planning could "ensure a proper balance between fundamental and applied research." Having been freed from the "destructive consequences of capitalism," science could become the "chief agent for social change."[7] Bernal's book was widely praised. Even so, in its single-minded fervour to promote the Soviet Union as a scientific paradise, and its equal willingness to condemn all activities of the scientific establishment in capitalist countries, his book was more a polemic than a thoughtful analysis. For instance, Bernal argued that aid to scientific refugees was a sop to make members of the establishment feel good, when in reality they were doing nothing to fight fascism.[8]

Martin Ruhemann was one of Bernal's key sources of information on Soviet science, so much so that he wrote an appendix for the book in which he outlined the organization of science in the country. He echoed Bernal's laudatory views about the centrality of science in the workers' paradise, pointing out that science had always served the interests of the ruling classes and that in the Soviet Union, "the ruling classes are the workers and peasants." Thus, Soviet science served practically the entire population, and because of this "there are no fears of a technocracy detrimental to the community at large." Martin, like his wife, nowhere acknowledged the fate that had befallen so many of their friends and colleagues in Kharkiv and elsewhere in the Soviet Union.[9]

Bernal took pains to avoid discussing the widespread persecution of Soviet and foreign scientists during the Great Terror. Understandably, this omission caught the ire of Michael Polanyi, Eva Zeisel's uncle, in a review of the book published in 1939. Polanyi *knew* all the left-leaning physicists who had fled Berlin for the Soviet Union after Hitler came to power. Moreover, he had heard Eva's account of her torture, he was well aware that both Weissberg and Fritz Houtermans had disappeared into the Soviet prison system, and he had received Saul Levy's desperate plea for help after his forced departure from Moscow. Polanyi shared none of Bernal's delusions about scientific freedom in Stalin's Soviet Union. He rightfully viewed Bernal's work as nothing more than Marxist propaganda; he "objected strongly to Bernal's relentless criticism of capitalist institutions and constant praising of Soviet Russia."[10] The Soviet system, rather than a model to be emulated, was "a more intelligent and more complete philosophy of oppression than is either Italian or German fascism." By ignoring the widespread arrests of scientists in the Soviet Union, Bernal had ignored the destruction of the very scientific freedom he was advocating as so crucial for science to flourish. Polanyi bitterly condemned Bernal's hypocrisy:

> Dragooned into the lip service of a preposterous orthodoxy, harried by the crazy suspicions of omnipotent officials, arbitrarily imprisoned or in constant danger of such imprisonment, the scientist in Soviet Russia is told,

from England, that the liberty which he enjoys can only be appreciated by living it. Since the terms of this liberty prevent him from answering his British colleagues, I have taken it upon me to point out the anomaly of the situation.[11]

Barbara Ruhemann abandoned physics after her departure from Kharkiv. It is not certain why, but it freed up her time for Communist Party work and for pursuing a new interest in ethnography and anthropology. At some point, it is not clear when, she may have earned a degree in anthropology. Not much is known about either Barbara or Martin in their first two years in Britain, although they likely lived in London at first with Martin's parents, themselves refugees from Germany. In March 1940 the Ruhemanns came to the attention of the British security services, the police Special Branch and MI5. For at least the next twenty years, Barbara's activities would be monitored by the security services.

For Communists in the West, the period between the signing of the German–Soviet Non-Aggression Pact in August 1939 and the German invasion of the Soviet Union in June 1941 was one of great confusion that challenged many people's loyalty to the party. The Ruhemanns were not among the doubters. From November 1939 to February 1940, Barbara lived in Somerset, where she came to the attention of the local police for distributing Communist literature. Barbara wanted to stay close to the couple's son Stephen, who had been evacuated from London at the start of the war and was hospitalized soon after for scarlet fever. The police observed that in hospital, Stephen "had received definite Communistic teachings; he refused to give any information regarding his parents."[12]

In June 1940 the Ruhemanns became the subject of an extraordinary report sent to the security services by Sir Graham Greene, a retired senior civil servant. Greene had known Martin's parents in Cambridge before the First World War and had met Barbara at an event at the Royal Anthropological Institute, which she had recently joined. Also attending the meeting was a Dr Lingren, an old friend of Barbara and Martin, who had met them in Berlin in 1926, before the couple was married. Lingren informed Graham:

Martin Ruheman [sic] has always impressed those who knew him as an honourable, steady, typically middle-class-intellectual man. Since his marriage he has seemed to be strangely under the power of, and changed by, his wife. His wife gives the impression of "inhuman," cold fanatic, single-mindedness about any interest in the hour. However it is quite possible that recent developments in Soviet Russia have disillusioned them both, and that they have no interest in Soviet aims and activities, or not in a political sense.[13]

Lingren was wrong about the couple being disillusioned by the Soviet alliance with Germany. Later that same year, Martin published an article that lauded Soviet science, very much in the style of Bernal's book. Again, he made no mention of the purges. In Britain, he wrote, "scientists are divorced from community and form a little group of their own," and the "opinion that science is intended for the welfare of the community is not widespread in scientific circles." In the Soviet Union, by contrast, barriers between scientists and workers had been broken down for the common good. This "penetration of the scientific approach into the cornfield and the factory, the bathroom and the kitchen is ... essential for development of society in which each shall work according to his ability and receive according to his needs."[14]

In June 1941 the Soviet Union became Britain's ally and it once again became acceptable to promote publicly the "wonders" of the Soviet system. In August, Barbara was a speaker, along with Bernal, Needham, Haldane, and George Derwent Thomson, at a two-day seminar on science and Marxist philosophy.[15] By this point, Barbara was working as a secretary at Communist Party headquarters and was a highly active member of her local party branch.[16] Two years later she began working as a translator for the Soviet government's TASS news agency, a position she would hold until the office was shut down in the early 1950s. Barbara fully immersed herself in the British Communist Party. She lived with fellow party members in a house painted bright red, located next door to the TASS offices. She spent all of her spare time helping the party achieve its goals, particularly in organizing lectures on Soviet science, and socializing with students from African colonies, whom she encouraged to learn about Marxism.

Martin and Barbara separated sometime between 1940 and 1942. Some sources suggest that Barbara's increasing radicalism was a major factor. After the war, Alexander Weissberg met with Martin, who accused him of having had an affair with Barbara while they were all in Kharkiv. Weissberg vehemently denied any romantic involvement with Barbara. The couple were finally divorced in 1947, and Martin remarried soon after. Whatever the reason for the break-up of that marriage, there is no indication that Martin actively supported the Soviets or the Communist Party after September 1942. Neither is there firm evidence that he ever joined the British party. As late as 1948, however, he could not bring himself to completely disavow either Stalin or the Soviet state. Weissberg, when he first visited his old friend in Manchester, was surprised to find him with a new wife and baby. Weissberg was even more shocked that Martin was still unwilling to accept that Stalin had murdered thousands of innocent people, including the entire Old Guard of Bolshevism during the purges. When he pressed Martin to accept this uncomfortable truth, all his friend could do was exclaim: "Alex, I find the whole problem extraordinarily difficult. What am I to do? I can't bring myself to believe that all my conscious political life, the whole movement into which you brought me, was an error. There are two camps in the world today, the Soviet Union and the United States. Am I to side with the American imperialists?"[17]

Martin remained unwilling to renounce his beliefs but willing to leave politics behind. He took the scientific and industrial knowledge he had gained in the Soviet Union and parlayed it into a successful career in Britain and Italy as a chemical engineer and applied physicist. In the early 1940s he joined the design team for Petrocarbon Limited, which built the first petrochemical complex in Europe at Carrington, near Manchester. In the early 1950s he moved to Italy, where he designed and built oxygen and nitrogen separation plants. He returned to Britain in 1957 to take a lead role in Petrocarbon's development of the first pure nitrogen plant in Britain. Eventually he became the company's executive director. He became known as one of the "founding fathers … in the arcane world of cryogenics."[18] In 1968, Martin was made visiting professor of chemical engineering at Bradford University. Barbara and Martin's son Stephen was a lecturer in the same department. Four years later, Martin made a triumphant return to Berlin on a lecture tour. While he was there, a laboratory at the Technical University was named after him. He died in 1993, aged ninety.

Barbara continued her party work after the war. In 1950 a former Communist briefed MI5 on her and other members' activities. His report described Barbara as "the most fanatical Communist" he had ever met, someone who "would stop at nothing to aide the USSR." She sustained her work with African students and had "numerous connections with subversive movements in the Gold Coast and elsewhere."[19] This description of Barbara as an extremist among the party's ideologues closely mirrors the views of many party members. Secretly recorded conversations held at party headquarters show she was regularly the subject of discussions about her disruptive behaviour on committees. She had an "unnecessarily sharp and aggressive ways of putting things" as well as a strong tendency to engage in "abstract and dreary harangues." One senior party official felt it would be impossible to get Barbara to change, since she reacted to any criticism "like water off a duck's back," and trying would only make her behaviour even worse. He described her behaviour as an unfortunate "combination of the Germanic type and the 'spinsterish sort.'"[20]

Despite these severe personality clashes, no one denied that Barbara had made significant contributions to the party's efforts to promote the decolonization of Africa. It is impossible to determine whether efforts to promote revolution in Africa actually had any effect on the collapse of the British Empire on that continent. Barbara's academic work in anthropology was not particularly successful; she wrote a mere handful of articles on kinship and marriage. She also acted as one of the English translators of several volumes of the collected works of Marx and Engels.[21] She died in April 1976, age seventy-one.

Faithful Communists in the German Democratic Republic

If Barbara Ruhemann was unable to foment revolution, three other refugee scientists – Gerhard Harig, Wolfgang Steinitz, and Fritz Lange – played significant roles during the early Cold War in supporting Communist domination of the German Democratic Republic (GDR), or East Germany. Harig, the physicist turned historian of science, had survived seven years in the Buchenwald concentration camp. His experiences there had reinforced his faith in Communist ideology. He wrote to a friend in November 1946: "It seems to me that you don't know where to put me and what you should think of me. I want to pour you pure wine right from the start and tell you that I am still a convinced Marxist and Communist. I don't want to be an independent private person, but an active employee for a new social order and an active supporter of a new worldview."[22] After his liberation, Harig co-authored three reports on the history of Buchenwald for the KPD, using records the SS had been unable to destroy.

From Buchenwald, Harig returned to his home city of Leipzig in the Soviet zone of occupation. He headed the statistical office, responsible for creating electoral lists for the first post–Nazi era elections. He soon established himself as a leading intellectual in the East German Communist Party (Sozialistische Einheitspartei Deutschlands – SED). From 1946 to 1948 he advised the SED's executive on Marxist philosophy. In 1947 he was made a professor in the history of science and technology at the University of Leipzig. One year later he was named the director of the new Franz Mehring Institute and head of the Department of Dialectical Materialism. The Franz Mehring Institute had been created by order of the Soviet Military Administration in Germany to undertake the training of teachers in Marxist-Leninist philosophy.[23]

Throughout the postwar years, Harig worked to be reunited with his wife Katharina and their son Georg. Harig had last had a letter from Katherina in April 1941, just months before the German invasion of the

Soviet Union. In 1945 he approached the Soviet authorities hoping to ascertain whether she was still alive. The Soviets noted his request but did nothing. Only when he was in Berlin did Soviet officials finally confirm that his family had been evacuated to the east before the German invasion and that both were still alive. Meanwhile, in the Soviet Union, Katharina was able to send letters to her husband. Finally, at the end of April 1948, the couple and their thirteen-year-old son were reunited, more than a decade after he had last seen them.[24]

In 1950, Harig took a leave of absence from the university to head the Department of Higher Education and Science in the Ministry of Education in the newly established East German government. His department soon became a separate ministry, and he was appointed a member of the Council of Ministers. Under his leadership, the study of Marxist-Leninism became a compulsory subject for all students. Anyone who opposed the Communist state was ruthlessly suppressed. On 30 May 1952, he travelled to the university in Halle to personally purge "criminal and hostile elements." He expelled eighty students, targeting those who spoke out against the regime, engaged in "anti-Soviet rabble-rousing," had close connections to West Germany, or were members of the bourgeoisie. Harig and his staff used the results from courses in dialectical materialism and Russian, and information from local party members, to identify their targets. He gave these students letters stating simply that they had been expelled. Any further explanations provided to individuals were made orally by university officials. Harig took pains that no written explanations were provided to the victims of the purge.[25]

The purge was soon extended beyond students to include suspect faculty.[26] Harig, no doubt, did not see the bitter irony that one of the victims of the mass dismissal of scholars by both Hitler and Stalin was now playing a key role in a similar expulsion under the Communist regime of East Germany. Harig also played an important role in the creation of a carefully constructed official East German historical mythology of the German people's solidarity with the Soviet Union in their great battle against fascism. In January 1952, Harig discussed the goals of the new Museum of German History under construction in East Berlin: "[It] shall utilize the struggle of the progressive forces of the past for our present-day national struggle ... Today, faced with attempts by the American warmongers to extinguish the national traditions and characteristics of the German people, it is more important than ever to rewrite German history, free of falsifications and distortions ... For in the maturation process of our national consciousness, knowledge and mastery of history play a decisive role."[27]

Harig did, however, try to limit the party hardliners' more extreme demands on universities. Communist dissidents, including Professor Wolfgang Harich, who supported democratic socialism but opposed the more extreme elements of Stalinism thrived while Harig was minister. It remains unclear whether his moderation was a result of his growing belief in diverse socialist political opinions or a desperate attempt to stem the mass exodus of university faculty to the West. Between 1952 and 1956, some 570 faculty fled the country. This brain drain crippled Harig's ability to grow the East German university system.[28]

Harig's zeal for the Communist party was not enough for him to remain in government. After the Hungarian Uprising in 1956, a series of protests broke out at East German universities aimed at the requisite courses in Russian and dialectic materialism. Harig tried to moderate the students' demands by temporarily suspending these courses in specific programs until better instructors became available to teach them. His efforts to appease the students did not sit well with the increasingly reactionary East German government, especially with Walter Ulbricht, the First Secretary of the SED Central Committee. Ulbricht considered Harig too soft on the state's opponents. In March 1957 he was replaced "by hardliner Wilhelm Girnus and returned to teaching."[29] Harig was lucky to get such a soft landing; when Harich was arrested in 1956, he had been sentenced to ten years in prison for "establishment of a conspiratorial counter-revolutionary group."[30]

Harig returned to the university, by then renamed Karl Marx University, and was appointed professor for history of the natural sciences and director of the Karl-Sudhoff-Institute for the History of Science, Technology, and Medicine. He held a number of senior university administrative positions and continued to be active in politics, albeit at the local level. In terms of scholarship, the last decade of his life was among his most productive. He helped found the *NTM Journal of the History of Science, Technology and Medicine*, which remains to this day a major German-language journal in the field. Not surprisingly, he focused strongly on applying the history of science to its function in society. He continued to write extensively about science and dialectic materialism and Soviet science. He wrote or edited a number of works on Alexander von Humboldt, the eighteenth- and early nineteenth-century scientist, considered an icon of German secular philosophy. James Howell explains: "East German authors and politicians recruited and reimagined Humboldt as a socialist scientist and a champion of the oppressed, in order to provide scientific and cultural legitimacy to the state's official ideology and political agenda."[31]

One of Harig's last contributions to the history of science was an "amazingly early" critical analysis of Thomas Kuhn's theory of scientific revolutions. Writing in 1966, before Kuhn's book had become widely accepted, Harig criticized Kuhn for treating scientific revolutions as replacing internal scientific paradigms with new ones. He argued that "only if the paradigm and the social function of science change at the same time will there be a scientific revolution." His motivation for this critique was Marxist, but it was also written to justify the East German government's efforts to utilize science and scientific planning to modernize society. This early-1960s social development program relied on science as the primary productive force, and Harig argued that it marked another step forward in the development of the social function of science similar in importance to the scientific revolution of the seventeenth century.[32] In 1966, Harig died quite suddenly of a heart attack, at age sixty-four.

In 1946, Wolfgang Steinitz and his brother-in-law, the physicist Hans Jürgen Cohn-Peters, both left Sweden and returned to the Soviet zone of occupied Germany. Cohn-Peters became a university administrator; Steinitz would become one of the most important and politically engaged scholars in East Germany, particularly in his work on *Volkskunde* (folklore). A staunch advocate of socialist and Marxist ideology, Steinitz was also a rebel who championed the creation of a diverse, international, interdisciplinary field, which included engaging East German scholars who were not members of the SDP or "staunch Marxists," as well as folklorists in the Western world.[33] Steinitz returned to Berlin, where he helped rebuild the German Academy of Sciences (Deutsche Akademie der Wissenschaften; DAW) and Berlin's university, which in 1949 was renamed Humboldt University. In 1946 Steinitz was appointed Professor of Finno-Ugrian languages; however, he also played an important role in promoting the teaching of Russian as the second language of instruction in all East German schools from elementary to university.[34]

In Sweden, Steinitz had written a textbook for teaching Russian to German-speaking students. That book, published shortly after his return to Berlin, was for many years *the* textbook of choice because students found it easy to comprehend. Widely used in East Germany, in Scandinavia, and, eventually, in West Germany, the book appeared in eighteen editions over nineteen years (1945–64). The textbook was "a simple and understandable explanation of the phonetics and grammar (which was based on a solid scientific basis), with concise exercises and reading texts of appropriate levels of difficulty, derived from life and the structure of the new society in the Soviet Union." Steinitz's book thus delivered two important elements to help legitimize the soon-to-emerge GDR: It was a useful tool for teaching the language of the new dominant partner of the German Communist state; and, by exalting a utopian vision of life in the Soviet Union, it provided a model for what could be accomplished in a socialist society. In that regard, Steinitz revealed himself an unrepentant Stalinist. Despite his family's experiences in Leningrad that had led to their flight to Sweden, he presented the same romantic vision of Soviet society that had been held by many Western Communists before the purges.[35]

By training Steinitz was a linguist, specializing in Finno-Ugric languages. After the war, however, he began switching his focus to the study of German folk traditions, particularly folk music. In 1951 he was appointed to head the DAW's Folklore Commission (Kommission fur Volkskunde), renamed in 1953 the Institute for German Folklore. Steinitz transformed the study of German folk culture by insisting it be part of a new "socialist consciousness." In doing so, he both affirmed his long-held ideological beliefs and fulfilled the will of the new state, which was looking to justify its existence through "scientific" study of the past. Steinitz thus created a mutually advantageous relationship between the scholar and the state. He received considerable funding and the institutional structures necessary to carry out his work; the state, for its part, benefited from history being reconstructed to reflect the requirements of contemporary socialism.[36]

In the early 1950s, Steinitz called for GDR scholars to study and collect German folklore, including poetry, song, and art, and to search in it for "the evidence of the working people and their long fight against oppression." In linking folklore to the Marxist view of history, which saw the past as a continuous struggle of the proletariat against elites, Steinitz hoped to reject the corrupt folk studies of earlier eras, particularly those of the Nazis. To Steinitz, the proper study of *Volkskunde* would link the past struggles of peasants and workers to the building of a modern socialist state.[37]

Steinitz's own scholarship focused on protest songs of the German peasantry. He rejected the notion that the pre-industrial workers lived an idyllic life, maintaining that in fact they lived terribly hard lives. The songs he meticulously collected showed that the downtrodden used them to articulate their sorrows and to protest their oppression. His *German Folk Songs of a Democratic Nature from Six Centuries* (*Deutsche Volkslieder demokratischen Charakters aus sechs Jahrhunderten*), published in two volumes between 1952 and 1962, contained 180 songs from the Peasants' Revolt to the fight against the Nazis. The volumes included little-known songs lamenting the suffering of Silesian weavers, the travails of soldiers in the Thirty Years' War, and the revolutions of 1848. In the 1970s, Steinitz's collection became an important source of material for musicians during the revival of folk music in East and West Germany. His work was deeply respected on both side of the Iron Curtain – a rare feat for an ideologically driven GDR scholar.[38] As a result of his support for the goals of the GDR, in the early 1950s he became one of the GDR's most influential social scientists. In 1954 he was appointed vice-president of the German Academy of Sciences. That same year, the government appointed him to the Central Committee of the SED.

Despite his strong commitment to the cause of socialism, and his close links to the GDR's ruling elites, Steinitz began to have his doubts about blindly following Stalin-era policies. He was deeply shaken by Khrushchev's secret speech, in which the new Soviet leader outlined Stalin's abuses of power, including the purges of innocent Communists. The Soviets' violent suppression of the Hungarian Uprising led him to realize that socialist states were very far from respecting the will of the working classes. In 1955 and 1956, Steinitz spoke to the Central Committee of the SED at its annual meeting. In his second speech to the committee, he denounced political abuses of science, including the tendency to exaggerate the successes of Soviet science. As examples, he included the support given to Lysenko's by then discredited genetic theories and the gagging of scientists classified by the East German government as "bourgeois." Ulbricht took great umbrage at Steinitz's remarks, interjecting criticisms as Steinitz spoke. Later, the East German leader removed Steinitz's entire speech from the records of the meeting. Incriminating evidence against Steinitz was twice gathered for a potential prosecution, and a Stasi (secret police) file on him was opened, but he was never arrested. Perhaps this was because he was too public a figure in the Communists' recasting of German history. He was, however, dropped from the Central Committee in 1958, although he would retain his position at the Academy of Sciences for another five years.[39] In 1967, Steinitz died in Berlin from the effects of a stroke. He was sixty-two years old.

Fritz Lange, the sole academic refugee to remain in the Soviet Union, returned to East Germany only in 1959. After the war, Lange led a Soviet team that developed centrifuges for separating uranium-235 for use in atomic bombs. In 1952 he was allowed to transition to civilian research, teaching at several institutions in the Soviet Union. He apparently had wished to be repatriated much earlier, but the Soviet authorities restricted his movements, likely because of his involvement in nuclear research. In 1951 he was denied permission to return to East Germany to attend his mother's funeral. By 1958, Lange, aided by Jürgen Cohn-Peters, then an influential East German scientific administrator and Lange's former research assistant, had launched a campaign to return home. The East German government convinced the Soviets that Lange qualified as a political migrant, having made an "official offer for him to take part in the construction of a Communist society in the GDR."

After twenty-four years in the Soviet Union, Lange was a stranger to most of the German scientific community, but he was highly regarded politically among the Communists who ran the GDR's Academy of Sciences. He was named the director of the new Institute for Biophysics at the Institute for Medicine and Biology in a suburb of Berlin. The institute had a staff of more than seven hundred, including one hundred scientists. Lange, now in his sixties, had no experience in the field of biophysics, so there were no great breakthroughs under his leadership. He did, however, succeed in establishing biophysics as an important scientific discipline in East Germany. He "played a more important role in GDR science policy than in science itself." Lange retired after just five years at the institute. He died in East Berlin on 25 July 1987, at the age of eighty-eight.[40]

Socialists and the Quest for Social Justice

Barbara Ruhemann, Harig, and Steinitz remained convinced Stalinists even after their experiences in the Soviet Union and during the war. Other refugee scientists rejected this unyielding hardline ideology but retained a strong commitment to socialism and social justice. Profoundly influenced by their experiences in the Soviet Union, this group retained their lifelong belief in collective action to bring about a more equitable world. Edgar Lederer was perhaps the most open about remaining faithful to the principles of the Communist Manifesto, and even to the Soviet Union, in the first decade after the war. Lederer had not personally experienced the worst of the purges; he had left the Soviet Union at the urging of his wife, who noticed the changing political climate in the country even while Lederer himself remained oblivious to it. He later confessed: "It took me a very long time (in fact until Khrushchev's report denouncing Stalin's crimes) to wake up to reality." Deprived of inspiration from the Soviet Union, he turned to a new model that "seemed most promising," Maoism. Here too he would eventually be disillusioned and forced to admit: "Alas, I was not quite right here again."[41]

Whatever Lederer's disappointment with the various manifestations of Marxism, he was never a member of a Communist Party, nor was he particularly committed to forging a new socialist society. He did not become politically active until the 1960s, when he emerged as a leading scientific voice in Europe's anti–Vietnam War movement. He attended several meetings of the International War Crimes Tribunal (aka the Russell Tribunal), a non-government group organized in 1966 by the

British philosopher and Nobel laureate Bertrand Russell. The tribunal attempted to expose American war crimes in Vietnam; in doing so, it was sharply criticized for failing to also examine atrocities allegedly committed by the North Vietnamese and the Viet Cong. In 1969, Lederer organized an international meeting at Orsay against the Americans' use of chemical weapons in Vietnam.[42]

Lederer and a group of close left-wing friends, including the physicist Alfred Kastler, a Nobel laureate, and the mathematicians Laurent Schwartz and Henri Cartan, were frequent participants in street demonstrations in Paris in support of human rights. The same group organized boycotts of scientific meetings in Franco's Spain and in Greece during the military junta. He also helped develop research capacity in a variety of Third World countries. He made it a policy to always welcome students from Africa, India, and Southeast Asia, and he always managed to find funding for them. He made a special effort to help refugees, because, he explained, "having been a refugee myself, several times in my life, I quite naturally understood the plight of others and helped them whenever I could." These refugees came from South America and Spain as well as from behind the Iron Curtain.[43]

Like Lederer, Ernst and Mia Emsheimer remained committed socialists throughout their lives. The couple rejected Stalinism, which accounts for their failure to join Wolfgang Steinitz in East Germany, but they continued to believe strongly in Marxism. In the 1960s they actively supported the anti–Vietnam War movement in Sweden, "a rare standpoint to be taken at that time by the head of any institution in Sweden." Like a number of other scientific refugees from the Soviet Union, Ernst advocated for maintaining connections with Soviet scholars, who after all had themselves been victimized during the purges.[44] The Emsheimers' unswerving commitment to social justice profoundly influenced those around them. Ernst's most important student, Jan Ling, was greatly influenced by the Ernsheimers, his mentors. He involved himself in Vietnam's solidarity movement, and "his politically 'uncool' activities as president of the Swedish–Soviet Friendship Association (Vanskapsforbundet Sverige-Sovjet) drove home the point that it was impossible to separate the politics and ethics of one's own work as an artist or intellectual from that of politics and ethics on a much larger scale."[45]

Charlotte Schlesinger set out to find a community and a scholarly environment that reflected her commitment to socialist cooperation and equality and where she could, to some degree, relive the avant-garde culture of late Weimar Berlin. After teaching at Foxhollow School throughout the war, in 1946 Schlesinger joined the faculty of Black Mountain College in the Blue Ridge Mountains of North Carolina. Founded in

1933, the school promoted radical ideas in educational philosophy and in governance. The college offered a relatively unstructured curriculum that emphasized interdisciplinary study and centred on the arts. Students were encouraged to study at their own pace and to pursue their own interests. They could take an unlimited number of elective courses. The only formal requirement was that students choose a mentor among the faculty. No formal course credits were given, although students could apply to have external professors grade and certify their performance.

Faculty and their families were expected to eat in the communal dining hall with the students and staff. In addition to room and board, the faculty received only a small stipend. The college was a mecca for refugee artists and scholars; in 1940, twenty-eight were on the faculty or staff.[46] The first refugee artists to arrive at Black Mountain were Josef Albers, an artist and art educator, and his Jewish wife, Anneliese (Anni) Fleischmann, a weaver. Albers had fled Germany after the Nazis closed down the Bauhaus art schools in April 1933. For Albers, art education included music, theatre, drama, literature, photography, drawing, painting, and design.[47] It is very likely that it was Albers, the rector of the college for a short time in 1946, who hired Schlesinger to teach music and piano.

Schlesinger arrived in September, just as the college was scrambling to complete a major expansion. The student body almost doubled in size to ninety-two students, caused by the influx of returning soldiers as a result of the GI Bill's education benefits. In her first year at Black Mountain, she taught introductory music, harmony, and chorus and co-taught counterpoint and piano with Edward Lowinsky. Lowinsky, a musicologist specializing in Renaissance music, was, like Schlesinger, a German refugee. Schlesinger's role was not to teach a few students specializing in music, but rather to provide an aesthetic education to all students in order to introduce them to music as a force shaping the personality of each individual.[48] Schlesinger quickly immersed herself in life at the college. At the first concert of the new term, she and Lowinsky, "at one piano, gave brilliant performances of two sparkling Mozart piano sonatas for four-hands." Schlesinger also directed the college's chorus, which comprised more than half the student body as well as many faculty. The chorus's first performance included chorales by Bach and an excerpt from Handel's *Messiah.*[49] Schlesinger's stay at Black Mountain College was brief. It had split into two rival factions, divided over management, funding, and curriculum. When a group of "reformers" in the faculty forced the resignation of Ted Dreier, the long-time rector, Albers and most of the rest of the faculty of the arts program, including Schlesinger, resigned.[50]

After leaving Black Mountain, Schlesinger took a position at the Wilson School of Music in Yakima, Washington, where she taught music to both young pupils and adults. She struggled with failing eyesight and feared she would go blind. In 1957, she was diagnosed with cancer; five years later she was forced to leave the United States to live with her brother Hans in England. She suffered through years of serious illness, finally dying in 1975, at the age of sixty-six.

When she died, Schlesinger and her work were almost completely forgotten. It is unclear whether, after leaving Berlin, she ever had an opportunity to compose music again. Almost all of her music was lost when she travelled to England in 1963. Believing she was only going for a brief vacation to recover, she left most of her manuscripts with American friends. Scholars have been unable to locate those friends. Of the ten major compositions she is known to have produced, four are missing or lost. A handful of Schlesinger's works were located by Christine Rhode-Jüchtern, a German musicologist, in an old suitcase left behind in her brother's house when she died.[51]

In May 2009, at the Sophie Drinker Institute for Musicological Research on Women and Gender in Bremen, Germany, a symposium was held marking Charlotte Schlesinger's hundredth birthday. At the symposium, scholars discussed the issues that faced woman refugee composers, including "to what extent escape, expulsion and exile experience combined for many people during the Second World War with an existential shock at their own life plans and their view of themselves." Other papers examined why so many women composers, like Schlesinger, were almost erased from history. The symposium ended with a performance of songs by Schlesinger, some of which may have been performed for the first time.[52] Since the symposium, performances of Schlesinger's music have become more frequent. In 2019, three of her songs were recorded on an album, "Forbidden Sounds: 'Degenerate' Women Composers in the 20th Century."[53]

The Anti-Communists in the Cold War

Lederer, Emsheimer, and Schlesinger continued to keep faith with the values of socialism contained in Marxism, but they also abjured the extremism of Stalinism. Not surprisingly, four other scholars, three of whom had experienced the horrors of the Great Terror and the Shoah, became outspoken critics of totalitarianism, especially all forms of Communism. This small group of anti-Communists had a profound influence, far greater than their small numbers would suggest, in shaping the response of western liberal democracies to the Soviet Union in the postwar world. Two of the four, Alexander Weissberg and Fritz Houtermans, had been devoted members of the KPD and supporters of the Soviet Union before migrating to Ukraine. In the early 1930s, the other two, Laszlo Tisza and Hermann Bouchardt, had both been fellow-travellers, not formal party members but closely associated with and supportive of the goals of international Communism.

On 18 June 1937, Hermann Borchardt arrived in New York, physically and psychologically damaged by his ordeals in Stalin's Soviet Union and Hitler's Germany. Having been forced to flee Germany in 1933 after the collapse of the Weimar Republic, he had learned to mistrust democracy. After seeing the Soviet police state from the inside, and being expelled from Minsk in 1936 with hardly a day's notice, he had come to loathe the delusion of a Communist utopia. On his return to Germany, he had experienced the horrors of the Nazi regime. Terrified of arrest, he and his wife had contemplated suicide. When the Gestapo did incarcerate Borchardt in concentration camps, the guards physically and mentally tortured him; his fellow inmates, especially Communist gangs, beat him. Upon his release, he was forced to flee for his life. Almost the final indignity was that when he arrived at Ellis Island, US immigration officials noticed his mutilated hand and almost refused to allow him to land. Borchardt was unable to hear the officials' demands for him to explain the cause of his injuries, and even if could have heard them, he was incapable of understanding them since he did not speak English. Only the intervention of Erich Cohn, a wealthy noodle maker and art collector, and friend Georg Grosz, convinced the immigration authorities to allow Borchardt to proceed.[54]

Borchardt and his family settled in a small apartment on the Upper West Side of Manhattan. Hermann's physical injuries, and his difficulty in learning spoken English caused by his profound deafness, prevented him working on a regular basis. The family would not have survived had they relied on Borchardt's few paid teaching assignments and the rare fees from his writing. Instead, the family eked out an existence, relying mainly on donations from friends, small grants from refugee aid committees, and his wife's wage labour.[55] Finally free from persecution for the first time since his initial flight from Germany in April 1933, Borchardt was determined to inform the world about the realities of the Soviet Union and Germany. In short order, he wrote a play and two autobiographical books.

His first book, an account of his experiences in Minsk, was a caustic, hard-hitting exposé of life in a police state. After finishing one draft, he put this book aside; only a few sections of it would ever be published. He focused his attention, instead, on his second work, in which he recounted his experiences in Germany and in the concentration camps. In June 1938 he submitted the book under a pseudonym to a competition organized by the Guild for German Cultural Freedom. A report written for the guild by an "expert" outlined the objections to this overly dark account of life in Nazi Germany:

> Incredible stories are told – yet somehow one believes every word – for this is in fact a historical document. But, though one feels deep sympathy for the victims ... somehow this book is hateful and its author fails to inspire the reader's sympathy. It seems as if this experience had not taught the author anything – as if he would go on living his life in scorn and hate for all his fellow-beings and because of this unsocial and hateful attitude the book leaves one with a nasty taste in one's mouth.

Christoph Hesse, the German cultural historian, who found this extraordinary review, heaped well-deserved scorn upon it: "The expert did not reveal what this experience should have taught the author. The opinion he had only hinted at, and which is still widespread to this day is that the concentration camp was a kind of reformatory."[56]

Borchardt refused to portray his fellow inmates as heroic victims, instead presenting many as active collaborators with the concentration camps' guards. This made his work immensely unpopular not just with the guild but with many members of the expatriate German community. At a public reading of parts of his book, two of Borchardt's friends had to physically protect him against the violent reactions of refugees in the audience. Borchardt's work was considered too heretical, too cynical, and, perhaps, too brutally honest to be published; hence only fragments of the manuscript survive.[57]

Borchardt's plays, *Liberation of Pastor Müller* and *The Brethren of Halberstadt*, fared little better than his books. *Liberation of Pastor Müller* was about Christian resistance to the Nazis and was loosely based on the experiences of a Pastor Müller from Ahlbeck, where his wife came from. Borchardt showed a synopsis of the play to Ernst Toller, the famous German Communist playwright, himself a recently arrived refugee from Germany, hoping to sell the play to him. When Toller demanded changes to the play, Borchardt instead wrote *The Brethren of Halberstadt* for Toller and kept the original play for himself. Toller did not buy the play. One year later, in 1939, however, Toller published his own play, *Pastor Hall*, which bore striking similarities to Borchardt's work. Borchardt had been unable to find a publisher for his plays and was enraged when Toller not only was able to publish his own but also sold a screenplay for a film version, to be made in Britain in 1940. Borchardt demanded compensation, and Toller obliged with a token payment. Borchardt regarded this compensation as totally inadequate, especially since Toller refused to publicly recognize Borchardt's contribution.[58] An unsuccessful lawsuit ensued, which he continued to pursue even after Toller committed suicide in 1940. Borchardt's son Frank, writing about the whole sordid affair, remarked: "The entire Toller incident was deeply embarrassing to the family, an episode in the struggle for existence among exiles in New York, especially those who could not possibly adapt to American circumstances, that is, most."[59]

Borchardt then began labouring at his greatest work, *The Conspiracy of the Carpenters: Historical Accounting of a Ruling Class*, in which he set out to create a "mirror world" where he could bring into play his complex political philosophy shaped by years of exile, flight, and torture. *Conspiracy* is a dystopian novel about a world shattered by the emergence of Dr Urban, a populist demagogue, and his fanatical followers, the Urbanites. Urban is, of course, modelled on Hitler; in Borchardt's fictional world, however, unlike in Germany, Urban is ultimately defeated by the wily manoeuvrings of the elderly but experienced industrialist and politician President Adam Faust, "the Fox of God," ably aided by his supporters, "the Demigods," and the members of the League of the Carpenters. The league is a revival of a medieval religious guild, willing to use violence, including murder, arrest, and detention in re-education camps, to stop the Urbanites. The goal of Faust and his Demigods is to encourage Urban to launch a civil war and thereby bring about a necessary "bloody purging of the nation." This novel extols the virtues of conservative Christians, both Catholic and Protestant, and industrial and religious elites, in thwarting fascism.

Borchardt included every possible faction in this sweeping book. The Communists, ineffective and dishonest, are portrayed as "The Iron Phalanx of Intelligence glittering with all the progressive party colors including the blood red of revolution." Borchardt painted middle-of-the-road bourgeois politicians as ineffective fence-sitters. He also attacked every aspect of modernity, including the Enlightenment, democracy, science, mass industrial production, technology, and secularism. The book as written was more than 1,000 pages long, and in order to find a publisher for an English translation it was necessary to pare it down to just over six hundred pages. Even in this abridged form, the novel was immensely complex, on the scale of Tolstoy's *War and Peace*. The book had 142 major characters, divided into seven major factions. Two indices at the end of the book helped the reader keep track of the huge cast.

The Conspiracy of the Carpenters was published in 1943 by Simon and Schuster and was widely reviewed in almost every major literary forum in the United States. Seldom does one read such a diversity of opinions about a single book, not just between reviewers, but also within the individual reviews themselves. One reviewer exclaimed: "I believe that 'The Conspiracy of the Carpenters' is wretchedly inept as fiction and thoroughly dangerous as a political tract." He went on the say that while the novel was "wonderfully bad," it contained "flashes of brilliance" which showed that Borchardt had "thought long and deeply on many matters about which he can be eloquent, provocative, contentious, nobly idealistic or mistaken."[60] Klauss Mann, the refugee writer and dissident, reviewed the book for the *Chicago Sun*'s *Book Week*. He called Borchardt "a vital original talent," and the book "truly an amazing work, bold and argumentative – and unique." It was "one of the most astonishing experiments in modern literature, incomparable in its particular accentuation and mannerism and yet typically Teutonic in its unrestrained moodiness and pedagogical claims." He concluded his review by calling *Conspiracy* "one of the most extravagantly reactionary books written since the time of Metternich."[61]

Despite almost universal praise for the powerful political ideas in the book, it was doomed by the almost equally unanimous criticism of its complexity, poor style, and reactionary conclusion. American conservatives were not interested in a book written by a foreign Jewish convert to Lutheranism, who increasingly leaned toward supporting Catholic theology, although he only formally converted in 1949. Sales of the book were poor, and Borchardt received no more than his initial advance cheque. His greatest work was all but forgotten before the end of the year. He did not receive the popular success he so craved; however, his increasingly reactionary views found an outlet in the fervently anti-Communist Catholic press. He wrote a series of anti-Communist and anti-fascist articles and pamphlets, all of which were published by the right-of-centre Catholic Information Society. He found a receptive audience in his arguments that there was little to differentiate between the Nazis' fascism and Stalin's Communism, since both "demanded everything, even the soul." Individuals in the modern total state "are permitted to have no will or conscience of their own, no duties to man or God except those ordained by the new idol, the state."[62]

It is doubtful that Borchardt had much influence as an anti-Communist warrior; his audience were already converts to the cause. But he did have one more opportunity to sway political opinion, when for a short time he acted as an adviser to Claire Boothe Luce, the writer, reporter, actress, wife of the publisher Henry Luce, and congresswoman, on issues regarding China. Boothe Luce, like Borchardt, was going through her own conversion to conservative Catholicism, and she saw a kindred spirit in the German intellectual. The two had a falling out over Borchardt's strident warnings that failure to support the Nationalists at all costs, and a Communist victory, would cause the West to become embroiled in a war in Asia that would last for generations.[63]

Borchardt was well aware of his lack of success in combating the forces that had so negatively impacted his life. Just before his death, he wrote that he had "gone from failure to failure so that it can no longer be described."[64] Among his small circle of friends, mainly in the German intellectual diaspora, including Brecht, the actor Ernst Josef Aufricht, and, of course, Georg Grosz, Borchardt's intellect continued to be held in high regard. He died of a heart attack in January 1951, at the age of sixty-one. Grosz paid tribute to his friend in a poem:

> In Germany, there lived a little man, Borchardthans, that was his name,
> They hired him as a teacher,
> He lived like a philistine.
> But secretly in his chamber he described the whole world of misery.[65]

Like Schlesinger's, Borchardt's life's work was almost completely forgotten. Most of his writing was never published, and much of it was lost. Even his great novel appeared in only the one edition, in an abridged English translation. Finally, in 2004, Stefan Weidle, a German publisher, was able to purchase the full manuscript from Hermann's son Frank for the German Exile Archives, and to publish this "last great unpublished work of German exile." *Die Verschworung der Zimmerleute* finally appeared in its entirety, all one thousand pages in two volumes, for the first time that same year, sixty-two years after it was completed. It still has not been widely read, and it is even less understood.[66] In 2018 a team of scholars in Germany commenced a multivolume project to publish the selected letters and works of Borchardt. He is no longer the forgotten intellectual of Weimar Germany.[67]

Borchardt's attempts to fight Communism had little more measurable effect than Don Quixote's efforts tilting at windmills. By contrast, Alexander Weissberg played a key role in the Cold War intellectual struggle against Communism and the Soviet Union. Weissberg and Eva Striker's experiences during the Great Terror strongly influenced their friend Arthur Koestler to renounce Communism and to write *Darkness at Noon*, one of the most important anti-Communist novels of the period. Koestler's final conversion to Communism had occurred after his political discussions with Weissberg in Eva Striker's Berlin apartment. He had visited the couple at UFTI, staying at Weissberg's apartment and socializing with many of the physicists on the staff of the institute. He remained a loyal member of the party, even risking his life to spy for it against Franco during the Spanish Civil War. Franco's forces captured Koestler, swiftly convicted him of espionage, and sentenced him to death. Koestler spent three months in prison, expecting any day to be shot. But he was released after a prisoner exchange. His experiences in Spain had shaken his fanatical faith in the party, but he was not yet ready to break from it. A subsequent series of events, including the show trials in Moscow, further eroded his beliefs. Wrestling with his conscience and frightened of being ostracized by many of his friends if he made a complete break, Koestler found himself paralysed with uncertainty about how he should proceed. Eva Striker's sudden arrival in the West was a key event that finally drove Koestler to publicly resign from the Communist Party.

Striker's account of her arrest and treatment in prison, and her knowledge of their many friends, including Weissberg, who had also been arrested on false allegations, transformed Koestler's views. He and Georg Placzek wrote a letter of support for Weissberg and Houtermans to Stalin. They then arranged for Irène and Frédéric Joliot-Curie and Jean Perrin, leading French left-wing scientists, to sign it. The letter stated that the two imprisoned scientists were faithful and loyal friends of Communism and the Soviet Union and pleaded for their release from jail. Koestler, however, was not surprised when nothing came of the letter.

In the spring of 1938, shortly after arranging for the letter in support of Weissberg, Koestler resigned from the party. It was a gut-wrenching decision. He later explained: "I was twenty-six when I joined the Communist Party, and thirty-three when I left it. The years between had been decisive years, both by the season of life which they filled, and the way they filled it with a single-minded purpose. Never before nor after had life been so brimful of meaning as during these seven years. They had the superiority of a beautiful error over a shabby truth."[68] In his letter of resignation, Koestler condemned the party's moral degeneration but continued to pledge his total support for the Soviet Union. His final epiphany that the degeneration was not just in the party, but also in the state, came only after the signing of the German–Soviet Non-Aggression Pact in the summer of 1939.

Having broken with his Communist delusions, Koestler set out to write a novel that would expose the means by which those arrested by the NKVD were broken, through interrogation and torture, so that they publicly acknowledged guilt to crimes they had not committed. His story of the fictional revolutionary leader Nicholas Salmanovitch Rubashov is based in part on his own experiences in Spain and also, in large measure, on others' first-hand accounts of imprisonment and interrogation in the Soviet Union. The most important witness to NKVD methods was no doubt Eva Striker.[69] Koestler began writing *Darkness at Noon* around the time of the Munich Accord; he finished it in April 1940. Koestler had no knowledge of Weissberg's experiences in prison, but he likely knew that early in 1940 the NKVD had turned his friend over to the Gestapo. While there is no firm evidence that Weissberg's fate helped shape *Darkness at Noon*, Koestler does write that he found that while writing the book, he gradually began to understand there was little difference between Hitler's and Stalin's regimes. This sameness became a recurring theme in the novel.[70]

When first published in Britain, *Darkness at Noon* sold poorly. In the United States, though, it became a bestseller after it was chosen as a Book of the Month Club selection. In the 1950s the book was transformed into an award-winning play. Its most significance reception was in postwar France, where it was published under the title *Le Zéro et l'Infini* (Zero and Infinity), in reference to a passage in the book where one's individual value in the social equation is said to be both nothing and infinite. In France at the end of the war, the Communists were the most powerful political party, and they were demanding a significant role in government, if not outright control. For those who opposed the Communists, Koestler's novel became an important tool for exposing the true nature of the Soviet state. Despite strong efforts by the Communists to suppress *Le Zéro et l'Infini*, more than 400,000 copies of the book were sold, helping sway public opinion against the Communists.[71] By the time the novel was published in France, Weissberg, having survived both the Soviets and the Nazis, had arrived in Paris. He was determined to establish himself in business, having proven himself an adept businessman both in the Soviet Union and in wartime Poland. He reconnected with Koestler, and the two excommunicated Communists quickly re-established their close friendship. Koestler urged Weissberg to write an account of his experiences in the Soviet Union.

Before Weissberg could publish his memoir, however, he was called as a key witness in a libel trial that pitted David Rousset, a French intellectual, former Trotskyite, and concentration camp survivor, on one side, against Pierre Daix, a French Communist and also a survivor of the Holocaust, and the Communist journal *Les Lettres françaises*, on the other. Rousset had made a considerable name for himself in postwar France by writing some of the first historical accounts of the Holocaust. In the early Cold War, Rousset, working with other left-wing French intellectuals disenchanted with the Stalinist French Communist Party (they included Jean-Paul Sartre), tried to found a new, independent French socialist movement, the Rassemblement démocratique révolutionnaire (RDR). Rousset wanted the RDR to be a social democratic party that rejected both capitalism and Stalinism. When it failed to grow into a significant political organization, he turned his energies toward developing a movement of Holocaust survivors who would fight for human rights by exposing the existence of concentration camps throughout the world. He wanted this new organization of survivors to turn their attention first to revealing that Soviet prisons and work camps were no less evil than Nazi concentration camps.

In 1949, on the front page of the conservative newspaper *Le Figaro litté-raire*, Rousset published a plea for assistance titled "Help the Deportees in the Soviet Camps: An Appeal from David Rousset to the Former Deportees of the Nazi Camps." Rousset called on fellow survivors to join him in aiding those who were suffering horrors in the Soviet camps, similar to those they had experienced at the hands of the Germans. In drawing this precise parallel, Emma Kuby has pointed out, Rousset downplayed the Nazi extermination camps.[72] The French Communist Party was already under strong attack from critics like Koestler because of its refusal to acknowledge Stalin's crimes and its insistence on maintaining close ties with Moscow. With considerable justification, many French Communists believed that some of the attacks on Stalin had been financed by American and British intelligence services and were designed to discredit the party. Rousset remained independent of these external groups, but since he had been a Trotskyite, albeit now a lapsed one, some Communists viewed him as the more potent threat to the party.

A group of Communists prepared a rebuttal for publication in *Les Lettres françaises*. The rebuttal was signed by Daix, who was himself a camp survivor and who, it was hoped, would counter much of the public sympathy given to Rousset. The rebuttal denied the existence of "concentration camps" in the Soviet Union and accused Rousset of lying about them. Rousset promptly launch a defamation lawsuit against both the journal and Daix. The lawyer for the Communists had only one strategy – to deny the existence of widespread imprisonment and forced labour in the Soviet Union. Rousset and his lawyers countered by calling a group of survivors of the purges to attest to the accuracy of Rousset's assertion that Stalin too had created a huge network of concentration camps and prisons. Daix's lawyers, knowing it would be difficult to impeach each witness's testimony, countered by trying to impeach the witnesses' credibility.[73]

Weissberg was the second of fifteen eyewitnesses called by Rousset's legal team to testify. His testimony was the most dramatic because it exposed the weakness of the defence's position. Weissberg, testifying through an interpreter in German, began by stating that he believed that as many as 8 to 10 million people had been incarcerated in Stalin's camps after the purges began in 1936. At this point, Claude Morgan, editor of *Les*

Image 16. David Rousset (right) and Alexander Weissberg in a staged press
photo re-enacting a dramatic moment from the 1949 trial. Weissberg appears
to be standing in the dock answering Rousset's questions. In truth, Rousset's
lawyers asked all the questions. (From the author's collection.)

Lettres françaises, threw the court into turmoil by interrupting Weissberg's
testimony, arguing that it was sickening that a French court would allow a
German to participate in a trial against the Soviet Union.

    After a brief adjournment, the president of the court warned the defen-
dants in the strongest possible terms about any further interruptions.
Then Theo Bernard, one of Rousset's attorneys, rose and addressed the
court: "The unspeakable attack of which the witness has just been the
victim on the part of one of the defendants, I would like to indicate in a
word that the one who is at the stand is an Austrian Jew whose father, wife
and two brothers were murdered by the Nazis, who himself was detained
for years by the Nazis, who participated in the Warsaw Uprising."

In a dramatic moment rarely seen outside of fictional courtrooms, Bernard next read into evidence the entire letter signed by Irène and Frédéric Joliot-Curie and Jean Perrin in June 1938 and sent to the Soviet Union attesting to Weissberg and Houtermans's loyalty to the party and to the Soviet Union, proclaiming their complete innocence, and pleading for the release of Weissberg and Houtermans. After reading the letter, Bernard proclaimed: "Here, gentlemen, is the man who comes, because he speaks German, to be the object of an unexpected and hateful attack on the part of one of the defendants and in terms which, I say, and Claude Morgan understands what I'm saying, are unworthy of someone who claims to be Communist." No one seemed aware that Koestler and Placzek had actually penned the letter and that none of the French scientists had ever met Weissberg.

The reading of the letter, signed by the three most important French Communist scientists, and Bernard's statement of fact that the witness was not German, but an Austrian Jew and a victim of the Holocaust, was the turning point of the case. Realizing that their position was desperate, the defence lawyers interrupted the proceedings, demanding the right to the floor. When denied this, they said it implied bias on the part of the court. When once again order was restored, and Weissberg was able to continue his testimony, he directly addressed one of the defence counsels to remind him that he had once been a Communist. At this point, someone in the court shouted at Weissberg, "Renegade!"

Weissberg was then asked by the defence where he had found the figure of 10 million people detained in Stalin's camps, since it was identical to the one put forward by Goebbels. Weissberg said it was his own calculation based on his experiences in the Soviet Union, and, moreover, that he had not had the opportunity to read the propaganda of Mr Goebbels, because at the time he was fighting against Hitler, a struggle that continued as long as Hitler lasted. Weissberg continued to outline his arrest, torture, and treatment in the Soviet prison system. He explained to the court that he had met thousands of people in the jails, none of whom had any idea why they had been arrested, and that almost all of them had been forced to make false confessions that implicated even more innocent persons.[74]

Weissberg's testimony was widely reported throughout France. The trial ended in early 1950 with a verdict in favour of Rousset. The judgment was a token amount, although the defendants were required to cover the legal costs. It is difficult to measure the importance of this trial, although the testimony did persuade several prominent French left-wing intellectuals, including Sartre, to acknowledge the veracity of the accounts of widespread abuses of human rights by Stalin's regime. In France, the trial was just one among several events and publications that gradually wore down broad public support for the Communist Party. The refusal of many French Communists to break with Moscow, and their continued denial of Stalin's crimes despite growing evidence of the purges and the Gulag, further eroded their credibility. Surprisingly, outside of France, the trial was not followed in any great detail. Not a single major American news organization covered the sensational events.[75]

Weissberg's account of his experiences in the Soviet Union appeared in 1951. Published originally in German under the title *Witches Sabbath: Russia in the Melting Pot of the Purges* (*Hexensabbat: Rußland im Schmelztiegel der Säuberungen*), it appeared in English in the United States as *The Accused*, and in Britain as *Conspiracy of Silence*. All three titles capture part of the complex narrative, which is non-linear, bouncing among the past, the present, and the future. Koestler wrote an extensive foreword in which he described Weissberg's book as "a rambling, sprawling, spouting whale of a book. The organization of the material is, from the craftsman's point of view, atrocious." Koestler, however, urged the reader to persevere:

> The book will grow on him steadily, entrance him more and more, and slowly carry him off his feet like a great muddy stream, until he finally realizes that he has become an eyewitness to a great saga of our time. It was a catastrophe like the Black Death, a witches' sabbath of human reason and the first full-sized example of the hitherto inconceivable ravages which modern despotism is capable of inflicting on the bodies and souls of its subjects.[76]

Weissberg's book was very well received by reviewers.[77] In the *New York Times*, David Dallin, a Russian refugee from the revolution, hailed Weissberg's account as something more than other accounts of the purges. It was "a landmark, a monument, and an inexhaustible source of penetrating insights into the souls of the men who confess and of those that make them confess."[78] Notwithstanding the critical acclaim, Weissberg's book never achieved anything like the success or influence of Koestler's fictional *Darkness at Noon*. Though gripping, Weissberg's book was just one among a growing body of eyewitness literature exposing the horrors of Stalin's Soviet Union. Moreover, neither Weissberg's story nor any other could sway those who remained disinclined to believe that Stalin's regime should be equated with the Nazi horrors. In a contemporary review of *The Accused*, Mark Graubard, the American historian of science, lamented:

> If Weissberg's account depicts the scene on the Communist side, the following episode epitomizes the reaction of the intelligentsia in America. A

leading American scientist, educator and top government adviser was asked whether he had read the book. "I do not listen to or credit turn-coats," was his prompt and virtuous answer. Clearly, the Kremlin is not alone in mocking and destroying the time-honored institution of the international brotherhood of scientists and scholars.[79]

A similar fate befell Fritz Houtermans's single effort to expose the horrors of the Soviet Union. *Russian Purge and the Extraction of Confession*, which he co-wrote with Konstantin Shteppa under the pseudonyms F. Beck and W. Godin, was published almost simultaneously with Weissberg's, something that is certainly not a coincidence. Koestler very likely had a role in its publication. In his foreword to Weissberg's study, Koestler revealed that he knew the real identities of Beck and Godin and recommended that people read their book as well because it provided a better analysis of the causes of the purges than Weissberg's effort.

While neither Weissberg's book nor Houtermans's was commercially successful, they became influential sources in the scholarly examination of the horrors unleashed by Stalin in the Soviet Union of the 1930s. Both books are still often cited by historians writing about what took place inside NKVD prisons. The flurry of revelations from the former Soviet archives after the collapse of the Soviet Union has done little to diminish the value of these books, given that the NKVD rarely recorded its methods of coercing confessions from its innocent victims.[80]

The two books were also closely linked to an American program that sought to win over the European intellectual left through a cultural Cold War. In the late 1940s, the US government, especially the State Department, the US Information Agency (USIA), and the Central Intelligence Agency, began financially supporting anti-Communist publications and organizations. Koestler was a founder of the CIA-supported Congress for Cultural Freedom (CCF), and he gave the keynote speech at the opening session of the CCF in Berlin in June 1950.

Melvin Lasky, the key organizer of the CCF's Berlin session, wrote a lengthy review of *Russian Purges*, calling it "first rate journalism and intelligent social analysis."[81] A USIA program of translating anti-Communist works made the book available in Chinese.[82] Through Koestler, Weissberg became involved with the CCF. In 1953 Weissberg took the lead in organizing the CCF's Committee on Science and Freedom (SFS). He realized that he had little prestige as a scientist, having last been an active scientist before his arrest in 1937, so he approached Michael Polanyi at the University of Manchester to head SFS. For Weissberg, Polanyi was the perfect candidate: a highly regarded chemist, and an outspoken and unyielding critic of Communism and the Soviet Union, as well as a pioneer in the social study of science. The meeting between Weissberg and his first wife's uncle may well have been tense; a year earlier Polanyi had written "a scathing review" of Weissberg's book in the *Manchester Guardian*. Polanyi had attacked Weissberg personally, asking why before his own arrest he had continued to support the Soviet government after witnessing the arrest of so many of Stalin's opponents and the disastrous famine that had followed the enforced farm collectivization. Polanyi concluded that *The Conspiracy of Silence* was a "standard biography of Modern Destructive Man at his most naïve and amiable best."[83]

Weissberg no doubt accepted the reproach; his book was filled with similar self-criticism. Polanyi was willing to be persuaded to take on the leadership of the SFS, and he used his contacts with the Rockefeller Foundation to secure half of the funding for the first SFS congress. Polanyi, with his well-established anti-Communist credentials, was able to persuade Warren Weaver, director of the foundation's Division of Natural Sciences, that the SFS would not be dominated by scientists considered "pinko, emotional, or naive." Weaver was concerned because the US Congress was investigating claims that major American foundations were supporting socialism and undermining American national interests. The other half of the funding for the SFS congress was supplied by the CIA-funded Fairfield Foundation. In July 1953, the congress was held in Hamburg, with Polanyi and Weissberg serving as co-chairs.[84]

The co-chairmanship of the Hamburg Congress marked the height of Weissberg's involvement with the CCF. Koestler had already been squeezed out of the leadership as the organization shifted its focus away from militant anti-Communism. The new approach was "to challenge the Soviet block at a level on which it was deemed to be most vulnerable – that of the arts, the sciences, and the humanities," targeting the intellectual elites with a "more sophisticated anti-Communism and pro-Americanism."[85] Weissberg had no place in the new CCF, but in recruiting Polanyi, he had brought to the forefront one of the great pioneers of science studies, who provided a different ideological basis than those provided by Marxist scholars like Harig and Steinitz.[86]

In 1953, Weissberg wrote an article on the political situation in the Soviet Union after the death of Stalin for the British right-wing journal *The Time and Tide*. The article was republished in the American anti-Communist magazine *The Freeman*. It is an insightful piece that predicts an inevitable power struggle within the Politburo. Optimistically, Weissberg predicted that in this struggle, "the solid armor which enclosed and constricted the life of a great and gifted people has been cracked. Now at last the Russian people have some hope of freeing themselves." It was wishful thinking, but Weissberg here revealed his continuing admiration for the Soviet people, his hope that they could throw off the yoke of their oppressors, and declared he did not blame them for his horrific ordeal.[87] This article marked Weissberg's final contribution to the cultural Cold War. Weissberg would continue to write, but on topics unrelated to his experiences in the Soviet Union. He never began work on the long-awaited sequel to his memoirs, which would have covered his extraordinary tale of survival in occupied Poland. Instead, he focused his attention on his successful business career. On 4 April 1964, Alexander Weissberg died in Paris at the age of sixty-three.

Laszlo Tisza did not directly engage in fighting Communism, but his experiences in the Soviet Union had a profound influence on one of the most controversial events in American science during the Cold War. Historian Gennady Gorelik has traced the roots of the anti-Communism of Edward Teller, the Hungarian-American physicist who was a principal designer of the first American atomic bomb and, later, the first hydrogen bomb. The public saw Teller as a Cold War warrior with an extreme, right-wing hatred of the Soviet Union. This hatred motivated him to play a key role in the hearing before the Atomic Energy Commission in 1954 that denied Robert Oppenheimer's application to have his security clearance reinstated. Oppenheimer's treatment has been widely considered a great injustice and a tragedy, since he never recovered from the humiliation. Teller was so out of step with the ideological views of most American scientists, who continued to believe that friendly engagement with the Soviet Union was possible, that he was vilified both by scientists and in American popular culture. Teller may have been the model for Stanley Kubrick's Dr Strangelove.

Teller long hid the roots of his fervent anti-Communism, but in the late 1990s, he revealed that much of it stemmed from the influence of his high school friend Tisza. It was, after all, Teller who had arranged for Tisza to work with Landau at UFTI. Teller had become friends with Landau at the Bohr Institute in 1930. Tisza provided Teller with a reliable, eyewitness account of the events that unfolded in Kharkiv during the Great Terror. Gorelik explains: "That a first-rate scientific institution had been destroyed, and that his socialist friend and world-class physicist Lev Landau had been imprisoned, told Teller more about the Soviet regime than all of the political events that the media had publicized."[88] Teller long hid Tisza's role in shaping his ideology because he did not wish his friend to share the burden he had suffered from being ostracized by most other American physicists.

It is a tragedy that Oppenheimer, who was guilty of little more than hiding an earlier flirtation with Communism, suffered as a result of the naivety he shared with so many American intellectuals about the Soviet

Union. Teller did not believe Oppenheimer would knowingly betray the United States; his main reservation concerned Oppenheimer's "wisdom and judgement" since 1945. Teller raised his apprehensions about Oppenheimer with the AEC, specifically criticizing him for his desire to curtail the development of the hydrogen bomb on moral grounds. Teller believed quite rightly that the Soviets would have no such moral scruples.

Oppenheimer certainly was aware of the purge of foreign and Soviet scientists during the late 1930s. He had entered the story of the Ensnared several times, and, like Teller, he had gathered details from eyewitnesses of events in Kharkiv, most notably from Charlotte Houtermans. He had gone to university with the Houtermans in Gottingen, fallen in love with Charlotte there, and sponsored her immigration to the United States after Fritz's arrest left her destitute in London. On 4 March 1954, Oppenheimer delivered a letter defending his actions to the AEC. In it, he explained that the treatment of scientists during the Great Terror was front and centre in his education, which made him realize that Stalin was a tyrant and that the Soviet system was evil. In 1938, George Placzek, Victor Weisskopf, and Marcel Schein had visited Oppenheimer at his ranch in New Mexico. All three physicists had "actually" lived in the Soviet Union in the 1930s. "What they reported," wrote Oppenheimer, "seemed to me so solid, so unfanatical, so true, that it made a great impression; and it presented Russia, even when seen from their limited experience, as a land of purge and terror, of ludicrously bad management and of a long-suffering people." Despite Oppenheimer's new grasp of the true nature of the Soviet regime, he explained: "At that time I did not fully understand – as in time I came to understand – how completely the Communist Party in this country was under the control of Russia." This realization only came about after the Nazi–Soviet Pact, when American Communists began to advocate for disengagement and neutrality in the war against Germany.[89]

Oppenheimer made no mention of the Houtermans in his account to the AEC. Charlotte had arrived in the United States in early 1939, well after the meeting with Placzek, Weisskopf, and Schein. Her knowledge of the Great Terror, including her account of her terrifying flight from Moscow after Fritz's arrest, was far more chilling than even Schein could have related. Oppenheimer may have been trying to shield his friend from the commission's scrutiny; that said, his omission also protected him from potential questions about his association with Communists while he was at Gottingen in the late 1920s, long before he ever admitted to associating with American party members.

There is reason to doubt Oppenheimer's assertions that he realized the Soviet state could not be trusted; in fact, there is more evidence that he continued to have faith in the possibility of some sort of international agreement, one that would preclude both the United States and the Soviet Union from developing the hydrogen bomb. Even after Charlotte's arrival in New York in 1939, there is no evidence that Oppenheimer took any action to intervene on behalf of his friends and colleagues languishing in Soviet prisons. Instead, he was firmly in the camp of American left-of-centre intellectual thought, which wishfully continued to believe in the basic goodwill of the Soviet Union. Those who believed in that ideology hailed Oppenheimer as a heroic martyr and condemned Teller as a right-wing extremist.

Just after the collapse of the Soviet Union in 1990, Vitaly Ginzburg, a pioneer of the Soviet hydrogen bomb, was shocked by the prevailing view of the Oppenheimer hearing – a view that, decades later, was still held by American scientists and the political left:

> While I visited the US [in the early 1990s] I saw a film on J. R. Oppenheimer and felt bitterness. All the good guys, like Oppenheimer's brother and others, were lefties, and all of them believed that the U.S. should not make an atomic weapon, that Russians were good guys and so on. At the same time the bad guys, including security officers, understood the situation. Now we know this. How Stalin might be given superiority?! This scoundrel – I am convinced – would not hesitate to strike the West.[90]

Conclusion

The handful of German and Central European scholars who had fled to the Soviet Union in the 1930s had a disproportionate impact during the Cold War. Linking together almost all of those who were later active in Cold War politics was that they had been sympathetic to Communism before going to Soviet Union. Given that they had survived both Hitler's Germany and Stalin's Soviet Union, it is not surprising they responded to their experiences in diverse ways. The fact that many would continue to embrace Marxist ideals is a remarkable testament to some people's reliance on faith over experience. Yet except for Barbara Ruhemann, all of those who remained loyal to Communism would eventually question the excesses of Stalinism. It is understandable that a number of the Ensnared would continue to advocate for social justice or seek to build the communal utopia promised by Marxism. Likewise, it is comprehensible that some of the most zealous supporters of Marxism had been thoroughly disillusioned by their experiences in the Soviet Union and became outspoken critics of that country during the Cold War. By the mid-1950s the ensnared scientists and scholars who had survived the Second World

War had either re-established themselves in academic careers or had found other ways to make a living. Yet there remained behind one small group that continued to suffer directly from the forced migration to the Soviet Union – the families of those scientists who had perished in the Great Terror, who continued to be victimized by those horrific events for decades after the war.

13   The Long Ordeal

Fritz Lange may have been the only refugee scholar remaining in the Soviet Union after the Second World War, but the families of four of the Ensnared remained behind as well. Only one of the families remained by choice in the Soviet Union: the mathematician Stefan Cohn-Vossen's wife Friedel and son Richard stayed in Moscow after his death in 1936. By the summer of 1937, Friedel was living with the German Communist intellectual Alfred Kurella, and the couple were soon married. Later that year, Friedel helped Charlotte Schlesinger and Charlotte Houtermans escape the Soviet Union. Kurella, unlike many German Communists, including his brother Heinrich, survived the Great Terror. In 1941, soon after the war with Germany began, Friedel and Richard were evacuated from Moscow to the Abkhazian mountain village of Pschu in the Caucasus. Kurella worked with the Red Army as an editor of frontline newspapers, and later for the National Committee for Free Germany, and was not reunited with his family until 1946. Within two years, he had left Friedel and received permission to return to Moscow. In 1954 he was recruited by Walter Ulbricht to return to East Germany, where he became a member of the Politburo and a notorious Stalinist. Friedel lived the rest of her life in Soviet Georgia, dying in 1957. After his mother's death, Richard joined his stepfather in Berlin, where he became a noted filmmaker.

The Family of Fritz Noether

For the families of those murdered during the purges, their long ordeal continued for years – for some, until well after the Soviet Union had ceased to exist. The families of Konrad Weisselberg and Hans Hellmann remained trapped in the Soviet Union. Fritz Noether's sons were allowed to leave, but for most of their adult lives the fate of their father remained a cruel mystery, only solved by Soviet officialdom after the Cold War ended.

This study began with the plaque set up by Fritz Noether's sons soon after they received a May 1989 letter from the Soviet Embassy confirming their father's murder in 1941. The Soviet officials acknowledged that on 23 October 1938, Noether had been convicted "on groundless charges" of sabotage. On that basis, the Soviet government had voided Noether's twenty-five-year sentence and fully rehabilitated him. Fritz Noether could not, of course, have been rehabilitated. On 8 September 1941, he had been convicted of further equally false charges of "engaging in anti-Soviet agitation" and sentenced to death. Two day later, he was executed. In the letter, the Soviet diplomat concluded: "Please, accept my deepest sympathy although I understand that no words can alleviate your pain." Hermann and Gottfried (Friedl) Noether finally knew what had happened to their father, more than fifty years after they witnessed his arrest. Young men in 1937 – Hermann was twenty-four, Gottfried twenty-two – they had spent most of their adult lives agonizing over their family's horrific experiences in Hitler's Germany and Stalin's Soviet Union.

Hermann and Gottfried Noether had been given just one month to leave the Soviet Union after their father's arrest. In early 1938, they fled to Sweden, entering on temporary visas. In April, the Noethers registered for US visas; the 1938–39 quota from Sweden had long been filled, but they were selected to be numbers 1 and 2 on the next quota list. This new quota, however, would not commence until July 1, 1939. Hermann and Gottfried's Swedish visa was due to expire in July 1938, which meant they were in grave danger of being sent back to Germany.

Yet the same action by the Nazi government that likely sealed their father's fate – the stripping of his German citizenship – may have saved his sons' lives. On 5 May 1938, all three Noethers, Fritz, Hermann, and Gottfried, had become stateless persons. Because they were no longer German, the Swedish government could not send them back there, and furthermore, it was obligated under international law to offer them safe haven. The Noether sons remained in Sweden until they received their US visas, at which point they left almost immediately for England. Eight days later, they boarded the SS *Britannic* in Southampton, heading for New York. On board the ship were dozens of other Jewish refugees, the fortunate few who had made it to safety just weeks before the Second World War began in Europe.[1]

Fritz's Noether's sons went on to have remarkable lives. Both completed doctoral degrees. Herman, as he now spelled his name, earned a PhD in chemistry from Harvard. He was supported in part by a fund established by Harvard students to assist refugees from Germany. He went on to have a successful career as an industrial chemist. Gottfried served with US Army intelligence during the Second World War as one of the famous Ritchie Boys. The Ritchie Boys, German-Jewish refugees, gathered 60 per cent of the creditable intelligence on Germany by the US army. They were responsible for prisoner interrogations, radio intelligence, and counter-intelligence operations.[2] After the war, Gottfried received his PhD in mathematics from Columbia, continuing a family tradition in the field that went back four generations. He went on to have a notable career as a statistician, serving on the faculties of New York University, Boston University, and the University of Connecticut. Herman and Gottfried Noether both married and had children and grandchildren. Gottfried died in 1991, at the age of seventy-six, just a few years after learning the fate of his father. Herman died in 2007, at the age of ninety-four. The brothers had done more than simply survive Hitler and Stalin – they had triumphed over both.

The Family of Konrad Weisselberg

The Noethers were by no means the longest-suffering members of the families of the Ensnared. The families of both Konrad Weisselberg and Hans Hellman remained trapped in the Soviet Union long after the other survivors had managed to flee the country. The treatment of (Anna) Galia and Alexander Weisselberg and Viktoria and Hans Jr Hellmann are remarkably similar. Galia was declared "a wife of an enemy of the people," and she and her young son Alex were evicted from Alexander Weissberg's apartment, which left them homeless. Galia managed to survive through the kindness of two brave individuals who provided her and Alex with a place to stay and just enough food to survive. She changed her name, adopting her maiden name of Mykalo.

Galia and Alexander's threadbare lives were again disrupted, this time by the German invasion of the Soviet Union. Kharkiv was taken by the Germans in October 1941, and soon after, the civilians left behind in the city began to suffer severe depravations, including starvation. Galia warded off hunger by making occasional visits for food to relatives in a village outside of Kharkiv. On one of these trips, she fell ill, perhaps with typhus, which was rampant in the area, and disappeared for four or five months. Unsupervised, and desperate to survive, Alex became a street urchin and joined a gang of boys. Constantly hungry, the boys scavenged in garbage dumps for kitchen scraps. At one point, Alex and another boy found beans; they ate these without first cooking them, and became ill. Many years later, Alex would tell his son Artjom (Art) that the beans may have been found in the excrement of German soldiers. Starving, the gang forced the youngest members, including Alex, who was just six or seven, to beg German soldiers for food. The boys were aware that approaching the soldiers was risky: sometimes they would be shot, other times they would be given some crumbs of food, or if they were really lucky, some chocolate.

Somehow, both Galia and Alex survived the war. Galia found them a room in an abandoned flat in Kharkiv, which they shared with other families, including the kitchen and bathroom. After the war, Alex Weisselberg remained an outcast, "a child of an enemy of the state." There were many outcast children in the postwar Soviet Union, and Alex was more fortunate than most because he still had a mother – many others had lost both parents and were living in orphanages. Alex, however, was allowed to study and find employment. After Stalin's death in 1953, the sheer number of people in the "enemy of the state" category, coupled with an acute labour shortage after the massive losses of the purges and the war, forced the government to allow these people to get an education and work. Alex went to technical school and became a truck mechanic. After his training, he spent two years in the far north of Siberia as part of an obligatory work program. He then returned to Kharkiv, where he lived with his mother and continued to work as a mechanic. Alex attended the technical university at night and after three years graduated as an engineer. He had a very successful career in the plastics industry.

Only after Stalin's death, and Khrushchev's public acknowledgment of the Soviet dictator's crimes against his own people, was Galia Weisselberg provided with any information about her husband's fate. In 1959 she received official confirmation of her husband's execution in 1937, along with a notice of his rehabilitation. While she and Alexander were no longer stigmatized as "enemies of the state," efforts to contact Konrad Weisselberg's family remained fraught with danger. In 1964, Walter Wodak, one of Konrad's friends from Vienna, became the Austrian ambassador to the Soviet Union. Wodak somehow managed to find Galia's address and wrote her a letter asking her to visit him in Moscow. Galia was terrified even to reply, but Alex, in his mid-twenties, had dinner with Wodak at the embassy. After Alex returned to Kharkiv, the KGB, as the secret police were then called, interrogated him, demanding to know what the dinner had been about, and then tried to persuade him to spy for them. Alex refused, but the incident reminded him that life was still dangerous in the Soviet Union.

Alex married Nadia Kharlamov in 1968, taking his wife's surname to try to escape the rampant antisemitism that existed in the Soviet Union. He rarely talked about his experiences as a child, only broaching his horrific childhood to his son Art when the boy was eleven or twelve. When

Image 17. Galia and Alex Weisselberg in Kharkiv, 1949. They had survived
the purges and the German occupation of their city. Alex is thirteen years old.
(From the collection of Art Kharlamov.)

Art was fifteen, Alex took him to the site in Kharkiv where it is believed
Konrad was executed. Today that site is a memorial park, but it remains a
dreary place because the trees planted there regularly die and need to be
replaced. Art was told that the trees probably died because of the quick-
lime that had been spread over the bodies of the hundreds of people
who were buried there after being executed. Alex eventually established
some contact with Konrad's family in the West, although this communication was infrequent and hindered by the "dangers and impediment
of the cold war." Finally, fifty years after Konrad's execution, during the

more liberal reign of Gorbachev, Alex and Nadia met Konrad's nephew Kiffer Weisselberg and his wife Alison in a Moscow hotel room. With the aid of an interpreter, they talked for the entire day and well into the night. When asked by Kiffer what he could send Alex from the West, Kiffer expected Alex to ask for Levis. Instead, in the aftermath of the Chernobyl disaster, Alex asked for a Geiger counter so that he could check the radiation levels in Kharkiv.

After this meeting, the families communicated regularly. Most of Konrad's family had moved to Britain just before the war. In 1991 the Kharlamov family visited Alex's first cousins in England, and together they watched on television the extraordinary events of the collapse of the Soviet Union. Five years later, the KGB opened its archives and Alex was able to access his father's NKVD file. The chilling contents of the file, which revealed that his father had done nothing to warrant his arrest, let alone his execution, convinced Alex that his family should leave Ukraine for the West.

Alex had no papers proving his claim to Austrian citizenship. Ironically, however, the Austrian government agreed to accept the evidence of it contained in Konrad's NKVD file. In 2000, Alex and his family were granted Austrian citizenship, and one year later Art moved his family to England. Alex and Nadia followed in 2005; by then, Alex was sixty-nine years old. It was, explained Alex's cousin Mica Nava, a difficult transition: "They experienced the tough uprooted lives of migrants." The change was particularly difficult for Alex and Nadia. Nadia was a very social person, but she spoke no English. Alex spoke some English, but he had been a respected professional in Kharkiv and now, in his late sixties, he had to start life again in a strange land. But the family, especially Alex's children and grandchildren, soon settled, and within a few years they were thriving. Alex died in 2008, knowing that his family was safe. They had finally put the ghosts of the Great Terror behind them.[3]

The Family of Hans Hellmann

After her husband's arrest, Viktoria Hellmann was declared "a wife of an enemy of the state," the same as Galia Weisselberg. Also like Galia and Alex, Viktoria and her son Hans Jr were left homeless when the Karpov Institute reclaimed their apartment. Viktoria approached the NKVD in Moscow several times to inquire about Hans Sr's fate. When she was threatened with arrest if she persisted, she fled the city and found a job teaching German in a middle school in a small town about 120 kilometres west of the Soviet capital.

The German invasion greatly affected the lives of Viktoria and her son. In September 1941, Viktoria was arrested by the NKVD and charged with spreading "anti-Soviet propaganda." She also was accused of preparing to act as a translator for the approaching Wehrmacht. The charges, like all others in this book, were not just false but absurd. Viktoria was, after all, a Ukrainian Jew who had fled Germany in 1934 with her German husband. After a few months in a Moscow jail, Viktoria was sent to a camp in the Semipalatinsk region of Kazakhstan, some 3,400 kilometres west of the Soviet capital. Hans Jr., then eleven years old, had witnessed his mother's arrest, as he had his father's three years earlier. Hans Jr was taken by the NKVD to a special children's home; he escaped and fled to his aunt's house in Moscow. The family changed Hans's name to Genadij Minchin, his aunt's surname, to hide him from the authorities. He would retain that name for the next fifty years. Sadly, he would not see his mother again until 1956, and even then only briefly, as Viktoria was forced to remain near the camp in Kazakhstan.

Viktoria was completely rehabilitated in 1956. Soon after, she applied for word about her husband. She received an inaccurate death certificate stating that Hans Sr had died of peritonitis while in prison. Viktoria then requested a rehabilitation certificate for Hans Sr, which she received in 1957. In 1958 the Red Cross provided Viktoria with the address of Hans' Sr's sister in Hamburg, but she was denied permission to visit her. In 1960, Hans Jr, also through the Red Cross, was put in contact with his relatives in Germany. From that point on, he kept up regular correspondence with them.

Like Alex Weisselberg, Hans Jr was able to get an education. He became a mining engineer and worked in a various places in the Soviet Union before settling in Kharkiv. It is unlikely he ever met Alex, and one can only imagine what they might have said to each other if they had. Only in August 1989, just a few months after the Noethers were informed of Fritz's fate, did Viktoria and Hans Jr finally receive an accurate death certificate and a letter that explained to them that Hans Sr had been executed.

Viktoria stayed in the far east of the former Soviet Union and lived well into her nineties. She died in 1992. Han Jr obtained a German passport from the embassy in Moscow in 1988. Three years later, soon after the collapse of the Soviet Union, Hans Jr, his wife, and their youngest son immigrated to Germany. Hans Jr immediately reverted to his birth name. Upon his return to Germany, he was able to participate in a revival of interest in his father's life and achievements, which had been all but forgotten. With Hans Jr's assistance, a scholarly biography of Hellmann was co-authored by the journalist and historian Sabine Arnold, chemist Eugene Schwartz, and other scientists. Arnold was able to acquire a copy of Hellmann's NKVD file, which he shared with his son. Like all of these files, there was no actual evidence of any crime, but rather the almost obligatory false confession, not written in Hellmann's handwriting, and signed by the shaky hand of a desperate and broken man.

In 1999 the Karpov Institute in Moscow held a symposium to mark what would have been Hans Hellmann's ninety-fifth birthday. The event was attended by Russian and German scientists, and a portrait of Hellman, painted by one of Hans Jr's cousins, was unveiled to hang permanently in the institute. The portrait was based on a 1933 photograph showing Hellmann giving a lecture in Hanover. The event was a sincere effort to make amends for the horrible injustices done to Hellmann. One can only wonder in the age of Putin whether such an event could be held today.[4] That same year the German Working Group for Theoretical Chemistry established the semiannual Hans Hellmann Prize, to be awarded to young researchers in the field of theoretical chemistry.[5] Hans Jr, now well into his nineties, has attended every award dinner since the prize's inception. Also in 1999, the University of Veterinary Medicine in Hannover, where Hans taught until he was dismissed in 1934, began giving the Hans Hellmann Memorial Prize for "the best doctoral dissertation on a basic-oriented topic."[6]

In 2003, on the centenary of Hans Hellmann's birth, symposia were organized in Hellmann's honour in Bonn and in Hanover. In 2012, as a final tribute to Hellmann, his great textbook on quantum chemistry was published in a full scholarly edition in German. Hans Jr wrote a lengthy biographical introduction to the book in which he described the tragedy that had befallen his father and his family after they became caught in the jaws of totalitarianism.

Conclusion

The stories of the families of the victims of the Great Terror are a chilling reminder of the long-term consequences of the forced migration of German intellectuals and scholars caused by Hitler. For Konrad Weisselberg, and his children and grandchildren, their story of flight in search of refuge spanned almost the entire twentieth century and continued well into the twenty-first. The Weisselbergs' ordeal began in 1907, when Konrad was just two years old, when his parents fled Berlad in the easternmost province of the Austro-Hungarian Empire. It did not end until Alex Kharlamov, Konrad's sixty-nine-year only son, arrived in England in 2005. At least the families of Weisselberg, Noether, and Hellmann survived. There was no one left to mourn Herbert Murawkin after he was rehabilitated by the Soviet government in 1957.

Conclusion: The Ensnared and History

This story of the refugee academics ensnared between Hitler and Stalin has followed their lives through some of the most tumultuous events of the twentieth century. These scholars had been caught up in two of the greatest mass murders in history; they had also endured the chaos that engulfed Central Europe in the decades before Hitler came to power. Those who survived the Great Terror faced the perils of the Second World War and the Cold War as well. They all sought a haven for themselves and their families where they could resume their scientific and scholarly careers.

This study goes well beyond simplistic analysis of the motivations behind the refugee scholars who relocated to the Soviet Union. In this group there were certainly a large number of dedicated Communists, some of whom – including Fritz and Charlotte Houtermans, Herbert Murawkin, Wolfgang Steinitz, and, likely, Alexander Weissberg – spied for the Soviet Union before leaving Germany. Yet only a few of the Communists in this group, even the dedicated ones, moved directly to the socialist utopia after fleeing Germany. Most of the scientists went there as a desperate last resort. Thus, the displaced scholars were not typical of the foreigners who visited or migrated to the Soviet Union in the 1930s. While some of the scholars shared the "favourable predispositions and the associated selective perceptions" of most travellers to the Soviet Union, they held sharply more diverse views of Soviet life than many Western observers; they were also more discerning eyewitnesses to the rapid changes that rocked the country in the 1930s. The recruitment of these scientists was made possible by the relative openness of Soviet science to the West before 1935.[1] The émigrés experienced Soviet society at its best, benefiting from the massive state investment in science and education. As in other countries, the refugee academics made significant contributions to this renaissance in learning before it was all swept away.

The crushing of the Soviet scientific renaissance, and the flight of foreign experts who were part of it, has been widely studied only in terms of events at UFTI; in this book, for the first time, these events have been studied with regard to the whole of the Soviet Union. By arresting these refugee scholars or forcing them to flee, the Soviet Union erased the benefits that many other countries enjoyed from this massive migration of knowledge, a process some historians have labelled "Hitler's Gift."

Understanding the experiences of the foreign experts during Stalin's purges, which culminated in the Great Terror, helps clarify which non-Soviet nationals were targeted for arrest and execution. Most of those arrested, and all of those murdered in 1937 and 1938, were either members of a Communist Party or had become Soviet citizens. Fritz Noether, who fits into neither category, was scheduled to be handed over to the Germans in 1940; the Nazi government's refusal to accept him doomed him to be murdered by the NKVD in the opening months of the Great Patriotic War. Had Noether been turned over to the Germans, he likely would have shared the same fate as Siegfried Gilde. Most of the foreign scholars were either expelled or allowed to leave; this fact says a great deal about the nature of the persecution, which was aimed mainly at perceived internal enemies of the state and the party. The perverse paradox of Stalin's Soviet Union was that his victims were both those who tried to integrate into Soviet society and espoused Marxist ideology, and those who opposed his policies. Most of the refugee scholars fell outside both categories and were able to escape once they had exit visas.

The extant scholarship has tended to ignore the changing nature of academic migration between 1933 and 1940. Having been forced to escape from the Soviet Union, the survivors found themselves confronting a second wave of displaced scholars. If the patterns of migration of the Ensnared are typical, then we can see how events unfolded quite differently than in 1933. Britain, which initially had been the most welcoming of displaced scholars, was by 1938 more restrictive, especially for those unable to secure a permanent position. Before 1936, British universities had welcomed more refugees than any other country; two years later, they were reaching the saturation point. This is illustrated by Walter Zehden's rejection by the University of Birmingham, as well as Ernst Simonson and Marcel Schein's failure to find positions despite their SPSL fellowships. France, by contrast, due to the changing political climate, was more open to the refugees than it had been in 1933. Sweden too hosted a number of the Ensnared, but this hospitality seems to have been more of a geographic convenience for those fleeing the Soviet Union. The Swedes wearily tolerated the refugees rather than welcoming them. Other countries in Europe hosted only two of the re-migrants, both of whom had been compelled to seek refuge on the peripheries of the continent. Migration to the United States was open to a select few in the second wave, but friends, family, and professional connections were essential to securing a coveted US visa. Interestingly, some of those who helped the Ensnared were themselves former German academics who had established themselves in the United States by the late 1930s. Notably, the AEC offered these scholars little direct assistance.

The third wave of academic flight began with the start of the Second World War. The German conquests in the spring of 1940 generated a torrent of desperate people even while avenues of escape were shrinking. Laszlo Tisza was the only one of those trapped on the continent to escape to the United States. His migration was made possible by his Hungarian citizenship, his prudence in applying for a visa before leaving Budapest for Paris, and his friends in the United States, who included his schoolmate Edward Teller. Guido Beck took the only available route of escape from Vichy, even if it was only to a precarious existence in Portugal. Those left behind struggled to survive; not all would do so. More scholarship is required to explore the changing nature of the migration of scholars from Europe between 1933 and 1940.

This book also challenges some of the assumptions about the role of academic refugee assistance organizations, particularly the AEC and the SPSL. The historiography of these organizations tends to highlight their successes in awarding fellowships to hundreds of displaced university faculty, allowing them to find academic positions in the United States, Britain, and elsewhere. Yet the sole success among those scientists who left the Soviet Union is Herbert Fröhlich, who parlayed his early expulsion into a SPSL fellowship to Bristol, the start of a long and successful scientific career that culminated in election to the Royal Society. Fröhlich was one of two scientists expelled from the Soviet Union in 1935, at the tail end of the first wave of migration. His success reflects the growth of the SPSL's fellowship program.

This study does not undercut the positive view of the rescue organizations; rather, it builds on recent scholarship to provide a more nuanced perception of their accomplishments.[2] The SPSL and AEC simply could not offer fellowships to more than a small fraction of those in need; even so, they worked tirelessly to assist all who applied. The SPSL's inability to help Emanuel Wasser, a journeyman physicist expelled at the same time as Fröhlich, illustrates the limitations of the rescue organizations as well as the disastrous consequences that could befall those left behind. The SPSL and the AEC categorized applicants on the basis of many other factors besides their scholarly accomplishments. Those too young or too old were unlikely to get support, and both organizations had to avoid fomenting an antisemitic backlash. The profile of those who went to the Soviet Union reflects the *modus operandi* of the academic rescue organizations. The assistance that Ernst Simonson and Marcel Schein received from the SPSL to help them migrate to the United States after fleeing from the Soviet Union provides two important case studies of the SPSL's efforts in the late 1930s to place more scholars in the United States, without direct support from the AEC.

The experiences of the Ensnared challenge the notion that the refugee experience in general was not part of the Holocaust. All of those fleeing Hitler faced grave dangers and hardships in their search for safe harbour. For those who sought refuge in the Soviet Union, those dangers included arrest, torture, and execution by the state. Many refugees succeeded in escaping Hitler's Germany – and, in the case of those in this study, the Soviet Union as well – only to perish later at the hands of the Germans and their Allies during the Second World War. Of the seven Ensnared who were murdered, four were killed by the Soviets, three by the Germans. Whichever organization carried out the murders, all of these deaths were directly attributable to the Nazis' persecution of the Jews. As well, Stephan Cohn-Vossen died of natural causes shortly after arriving in Moscow. Ruth and Marina Gilde and (probably) Sara and Elizabeth Wasser were also killed in the Holocaust. The overall death rate of the Ensnared was 20 per cent, a high figure for a group of young and middle-aged European men. That figure is, of course, nowhere near the overall death rate of European Jews, but it is high enough to indicate that the story of these refugees is integral to the history of the Holocaust. And that death toll could easily have been much higher: Hermann Borchardt, Gerhard Harig, Heinrich Luftschitz, Helmuth Simons, and Alexander Weissberg all faced death in concentration camps, ghettos, or prisons. A number of the survivors, including Guido Beck, Heinrich Luftschitz, Helmuth Simons, and Alexander Weissberg, because they could not find a safe haven before the Second World War began, were unable to help their families escape the Holocaust. There are doubtless others who were forced to endure the murder of loved ones they could not save.

The Ensnared were not simply an integral part of the Holocaust, they were also part of the much longer history of the movement of displaced peoples in Europe in the first half of the twentieth century, the "forty year crisis."[3] An examination of the lives and careers of these scientists before they lost their livelihoods in 1933 reveals that a surprising number of displaced academics had been refugees, in some cases several times, before Hitler came to power. Prior to 1933, the movement of scholars, particularly of Jewish heritage, was not away from German universities but *toward* them. This is demonstrated by the very high number of the Ensnared who were born outside Germany. After the Second World War, it took years for some of these academics to find permanent places of refuge. Of course, some never did, or if they found security, it was without the ability to continue their academic careers. As the lives of Michael Sadowsky and Guido Beck exemplify, it was not uncommon for displaced people to experience multiple flights in what must have seemed a never-ending quest for a secure refuge. The long suffering of the families of Hans Hellman, Konrad Weisselberg, and Fritz Noether point to the need to expand the temporal focus of studies of the forced migration of European people well beyond the "forty year crisis." The Weisselberg family's experience with migration spanned almost an entire century: in 1907, when Konrad was just two years old, his family fled from anti-Jewish pogroms in Berlad, Austria-Hungary, and their ordeal only ended in 2005, when his sixty-nine-year-old son, Alexander, moved to Britain.

The pre-1933 experiences of Helmuth Simons, Hermann Borchardt, and Herman Muntz bring out another aspect of the forty-year crisis that has not received enough attention in the literature. For these scholars, the crisis was an internal one caused by the chaos that engulfed Weimar Germany between 1918 and the mid-1920s, preventing them from finding secure university positions. No doubt antisemitism played some role in this, but before 1933, it was less of a factor in Germany than elsewhere in Central Europe. Only in the Soviet Union could these three scholars be employed at a level commensurate with their qualifications and abilities. This book, therefore, is an important historical case study, one that furthers our understanding of the movements of refugees, neatly fitting into recent trends in the historiography, which has focused on the "agents, means, and structures of aid for refugee academics, the motives behind and consequences of aid efforts, and the obstacles and push factors that shaped the outcome of aid efforts currently of great importance for scholars working in this field – not least on account of their topicality."[4]

The intellectual legacy of the Ensnared has been largely overlooked. Bernal's *The Social Function of Science*, including the reaction to it, has been extensively studied since its publication in 1939; yet the great influence of Martin Ruhemann on Bernal has gone almost unnoticed, even though he wrote that book's appendix on Soviet science.[5] In fact, it is doubtful that many historians of Bernal even know who Ruhemann was, in part because he broke with the British scientific left after separating from Barbara. Having experienced Hitler's Germany and Stalin's Soviet Union first-hand, a few of these scholars exerted an impact on the Cold War far greater than their small numbers might suggest. That so many retained their faith in Communism and the Soviet Union after experiencing the Great Terror is a testament to human obstinacy. Harig and Steinitz were prominent intellectuals and scientific leaders in East Germany, and their work, unlike that of most scholars from the GDR, was highly regarded in the West. Except for Barbara Ruhemann, all of the Marxists eventually came to doubt, if not Communism itself, then Stalin's perversion of it. Even Fritz Lange, the only scholar allowed to remain unscathed in the Soviet Union, eventually fought for permission to return to scientific research back home in East Germany. It is far easier to understand why many of those who flirted with Communism before going to the Soviet Union ended up rejecting Stalinism even while retaining a passionate interest in social justice and socialism. Edgar Lederer and Ernst Emsheimer were passionate advocates for human rights while maintaining successful scientific careers. That a number of others became profoundly anti-Communist and anti-Soviet is even less surprising. Hermann Borchardt may have failed as a Cold War warrior, but Alexander Weissberg, in particular, did not. Weissberg's and Eva Striker's experiences profoundly influenced Arthur Koestler as he began writing *Darkness at Noon*, "one of the most important political novels of the twentieth century."[6] The books by Weissberg and Fritz Houtermans – among the earliest accounts of the Great Terror, the NKVD, and the Soviet prison system – threw down important challenges to the belief held by much of the European left that Stalin's Soviet Union was a proletarian paradise. Weissburg played a vital role in the early days of the Congress for Cultural Freedom and in the founding of the Committee on Science and Freedom, key organizations in the cultural and intellectual struggle against those who supported the Soviet Union. Tisza's impact on Edward Teller links the story of the Ensnared to one of the most controversial episodes in the dark history of American anti-Communism in the early Cold War.

Perhaps the most important contribution of *The Ensnared* is that it gives voice to some of those scholars whose memory Hitler and Stalin wished to erase from history. Only a few of them had an opportunity to fulfil their early promise as scholars and scientists. Occasionally, echoes of these scholars appear in the scientific literature. For instance, Fritz Duschinsky's scientific work serves as the basis for a significant body of literature on spectral analysis. It is unlikely that any of the scientists benefiting from Duschinsky's work actually know of his tragic fate.[7] Siegfried Gilde and his family were virtually expunged from the historical record, remembered only in an oral history interview by his niece and in brief mentions in a few studies on medical history. Gilde's role in the medical research conducted in the Warsaw Ghetto is revealed in *The Ensnared* for the first time. Even many of the survivors were all but forgotten, perhaps none more than Guido Beck, one of the least-known of the pioneers in quantum mechanics. The work of a few of the Ensnared, including Charlotte Schlesinger, Hans Hellmann, and Hermann Borchardt, has become part of a growing movement in Germany to rediscover the intellectuals, scientists, and artists whose work was almost lost as a result of the Nazi persecutions. It is in this rediscovery of these "disappeared" scholars that this study has contemporary relevance. The academic refugee crisis caused by Hitler and Stalin might be a thing of the past, but there are again entire generations of scholars in danger of becoming lost. By showing what was vanished in the past one may hope that the importance of rescuing displaced scholars may be recognized before they too are lost, before they too can fulfil their full intellectual promise.

The work of academic refugee organizations has continued unabated since the Second World War. The British organization, now renamed the Council for At-Risk Academics (CARA), continues its work today. The AEC was closed after the Second World War, but the Institute of International Education (IIE) has carried on providing aid to scholars fleeing persecution. In 2002, the IIE formed the Scholar Rescue Fund to oversee this vital work. The two organizations continue to work closely together, along with the Scholars at Risk Network (SAR), an international network of institutions and individuals established in 1999, now based at New York University, whose mission is to protect scholars and promote academic freedom. Today the refugees come from all over the globe, including Afghanistan, Turkey, Syria, Egypt, Venezuela, China, Cameroon, Yemen, Ukraine, and Iran. A better and more inclusive understanding of the academic refugee experience in the 1930s will improve our understanding of the barriers faced by those who are today fleeing persecution. There can be little doubt that this alone makes an account of the scholars ensnared in Hitler's and Stalin's web of hatred and violence an important story.

Notes

Introduction

1 The original German reads:

> *Prof. Dr. Fritz Alexander Noether*
> *7 Okt. 1884 Erlangen – 10. Sept. 1941 Orel*
> *Eisernes Kreuz 1914–1918*
> *Opfer Zweier Diktaturen*
> *1934 Aus Deutschland Wegen Rasse Vertieben*
> *1938 Von Den Sowjets Angeklagt Und Verurteilt*
> *1941 Hingerichtet 1988 Unschuldig Erklart*

2 Figures for academic refugees are from the SPSL's Annual Report for 1938, November 1938, SPSL 1/4. This was the last report undertaken until 1946. Figures contained in this report are not reliable and only account for those known to the society.

3 Strauss, "The Migration of the Academic Intellectuals," 67.

4 On dislocation and refugees in Europe before 1933, see Gatrell, *A Whole Empire Walking*; Kushner, *Journeys from the Abyss*; and Marrus, *The Unwanted*.

5 For Jews and antisemitism in German universities and research institutions before 1933, see Karin Orth, *Die NS-Vertreibung der jüdischen Gelehrten*, 27–61; and Volkov, *Germans, Jews, and Antisemites*, 224–47.

6 See, for example, Hollander, *Political Pilgrims*; Jones, "Silence, Exile and Cunning"; and Zamfira, "The Enthusiasm of Intellectuals."

7 M. Ruhemann, "Science and the Soviet Citizen."

8 Strauss, "The Migration of the Academic Intellectuals," 71.

9 For recent examples of scholarship on Soviet espionage activities in this period, see Murphy, "Soviet Espionage in France between the Wars"; and "First Decade of Soviet Espionage in America."

10 Works on Soviet science in this period include Demidov and Levshin, *The Case of Academician Nikolai Nikolaevich Luzin*; Ings, *Stalin and the Scientists*;

Kojevnikov, *Stalin's Great Science*; Pringle, *The Murder of Nikolai Vavilov*; and Rogacheva, *The Private World of Soviet Scientists*.

11 The brief history of the Ukrainian Physical Technical Institute (UFTI) as a world-class centre for physics, and its precipitous fall from grace, is the one part of the academic refugee experience in the Soviet Union that historians have explored extensively prior to this study. The story of the research centre was first told by Alexander Weissberg shortly after the Second World War, though his account focuses more on his own horrific experiences at the hands of the NKVD than on his scientific career at Kharkov. Since 1992, with the collapse of the Soviet Union and the creation of an independent Ukraine, at least six articles and books have been published about UFTI and the Soviet and foreign physicists who worked there. Works that have examined the history of UFTI, in chronological order, include Weissberg, *The Accused*; Khriplovich, "The Eventful Life of Fritz Houtermans"; Pavlenko, Raniuk, and Khramov, *"Delo" UFTI 1935–1938*; Amaldi, *The Adventurous Life of Friedrich Georg Houtermans*; Szapor, "Private Archives and Public Lives"; Frenkel, "Professor Friedrich Houtermans"; and Reinders, *The Life, Science and Times of Lev Vasilevich Shubnikov*.

12 For life in the Soviet Union, see for example, Hoffmann, *The Stalinist Era*; Ilic, *Soviet Women*; Johnston, *Being Soviet: Identity, Rumour, and Everyday Life under Stalin 1939–1953*; and Samuelson, *Tankograd*.

13 Schlügel, *Moscow 1937*, 388.

14 For studies of refugees and the Holocaust, see: Dwork and Van Pelt, *Flight from the Reich*; Kushner, *Journeys from the Abyss*; and Marrus, *The Unwanted*.

15 S.M. Harris, *The CIA and the Congress for Cultural Freedom*, 6. See also Kubby, "In the Shadow of The Concentration Camp"; Nye, *Michael Polanyi and His Generation*; and Wilford, *The CIA, the British Left, and the Cold War*.

16 See, for example, Alter, *Out of the Third Reich*; Ash and Sollner, *Forced Migration*; Daum, Lehmann, and Sheehan, *The Second Generation*; and Lehmann and Sheehan, *An Interrupted Past*.

17 Crawford, Ulmschneider, and Elsner, *Ark of Civilization*, 3.

18 Author to Laurel Leff, 2 March 2020; Leff to author, 6 March 2019; Leff, *Well Worth Saving*.

19 Orth, *Die NS-Vertreibung*.

20 See, for example, Bentwich, *The Rescue and Achievement of Refugee Scholars*; Beveridge, *A Defence of Free Learning*; Seabrook, *The Refuge and the Fortress*; Duggan and Drury, *The Rescue of Science and Learning*; Zimmerman, "Protests Butter No Parsnips."

21 See for example, Zimmerman, "Competitive Cooperation," 151–69; and Leff, *Well Worth Saving*.

1. Scholars and Scientists

1 For examinations of émigré scholars, see for instance, Ash and Sollner, "Introduction," in *Forced Migration and Scientific Change*, 6–9; and Strauss, "Some Demographic and Occupational Characteristics of Emigres included in the Dictionary," in *International Biographical Dictionary of Central European Emigres 1933–1945*, vol. 2: *The Arts, Sciences and Literature*,), 77–86.
2 Shifman, *Physics in a Mad World*, ch. 5.
3 SPSL 244/8, Lutftschitz to Walter Adams, SPSL, 29 April 1934. The photograph is enclosed in the file. It should be noted that of the several hundred application files I have read, this is the only one in which the applicant enclosed a photograph.
4 Fritz Alexander Ernst Noether, http://www-history.mcs.st-and.ac.uk /Biographies/Noether_Fritz.html. See also the entries for Emmy and Fritz Noether in *International Biographical Dictionary of Central European Emigres, 1933–1945*, vol. II.
5 Weindling, *Health, Race, and German Politics*, 328.
6 Hopwood, "Biology between University and Proletariat"; *International Biographical Dictionary of Central European Emigres 1933–1945*, vol. II, 368.
7 O'Connor and Robertson, "Stefan Emmanuilovich Cohn-Vossen."
8 Nussenzveig, "Guido Beck." Also see Passos Videira and Puig, "Introduction," in *Guido Beck: The Career*, 15–19; Havas, "The Life and Work of Guido Beck."
9 Khriplovich, "The Eventful Life of Fritz Houtermans," 29.
10 Khriplovich, "The Eventful Life of Fritz Houtermans," 31.
11 The best account of Houtermans's early work and life is in Shifman, *Physics in a Mad World*, chs. 2–4.
12 Very little has been written about Lange. See Hoffmann, "Fritz Lange."
13 Schmidt, "High Voltages."
14 See short biography of Hans G.A. Hellmann (1903–1938), http://www .tc.chemie.uni-siegen.de/hellmann/hellbioe.html. See also Jug et al., "Hans Hellmann."
15 Tully, "Diatomics-in-Molecules."
16 Pantelides and Di Ventra, "Hellmann–Feynman Theorem."
17 For more on the Hellmann–Feynman Theory, see Wallace, "An Introduction to the Hellmann-Feynman Theory."
18 Karl Jug et al., "Hans Hellmann," 15.
19 Potier, "Biography of Edgar Lederer."
20 Murawkin, *Massenspektra von Gläsern, Salzen und Metallen.*
21 Murawkin, "Beiträge zur Theorie und Konstruktion"; "Massenspektra von Gläsern, Salzen und Metallen," issues 3, 4, and 8.
22 Duschinsky's background can be found in his American Emergency Committee file, AEC 53/11, and his SPSL file, SPSL 480/1.

23 Siegfried Gilde, "Vergleichende Versuche."
24 Czechoslovakia (4, or 18 per cent) and Austria (4, or 18 per cent), and one came from Hungary. Four of the scholars where dual nationals: German-Russian, German-Polish, German-Austrian, and a German-British.
25 Lederer, "Adventures in Research."
26 Entretien avec Edgar Lederer. See also oral history interview with Guido Beck.
27 Sadowsky to Walter Adams, Academic Assistance Council, 10 April 1934, SPSL 284/5.
28 Frenkel, *Physics in a Mad World*, 526.
29 "Birlad."
30 Saltzman, "Is Science a Brotherhood?"
31 Simons biography, SPSL 265.
32 Borchardt to Isodore Lipschutz, 30 June 1944, and Borchardt, "Als deutscher Leher in der Sowjetunion" (As a German Teacher in the Soviet Union), both in Herausgegeben von Hermann Haarmann, *Hermann Borchardt Werke Band 1: Autobiographische Schriften*, ed. Christoph Hesse and Lukas Laier, 93–100.
33 F. Borchardt, "Hermann Borchardt," 121–2. Translation by author.
34 B. Ruhemann, "Die Kristallstrukturen von Krypton, Xenon, Jodwasserstoff und Bromwasserstoff," contains a brief biographical sketch. See also McRae, *Nuclear Dawn*, 25.
35 Shifman, *Standing Together*, 33.
36 Shifman, *Standing Together*, 35.
37 Her legal name was Sophie Charlotte Schlesinger.
38 Anna-Christine Rhode-Juchtern, "Charlotte Schlesinger."
39 Schlesinger, "Тетрадь, найденная на чердаке."
40 Academic Assistant Council Permanently Placed Abroad, USSR, 11 October 1934, found in Julius Schaxel File, AEC 111/31; Mills, *Biological Oceanography*, 109–14.
41 Hildebrandt, "Anatomy in the Third Reich"; Hoffmann, "Fritz Lange," 407.
42 Williams, "The Publications of Boris Davison."
43 Glad, "Hermann J. Muller's 1936 Letter to Stalin."

2. German Scientists in the Soviet Union before Hitler

1 Siegmund-Schultze, *Mathematics Fleeing from Nazi Germany*, 36–40.
2 Sorokina, "Within Two Tyrannies," 227.
3 Report of Informal Meeting at Oxford on Displaced University Teachers, 13 November 1937, SPSL 59.
4 Graziosi, "Foreign Workers in Soviet Russia."
5 Reinders, *The Life, Science and Times of Lev Vasilevich Shubnikov*, 79–80.

6 Ortiz and Pinkus, "Herman Muntz," 1.
7 Ortiz and Pinkus, "Herman Muntz," 22–31.
8 Graziosi, "Foreign Workers in Soviet Russia," 39.
9 Cesarani, *Arthur Koestler*, 78–9; Scammell, *Koestler*, 67, 77–9.
10 M. Ruhemann, "Kharkov in the Thirties."
11 The documents are quoted in Shifman, *Physics in a Mad World*, Corrections and Addenda, 2016, 1–3, https://www.academia.edu/19831333/Physics_in_a_Mad_World_Ed_M_Shifman_World_Scientific_Singapore_2015_Corrections_and_Addenda_Update_4_20_16?email_work_card=view-paper.
12 Interview of Victor Weisskopf by Thomas S. Kuhn and John L. Heilbron on 10 July 1963, Niels Bohr Library and Archives, American Institute of Physics, College Park, Maryland, www.aip.org/history-programs/niels-bohr-library/oral-histories/4944.
13 Ruhemann, "Kharkov in the Thirties."
14 Weissberg, Alexander, *The Accused*, loc. 1620–1623 of 10573, Kindle.
15 McRae, *Nuclear Dawn*, 17.
16 Ruhemann, "Kharkov in the Thirties."
17 Curriculum vitae of Prof. Dr Ernst Simonson, SPSL 419/4.
18 Simson to Hill, 9 November 1938, SPSL 419/4.
19 Simonson, "Die wissenschaftliche arbeit in Russland," 8.
20 Bernal, *The Social Function of Science*.
21 Crowther, *Soviet Science*, 145. Crowther's other book was *Industry and Education in Soviet Russia*.
22 Curriculum vitae, Dr Emanuel Wasser, Lodz, 13 May 1935, SPSL 342.
23 Michael Polanyi to Walter Adams, 14 October 1937, SPSL 334/1.
24 Bertolotti, *The History of the Laser*, 112–13.
25 Saul Levy, curriculum vitae, 1937, SPSL 334/1.
26 Walter Zehden, curriculumvitae, SPSL 343/6.
27 Rechenberg, "Ladensburg, Rudolf."

3. Scientists and Communists

1 Strauss, "The Migration of the Academic Intellectuals," 71.
2 Fisher, *The Communist Revolution*, 13.
3 Shifman, *Standing Together*, 48.
4 Email from Christopher Hesse, 24 May 2021.
5 F. Borchardt, "Hermann Borchardt," 121.
6 Lederer oral history interview.
7 Tisza, "Adventures, Part I."
8 Interview of Laszlo Tisza by Kostas Gavroglou.
9 Tisza, "Adventures, Part I," 71.

10 Solomon, "Introduction," in *Doing Medicine Together*, 3–34; Paul Weindling, "German Overtures to Russia, 1919–1925," in Solomon, *Doing Medicine Together*, 35–61.
11 Kojevnikov, *Stalin's Great Science*, 77.
12 Kojevnikov, *Stalin's Great Science*, 83. Also see Hall, "The Schooling of Lev Landau."
13 Tisza, "Adventures of a Theoretical Physicist," 209.
14 Ortiz and Pinkus, "Herman Muntz," 11.
15 Weindling, *Health, Race, and German Politics*, 328.
16 Reiß et al., "Introduction," 174.
17 Nick Hopwood, "Biology between University and Proletariat," 407; *International Biographical Dictionary of Central European Emigres*, vol. 2, 1026.
18 "Steinitz, Wolfgang," *Verfolgung und Auswanderung deutschsprachiger Sprachforscher 1933–1945*, http://zflprojekte.de/sprachforscher-im-exil/index.php/catalog /s/451-steinitz-wolfgang. See also Wolfgang Steinitz, http://www.fraenger.net /per_steinitz.html.
19 Leo, *Leben als Balance-Akt*, 111–12.
20 Murawkin, *Massenspektra von Gläsern, Salzen und Metallen*. See also Rykov, *Tesla protiv Eynshteyna*.
21 Lins, *Dangerous Language*, 218.
22 Murawkin's espionage activities are discussed in Primakov, *Survey of the History of Russian Foreign Intelligence*, vol. 3, ch. 42. This source contains a number of inaccuracies, particularly in regard to atomic research in the 1930s, but appears based on files in the archives of the KGB.
23 Lins, *Dangerous Language*, 218.
24 Rykov, *Tesla protiv Eynshteyna*, 86. Translation by author.
25 Interview with Art Kharlamov and Mica Nava, 4 December 2018; information on Weisselberg's fiancée provided by Mica Nava, July 2020. Weisselberg's NKVD file has been published in Gottvald and Shifman, *George Placzek*, Kindle 1952–2335; Nava, *Visceral Cosmopolitanism*, 139–41; *Dokumentationsarchiv des osterreichischen Widerstandes*, https://www.doew .at/erinnern/biographien/oesterreichische-stalin-opfer-bis-1945/stalin -opfer-w/weisselberg-konrad; and Weisselberg, *Beiträge zum Nachweis*.
26 O'Connor and Robertson, "Fritz Alexander Ernst Noether"; Noether, "Fritz Noether (1884–194?)." See also "Noether, Fritz Alexander Ernst," *Neue Deutsche Biographie* 19 (1999): 321–2.
27 Schiffer and Samelson, "Dedicated to the Memory of Stefan Bergman."
28 Jug et al., "Hans Hellmann," 414.
29 Mott, "Herbert Frohlich," 148–9.
30 E.W. Bagster Collins to E.R. Murrow, AEC, 25 January 1934, AEC 60/48.
31 Romberg as quoted in O'Connor and Robertson, "Werner Romberg."
32 Zinnemann was working with Professor Hugo Braun, University of Heidelberg, and Professor Max Neisser, University of Frankfurt. Zinnemann, interrogation transcript, NKVD File.

33 Der Magistrats-Personaldezerent to Zinnemann, 3 April 1933, Pam Zinnemann-Hope's papers.
34 Zinnemann-Hope, "My Father Tells About 1934," in *On Cigarette Papers*.
35 Zinnemann, interrogation transcript, NKVD File.
36 Pam Zinnemann-Hope, various poems, *On Cigarette Papers*.

4. Scientists in Flight to the Soviet Union

 1 Interview of Victor Weisskopf by Thomas S. Kuhn and John L. Heilbron on 10 July 1963, Niels Bohr Library and Archives.
 2 "A Scientist's Odyssey with Victor F. Weisskopf," *Conversations: A Video Series on International Affairs*, Institute of International Studies, University of California–Berkeley, UCTV, 1988, https://www.youtube.com/watch?v =utZY2vlgphc. See also interview of Victor Weisskopf by Thomas S. Kuhn and John L. Heilbron.
 3 Academic Assistance Council: Second Annual Report, 20 July 1935, SPSL 1/2.
 4 Duggan, AEC to Frederick Keppel, President of Carnegie Corporation, 15 April 1939, SPSL 155.
 5 Charlotte, as quoted in Amaldi, *The Adventurous Life of Friedrich Georg Houtermans*, 7.
 6 Shifman, *Standing Together*, 56; Schlesinger, "Тетрадь."
 7 Hoffmann, "Fritz Lange," 411.
 8 Shifman, *Standing Together*, 56.
 9 Schlesinger file, SPSL 290; Rhode-Juchtern, "Charlotte Schlesinger"; Schlesinger, "Тетрадь."
10 Munro to the General Secretary AAC, 11 June 1934; Simons to the AAC, 12 June 1934, SPSL 265.
11 Hugh-Jones, "Wickham Steed."
12 Brinson and Dov, *A Matter of Intelligence*, 57.
13 Brinson and Dov, *A Matter of Intelligence*, 58; Brinson, "The Strange Case."
14 Brinson and Dov, *A Matter of Intelligence*, 60.
15 Simons to the Secretary of the AAC, 22 October 1934, SPSL 265.
16 Weizmann to A.J. Makower, Jewish Professional Committee, 11 February 1935; Adams, General Secretary, SPSL, to Simons, 19 February 1936; J. Rigg, Cawthron Institute, Nelson, New Zealand, to Adams, 8 July 1936, SPSL 265; Dr Berna Simons, Barcelona (signed Mickey Mouse), to Hellmuth Simons, London, 20 February 1936, NAUK, KV 2/2001. Helmuth's relationship to Berna or Bertha Simons remains unclear.
17 Translation of Int. H 200 M., Simons to Pawlowsky, 6 March 1936, NAUK, KV 2/2001.
18 Translation of Int. H 200 M, To be given to Prof. Kein for information, n.d. but March 1936; Simons to Bertha Simons, found on same page as

previous document, also n.d. but March 1936; gist of Int. H 127 C, Dr Berna Simons to Hellmuth Simons, 20 February 1936, NAUK, KV 2/2001. For the Communist view of Zionism, see Franzén, "Communism Versus Zionism," 6–24.

19 Liepman, *Death from the Skies.*

20 Bentwich, *The Rescue and Achievement of Refuge Scholars,* 16.

21 For more on the situation in France, see Balibar, *Savant cherche refuge,* 56–8.

22 Borchardt, "As a German Teacher in the Soviet Union," in *Hermann Borchardt Werke Band 1,* 101.

23 Borchardt, "My Last Days in the Soviet Union," 44.

24 Borchardt, *I Was a Teacher in Soviet Russia,* 3.

25 Borchardt, *Communism and Fascism,* 3.

26 For Langevin and Communism, see Caute, *The Fellow-Travellers.*

27 Borchardt, "My Last Days," 44.

28 Borchardt, "As a German Teacher," 114–15.

29 Lederer oral history interview.

30 Gilde to the AAC, "Request for Granting of a Research-Scholarship," n.d. but late September or early October 1934, SPSL 491/1.

31 Skepper to A.V. Hill, 4 October 1935, SPSL 491/1.

32 Skepper to de Margerie, 2 March 1939, SPSL 491/1.

33 For the AAC's internal difficulties, see Zimmerman, "Protests Butter No Parsnips," 40–1.

34 For Prinsheim, see https://de.wikipedia.org/wiki/Peter_Pringsheim.

35 Note by Pringsheim, n.d. but likely early 1934, SPSL 480/3.

36 See the various correspondence between Duschinsky and the AAC in SPSL 480/3. His AEC file is 53/11. Duschinsky's appointment to Leningrad is discussed in his 1937 CV, found as part of his entry in the Central Database of Shoah Victims' Names, Yad Vashem, https://yvng.yadvashem .org/nameDetails.html?language=en&itemId=1772506&ind=32&win Id=-5625036566696593314.

37 O'Connor and Robertson, "Stefan Emmanuilovich Cohn-Vossen." See also Cohn-Vossen's files, SPSL 325/5; AEC 49/64.

38 Luftschitz, "Request for Work," SPSL 244/8.

39 Enclosure to 1.3, Review of *Die Rasumaenderungen Der Baustoffe* (The Volume Change in Building Material), 1932.

40 Skepper to Luftshitz, 19 October 1934, and 21 December 1934, SPSL 244/8.

41 Extracts from letters from Professor Adam Warschavsky, 5 May 1933, 15 June 1933, and 16 September 1933; Luftschitz, curriculum vitae, 1948, SPSL 244/8.

42 For more on the early years of the American Emergency Committee, see Zimmerman, "Competitive Cooperation."

43 Schein, *Becoming American,* Kindle, loc. 73–92 of 1476.

44 Schein CV, 8 March 1938, AEC 111/37.

45 Schmid, "A Learning Journey to Trstena," 7.

46 Beck, oral history interview.

47 Schein, *Becoming American*, loc.154–165 of 1476, Kindle.

48 Sadowsky CV, 10 April 1934, SPSL 284/5; Hilda Sadowsky, "Nothing New under the Sun," December 1981, George Sadowsky's private archive.

49 Sadowsky to the AAC, 4 April 1934, 25 October 1934, SPSL 284/5.

50 Sadowsky to the AAC, 17 December 1934, SPSL 284/5.

51 For an example of another of those back-and-beyond countries, see Kersfeld, *La migración judía en Ecuador*.

5. Living in Stalin's Soviet Union

1 Hans Helmann Jr's account can be found in *Einfuhrung in die Quantenchemie*.

2 Fritz Alexander Ernst Noether biography. O'Connor and Robertson, "Fritz Alexander Ernst Noether."

3 Kojevnikov, *Stalin's Great Science*, 92.

4 Zinnemann, NKVD interrogation file.

5 Scammel, *Koestler*, 97.

6 M. Ruhemann, "Kharkov in the Thirties," 78.

7 Weissberg, *The Accused*, Kindle, loc. 1655–658 of 10573.

8 Weissberg, *The Accused*, Kindle, loc. 1627–31 of 10573.

9 Borchardt, *The Soviet Caste System*, 3.

10 Amaldi, *The Adventurous Life of Friedrich Georg Houtermans*, 39.

11 Borchardt, "My Last Days in Soviet Russia," 41.

12 Weissberg, *The Accused*, Kindle, loc. 5639 of 10573.

13 On the housing crisis in the Soviet Union, see Fitzpatrick, *Everyday Stalinism*, 46–50.

14 Lederer oral history interview.

15 Leo, *Leben als Balance-Akt*, 112.

16 Leo, *Leben als Balance-Akt*, 119.

17 Professor Karl Weissenberg to Adams, 13 December 1937, SPSL 491.

18 Hans Helmann Jr account.

19 Borchardt, *I Was a Teacher in Soviet Russia*.

20 Borchardt, "My Last Days, 42; Amaldi, 37.

21 Schein, *Becoming American*, Kindle, loc. 165–78 of 1476.

22 Ortiz and Pinkus, "Herman Muntz," 11.

23 Lederer oral history interview.

24 Fritz Alexander Ernst Noether biography. O'Connor and Robertson, "Fritz Alexander Ernst Noether."

25 Schein, *Becoming American*, 200.

26 Fitzpatrick, *Everyday Stalinism*, 51.

27 Zinnemann and Zinnemann, "Incidence and Significance."
28 K. Zinnemann, "Ordered to the Donetz Basin," n.d. but likely 1942, from Pam Zimmerman-Hope's personal archive.
29 Shifman, *Standing Together*, 157.
30 Leo, *Leben als Balance-Akt*, 115, 122.
31 Tagg, "The Goteborg Connection," 225.
32 Leo, *Leben als Balance-Akt*, 114–15.
33 Leo, *Leben als Balance-Akt*, 124–5, Federal Foundation for the Evaluation of the SED dictatorship. Gerhard Harig acknowledged he had contact with the NKVD before his arrest. Harig as qtd in Bernhardt, "Gerhard Harig (1902–1966)," 15.
34 Bernhardt, "Gerhard Harig (1902–1966)," 15–17.
35 M. Ruhemann, "Kharkov in the Thirties," 81.
36 Tantscher, "Ernst Emsheimer." For the German expatriate community in Sweden during the Second World War, see Müssener, *Exil in Schweden*.
37 Borchardt, "My Last Days," 44.
38 F. Borchardt, "Hermann Borchardt," 122.
39 Works that have examined the history of UFTI, in chronological order: Khriplovich, "The Eventful Life of Fritz Houtermans"; Pavlenko, Raniuk, and Khramov, *"Delo" UFTI 1935–1938*; Amaldi, *The Adventurous Life of Friedrich Georg Houtermans*; Szapor, "Private Archives and Public Lives"; Frenkel, "Professor Friedrich Houtermans"; Reinders, *The Life, Science, and Times*.
40 Szapor, "Private Archives and Public Lives," 99.
41 Weissberg, *The Accused*, Kindle, loc. 3419 of 10573.
42 Weissberg, *The Accused*, Kindle, loc. 3611 of 10573.
43 Reinders, *The Life, Science, and Times*, 102.
44 Reinders, *The Life, Science, and Times*, 103.
45 Weissberg, *The Accused*, Kindle, loc. 1623 of 10573.
46 Weissberg, *The Accused*, Kindle, loc. 3436 of 10573.
47 Tisza, "Adventures, Part I: Europe," 81.
48 Shifman, *Physics in a Mad World*, Kindle, loc. 2262 of 8664.
49 Reinders, *The Life, Science, and Times*, 108–9.
50 Weissberg, *The Accused*, Kindle, loc. 3566 of 10573. Weissberg's account must be used with caution, since he places Korets's arrest and trial as occurring during Davidovich's directorship rather than just after.
51 Fitzpatrick, *Everyday Stalinism*, 7.

6. Refugee Scholarship in the Soviet Union

1 Ortiz and Pinkus, "Herman Muntz," 9–12.
2 Ernst Simonson, curriculum vitae, n.d. but likely November 1938, SPSL 419/4.

3 Svetlana Tantscher, "Ernst Emsheimer"; see also Tagg, "The Goteborg Connection," 224–5.
4 Dr. Walter Zehden, "Curriculum vitae," nd but 1938, SPSL 343/6. For information on Landsberg see http://eleven.co.il/jews-of-russia-and-ussr/jewish-contribution-to-culture-science-economy/12324/.
5 Zehden, "Curriculum vitae."
6 Crowther, *Soviet Science*, 145. Crowther's other book was *Industry and Education in Soviet Russia*. See also Walter Zehden, "Curriculum Vitae," SPSL 343.
7 Ginzburg, "Notes of an Amateur Astrophysicist," 7.
8 Saul Levy, curriculum vitae, n.d. but 1937, SPSL 334/1.
9 Weissberg, *The Accused*, Kindle, loc. 3944. Reinders states that Soviet physicist Dimitry Ivanenko was also instrumental in establishing the journal, but gives no further details on his role.
10 Reinders, *The Life, Science, and Times*, 88–9.
11 M. Ruhemann, "Kharkov in the Thirties," 78–81. See also Reinders, *The Life, Science, and Times*, 135–6.
12 Amaldi, *The Adventurous Life*, 68.
13 Tisza oral history interview, American Institute of Physics, https://www.aip.org/history-programs/niels-bohr-library/oral-histories/4915-1.
14 Tisza oral history interview.
15 Tisza oral history interview.
16 Tisza, "Adventures, Part I: Europe," 85.
17 Shifman, *Physics*, Kindle, loc. 2205 of 8664.
18 Reinders, *The Life, Science, and Times*, 197.
19 Khriplovich, "The Eventful Life," 32.
20 Weissberg, *The Accused*, Kindle, loc. 1127–32 of 10573.
21 Frenkel, "Professor Friedrich Houtermans," in Shifman, *Physics in a Mad World*, Kindle, loc. 2100–2200 of 8664.
22 Hoffmann, "Fritz Lange," 411.
23 Rykov, *Tesla vs Einstein*, 135.
24 The author wishes to thank Dr Milton Leittenberg for information on Pavlosky and on Soviet biochemical warfare research. See Leitenberg, Zilinskas, and Kuhn, *The Soviet Biological Weapons Program*; and Rimmington, *Stalin's Secret Weapon*.
25 Hellmann as qtd in W.H.E. Schwartz, D Andrae, et al., *Hans G.A. Hellmann (1903–1938)*, trans. Mark Smith and W.H.E. Schwartz, http://www.tc.chemie.uni-siegen.de/hellmann/hh-engl_with_figs.pdf, 14.
26 Hellmann, *Lebenslauf von Hans Hellmann*, 10.
27 Hellmann, *Lebenslauf von Hans Hellmann*, 11.
28 Schwartz et al., *Hans G.A. Hellmann (1903–1938)*, 19.
29 Schwartz et al., *Hans G.A. Hellmann (1903–1938)*, 20.
30 Schwartz et al., *Hans G.A. Hellmann (1903–1938)*, 14.

31 Reiß, "No evolution." Also see *Biographical Dictionary of Central European Emigres 1933–1945*, vol. 2: 1026.

32 Institute of the Peoples of the North, https://www.herzen.spb.ru/main/structure/inst/ins/1390565482.

33 Leo, *Leben als Balance-Akt*, 126–44.

34 Nussenzveig, "Guido Beck," 89–90.

35 Guido Beck, oral history interview.

36 Marcel Schein, curriculum vitae, 8 November 1937, AEC 111/37.

37 Sadowsky to the Emergency Committee, 16 October 1937, AEC 109/53.

38 Rhode-Jüchtern, "Charlotte Schlesinger."

39 Entretien avec Edgar Lederer, by Jean-François Picard.

40 H. Weiner, Central Office for Refugees, Medical Department to Simpson, SPSL, 12 December 1940, SPSL 365/11.

41 Qian Peng et al., "Excited State Radiationless Decay Process." See also Duschinsky's article, "The Importance of the Electron Spectrum." According to the Web of Science Database, this article has been cited 896 times.

42 Duschinsky, 1937 CV, Central Database of Shoah Victims' Names, Yad Vashem.

7. The Great Terror

1 McLoughlin and McDermott, "Rethinking Stalinist Terror," 6.

2 Harris, *The Great Fear*, 144.

3 Harris, *The Great Fear*, 120–1.

4 Firsov, "Dimitrov," 56–81.

5 Beck and Godin, *Russian Purge*, 119. This book was actually co-authored by Fritz Houtermans and Russian historian Konstantin Shteppa.

6 Mott, "Herbert Frohlich," 6.

7 Ortiz and Pinkus, "Herman Muntz," 11.

8 Wasser to Walter Adams, 20 May 1935, SPSL 342.

9 Mott, "Herbert Frohlich," 149.

10 Borchardt, *I Was a Teacher in Soviet Russia*, 4–5.

11 Borchardt, "My Last Days," 40–4.

12 Leo, *Leben als Balance-Akt*, 120, 139–40.

13 Leo, *Leben als Balance-Akt*, 164.

14 Koestler in the preface of *The Accused*, Kindle, loc. 1073 of 10573.

15 Weissberg, *The Accused*, Kindle, loc.1414 of 10573.

16 Levin, "Anatomy of a Public Campaign," 93.

17 As quoted in Levin, "Anatomy of a Public Campaign."

18 Josephson and Sorokin, "Physics Moves to the Provinces," 321.

19 Josephson and Sorokin, "Physics Moves to the Provinces," 322. Josephson and Sorokin misidentify Noether as Netter, perhaps a misspelling based on transliteration from Cyrillic to Latin. Email between author and Josepheson, 1 October 2018.

20 As quoted in Uri, *Dangerous Language*, 25.

21 Lins, *Dangerous Language*, 13.

22 Schlesinger, "Тетрадь." Schlesinger provides no specific date for Weller's arrest. Russian and Ukrainian sources only state that Weller was convicted and executed on 19 November 1937. See Веллер Григорий Максимович (1897), https://ru.openlist.wiki/ (1897).

23 Tagg, "The Goteborg Connection," 225.

24 Weissberg, *The Accused*, Kindle, loc. 1412–15 of 10573.

25 Tisza interview, 2009, 83–4.

26 Amaldi, *The Adventurous Life*, 40–1.

27 Shifman, *Physics in a Mad World*, Kindle, loc. 1276–87 of 8664.

28 Weissberg, *The Accused*, Kindle, loc. 1210 of 10573.

29 Weissberg, *The Accused*, Kindle, loc. 582–86 of 10573.

30 Weissberg, *The Accused*, Kindle, loc. 1291 of 10573.

31 Weissberg, *The Accused*, Kindle, loc. 1674–78 of 10573.

32 Weissberg, *The Accused*, Kindle, loc. 1679 of 10573.

33 Tisza, "Adventures, Part I: Europe," 84.

34 Sadowsky to AAC, 17 May 1937, SPSL 284/5.

35 Schlesinger, "Тетрадь."

36 Weisskopf to Walter Adams, SPSL, 11 July 1937, SPSL 339–41.

37 Schein, *Becoming American*, Kindle, loc. 211 of 1476.

38 Simonson to Hill, 9 November 1937, SPSL 419–4.

39 Zinnemann-Hope, "Arrested," in *On Cigarette Papers*, Kindle, loc. 384–91 of 1167.

40 Kurt and Irina [Irene] Zinnemann's NKVD File.

41 Reinders, *The Life, Science, and Times*, 171.

42 Lederer oral history interview.

43 Leo, *Leben als Balance-Akt*, 123–4. The quotes are the author's translation of Leo's German text.

44 Bernhardt, "Gerhard Harig," 15. Translation by author.

45 Walter Zehden, curriculum vitae, n.d. but 1938, SPSL 343. On Zheden's family, see Rudolf Peierls to Georg Placzek, 22 October 1938, in Gottvald and Shifman, *A Nuclear Physicist's Odyssey*, Kindle, loc. 2630 of 6271.

46 Schlesinger, "Тетрадь."

47 Ruth Gilde to Weissenberg, 30 November 1937; K. Weissenberg to Walter Adams, 13 December 1937, SPSL 491. Translation of Ruth's letter by author.

48 Fritz Alexander Ernst Noether biography.

49 Hellmann as qtd in W.H.E. Schwartz and D. Andrae, *Hans G.A. Hellmann*, 22. See also O'Connor and Robertson, "Fritz Alexander Ernst Noether."

50 Weissberg, *The Accused*, Kindle, loc. 1878–93 of 10573.

51 Weissberg, *The Accused*, Kindle, 7492 of 10573.

52 Pavlenko et al., *"Delo" UFTI 1935–1938*, 202.

53 Shifman, *Physics in a Mad World*, Kindle, loc. 524–30 of 8664.

8. Into Stalin's Frying Pan

1 Friedrich Georg Houtermans, "Chronological Report of my Life in Russian Prisons," in Shifman, *Physics in a Mad World*, Kindle, loc. 5462–652 of 8664.
2 Weissberg, *The Accused*, Kindle, loc. 4690–5345 of 10573.
3 Weissberg, *The Accused*, Kindle, loc. 5375 of 10573.
4 Pavlenko, N. Raniuk, and A. Khramov, *"Delo" UFTI 1935–1938*.
5 Reinders, *The Life, Science, and Times*, 155.
6 Reinders, *The Life, Science, and Times*, 155–60.
7 Houtermans, "Chronological Report of my Life in Russian Prisons," loc. 5500–5507 of 8664, Kindle.
8 Beck and Godin, 60.
9 Beck and Godin, 61.
10 Weissberg, loc. 7048 of 10573, Kindle; Zinnemann-Hope, *On Cigarette Papers*, loc. 441 of 1167, Kindle.
11 Houtermans, "Chronological Report," Kindle, loc. 5567 of 8664.
12 Weissberg, *The Accused*, Kindle, loc. 4798 of 10573.
13 Weissberg, *The Accused*, Kindle, loc. 2215 of 10573.
14 Zinnemann-Hope, *On Cigarette Papers*, Kindle, loc. 417–18 of 1167.
15 Zinnemann-Hope, *On Cigarette Papers*, Kindle, loc. 9873–913.
16 История эсперанто в России и Советском Союзе. Биографии репрессированных эсперантистов, архивные материалы (History of Esperanto in Russia and the Soviet Union. Biographies of repressed Esperantists, archival materials), http://historio.ru/murav.php. Author's translation.
17 История эсперанто в России и Советском Союзе; Lins, *Dangerous Language*, 13n37; Kuenzli, "Dem Stalinismus."
18 Gottvald and Shifman, *Placzek*, Kindle, loc. 2262–68 of 6271.
19 Schwarta, Andrae, et al., *Hellmann*, 22.
20 Hellmann, "Lebenslauf von Hans Hellmann," 12. Translation by author.
21 Schwarta, Andrae, et al., *Hellmann*, 23.
22 Fritz Noether, Index of Jews Whose Nationality Was Annulled by the Nazi State, 1935–44, National Archive and Records Service.
23 Karen Kettering, "Appendix A: Behind the Scenes," in Zeisel: *A Soviet Prison Memoir*, Kindle, loc. 2256 of 2902.
24 Zinnemann, NKVD file.
25 Mrs Houtermans to Blackett, 3 January 1938; Blackett to Russian Ambassador, 2 February 1938; Adam to Blackett, 15 February 1938, SPSL 330/8.
26 Szapor, "Private Archives and Public Lives," 104–5.
27 Bloch to Eva Zeisel, 9 June 1938, in Gottvald and Shifman, *Placzek*, Kindle, loc. 657–73 of 6271. See also George Placzek to Eva Zeisel, 15 May 1938, Kindle, loc. 1400–1416 of 6271.

28 Adams to Niels Bohr, 27 April 1938; Niels Bohr to Adams, 17 May 1938, SPSL 330/8.
29 Shifman, *Physics in a Mad World*, 365.
30 Charlotte Houtermans to Beria, 11 February 1939, SPSL 330/8.
31 Nancy Searle, SPSL to Blackett, 14 April 1939, SPSL 380/8.
32 Quote is from the announcement of the founding of the society found in *Nature* 139, 188–9 (30 January 1937). See also Simpson to Gardiner, 20 April 1939, SPSL 330/8.
33 Gardiner to Simpson, 2 June 1939, SPSL 330/8.
34 Gardiner to Simpson, 13 June 1939, SPSL 380/8.
35 Uwe Hoßfeld, "Schaxel, Julius," 597–8, https://www.deutsche-biographie.de/pnd118822977.html#ndbcontent.

9. From Stalin's Frying Pan into Hitler's Fire

 1 Tischler, *Flucht In Die Verfolgung*, 137.
 2 Tischler, *Flucht In Die Verfolgung*, 135–7. For the stripping of Noether and Gilde's German citizenship, see Germany, Index of Jews Whose German Nationality was Annulled by Nazi Regime, 1935–1944, National Archives and Records Administration (NARA), RG ARC ID: 569; no. T355.
 3 Andrei Parastaev, First Secretary Embassy of the USSR, Washington, to Dr H.D. Noether, 12 May 1989, in *Integral Equations and Operator Theory* 13 (1990): 305. See also Houtermans, "Chronological Report," Kindle, loc. 5606.
 4 Weissberg, *The Accused*, Kindle, loc. 10038–259 of 10573; "Chronological Report," Kindle, loc. 5575–621of 8664.
 5 For more on these exchanges see Tischler, *Flucht In Die Verfolgung*, 135–6.
 6 M. Gilde-Berent to Weissenberg, 13 February 1940, SPSL 491/1.
 7 Simpson to Bracey, 22 February 1940 (emphasis added). See also Weissenberg to SPSL, 17 February 1940, SPSL 491/1.
 8 Bracey to Simpson, 26 February 1940, SPSL 491/1.
 9 George Placzek to Eva Zeisel, 21 July 1940.
10 Gilde's identity papers were discovered by Michal Palacz in Siegfried Gilde's file in the Main Medical Library in Warsaw, Documents of the Warsaw–Białystok Chamber of Physicians.
11 Henry Fenigstein, "The Holocaust and I: Memoirs of a Survivor by Dr Henry Fenigstein as told to Saundra Collis," n.p., n.d. (1990s); pts 1 and 2, 136–7. Courtesy of Dr Maria Ciesielska, Lazarski University in Warsaw. Provided to the author by Michal Palacz. The founding of the Ghetto Medical School is discussed in Fenigstein, "The History of the 'Czyste': The Jewish Hospital in Warsaw from 1939 to 1943 during the Nazi Regime," xxii, unpublished manuscript, 1990; courtesy of Professor Claude Romney, University of Calgary. Provided to the author by Michal Palacz.

12 Fenigstein, "The History," xvii.
13 As qtd in Fenigstein, "The History," xvi.
14 Fenigstein, "The History," xvii.
15 Fenigstein, "The History," xix.
16 Fenigstein, "The History," xxxiv.
17 Fenigstein, "The History," xxxvii.
18 "Transport 46 from Drancy Camp, France, to Auschwitz Birkenau, Extermination Camp, Poland on 09/02/1943," https://deportation .yadvashem.org/index.html?language=en&itemId=5092618; Ruth Gilde entry, https://yvng.yadvashem.org/nameDetails.html?language=en&itemId =3177744&ind=1; Marina Gilde Entry, https://yvng.yadvashem.org/index .html?language=en&s_id=&s_lastName=Gilde&s_firstName=Marina&s _place=Paris&s_dateOfBirth=&cluster=true, Yad Vashem, Central Database of Shoah Victims' Names. Marina's inclusion in Transport 46 is not certain, but it seems more than likely she was murdered with her mother.
19 "The Stories of Six Righteous among the Nations in Auschwitz: Dr Ella Lingens," https://www.yadvashem.org/yv/en/exhibitions/righteous-auschwitz/lingens.asp.
20 Gowing, *Britain and Atomic Energy*, 23–5.
21 Khriplovich, "The Eventful Life," 35.
22 Fritz Reiche – Session III, American Institute of Physics Oral History Collection, https://www.aip.org/history-programs/niels-bohr-library/oral-histories/4841-3.
23 Fritz Reiche – Session III.
24 For Houtermans's message see Powers, *Heisenberg's War*, 103–9.
25 For the origins of the Manhattan Project, see Stanley Goldberg, "Inventing a Climate of Opinion"; and Zimmerman, "The Tizard Mission."
26 Shifman, *Standing Together in Troubled Times*, 85.
27 Amaldi, "The Adventurous Life," 79–81; Khriplovich, "The Eventful Life," 35–6; Shifman, *Physics in a Mad*, Kindle, 831–51 of 8664; excerpts of the letter from Houtermans to B. Touschek, July 1947, SPSL 330; Marie Rausch von Traubenberg to SPSL, 22 November 1945, SPSL 330.
28 Max Born to Ursell, 10 November 1947, SPSL 330.
29 Cammann was a fighter ace in the First World War and briefly saw further operational service in 1939–40, when he shot down one additional aircraft. The rest of his service in the Luftwaffe appears to have been in non-operational roles. Franks and Baily, Above the Lines, 92.
30 Marie Rausch von Traubenberg to the SPSL, 22 November 1945.
31 Born to Ursell, SPSL, 10 November 1947, SPSL 330.

10. From the Great Terror to the Shoah

1 Mott, "Herbert Frohlich," 161.
2 Adams to Wasser, 17 May 1935, SPSL 342. Emphasis added.

3 Wasser to Adams, 30 September 1935, SPSL 342.

4 Adams to Wasser, 11 October 1935, SPSL 342.

5 Wasser to the SPSL, 11 April 1938, SPSL 342.

6 Simpson to Wasser, 21 April 1938, SPSL 342.

7 Wasser to the SPSL, 10 September 1938, SPSL 342.

8 David Gleghorn Thomson to Wasser, 13 September 1938, SPSL 342.

9 F. Borchardt, "Herman Borchardt," 122. Translation by author.

10 Borchardt, *Communism and Fascism*, 5.

11 Leonard Fischl, "Unbekannte Briefe zwischen Borchardt und Grosz Verfemt und verfolgt: George Grosz und Hermann Borchardt verband eine enge Freundschaft. Eine Edition ihres Briefwechsels dokumentiert die unbekannte Beziehung," *Tagesspiegal*, https://www.tagesspiegel.de/themen /freie-universitaet-berlin/unbekannte-briefe-zwischen-borchardt-und-grosz -verfemt-und-verfolgt/23757594.html.

12 Sadowsky to AAC, 17 May 1937, SPSL 284/5; Simpson to John Whyte, Assistant Secretary, AEC, 4 June 1937; Whyte to Concha Romero James, Chief, Division of Intellectual Cooperation, Pan American Union, Washington, 29 July 1937, AEC 109/53.

13 Sadowsky to the AEC, 16 October 1937, AEC 109/53.

14 Harig, Military Government of Germany, Concentration Camp Inmates Questionnaire, 27 April 1945, Harig File, Arolsen Archives, https://collections .arolsen-archives.org/en/archive/6065068/?p=1&s=Harig&doc_id=6065082.

15 Zegenhagen, "Buchenwald Main Camp," 291.

16 German Embassy Moscow to E. Harig, Leipzig, 17 April 1943; declaration by G. Harig, 25 April 1939, Gerhard, Harig file. See also https://collections .arolsen-archives.org/en/search/person/6065068?s=Harig&t=0&p=1.

17 Harig, Military Government of Germany, Concentration Camp Inmates Questionnaire, 27 April 1945, Harig file.

18 Balibar, *Savant cherche refuge*, 58–64. CNRS was formed on 19 October 1939, when the National Center for Applied Scientific Research (CNRSA) and the High Council for Scientific Research were merged.

19 MIT News Release, "Laszlo Tisza, physics professor emeritus, 101," 16 April 2009, http://news.mit.edu/2009/obit-tisza-0416.

20 Tisza, "Adventures, Part II: America," *Physics in Perspective* 11 (2009): 120–2.

21 O'Connor and Robertson, "Stefan Bergman." See also Bergman's Petition for Naturalization in the Rhode Island, State and Federal Naturalization Records, 1802–1945, 5 November 1950, accessed through Ancestry.ca.

22 Betty Durry, Emergency Committee, to Walter Adams, SPSL, 12 May 1938; Beck to Walter Adams, 3 June 1940, SPSL 323.

23 Rapkine to Simpson, 9 May 1938, SPSL 323.

24 Beck oral history interview; Beck's survey form, 1958, SPSL 61.

25 Beck oral history interview; Beck to the General Secretary and Personal Data on my Mother, n.d., SPSL, 22 March 1940, SPSL 323.

26 Mizelle, "Peron's Atomic Plans" and "More about Peron's Atomic Plans"; see also "Beck Writes the Editor" and "Henry Wallace Replies," *New Republic*, 31 March 1947, 20–1.

27 For more on Beck's time in South America, see Nussenzveig and Videira Passos, *Guido Beck Symposium.*

28 Beck's survey form, 1958, SPSL 61; Beck to Ursell, SPSL, 24 June 1947, SPSL 323; Beck oral history interview; Mizelle, "Peron's Atomic Plans" and "More about Peron's Atom Plans." See also "Beck Writes the Editor" and "Henry Wallace Replies."

29 Lederer, oral history interview.

30 Majower to Simpson, SPSL, 2 September 1938, SPSL 265.

31 For Simons's son, see Szajkowski, *Jews and French Foreign Legion*, 161. For another account claiming that bribes had to be paid to leave Vichy France, see Arno Motulsky, "A German-Jewish Refugee," 1294. For more on refugees in Marseille, see Tortel, "Marseille."

32 Simons to the Institute of International Education, 22 September 1938, NYPL AEC 117/4.

33 Dr Andre Lwoff, Pasteur Institute, 28 February 1938, ACC 117/4.

34 Simons to Flynn, 20 August 1940, AEC 117/4.

35 Tortel, "Marseille," 367.

36 Dr HCR Simons, "Escape from Death," Wiener Library, 1656/3/9/60.

37 Wendland discussed his support of Simons with US army interrogators as proof of his good character. He claimed that until after the war, Simons had no idea who had provided the money. Translation of the interogation of Hans Wendland, 4 September 1946, Schweizerisches Bundesarchiv, E9500.239A#2003/48#79*, USA 3: NARA.

38 Allen Dulles, telegram from OSS Bern, 8 December 1943, in Petersen, *From Hitler's Doorstep*, 173.

39 Simons, curriculum vitae, 3 November 1933, AEC 117/4.

40 Croddy, Perez-Armendariz, and Hart, *Chemical and Biological Warfare*, 71.

41 Dr Emanuel Wasser, Jewish Transmigration Bureau Deposit Cards, 1939–54 (JDC), Roll 9.

42 Emanuel Wasser, Yad Vashem, Central Database of Shoah Victims' Names, http://yvng.yadvashem.org/nameDetails.html?language=en&itemId=3228250&ind=1.

43 Ursell to K. Weissenberg, 2 June 1947; Weissenberg to Ursell 5 June 1947; Ursell to E. Duschinsky, 14 June 1947, SPSL 480.

44 E. Duschinsky to Ursell, 24 January 1948, SPSL 480.

45 Ursell to E. Duschinsky, 27 January 1948; E. Duschinsky to Ursell, 28 February 1948, SPSL 480. See also Friedrich Duschinsky, Central Database

of Shoah Victims' Names, Yad Vashem, https://yvng.yadvashem.org
/nameDetails.html?language=en&itemId=5399671&ind=1.

46 H.A. Bethe to P.M.S. Blackett, 8 March 1938. See also R. Peierls to Blackett,
5 February 1938, SPSL 337/9.

47 O'Connor and Robertson, "Werner Romberg."

48 Lutschitz left a brief untitled and undated handwritten account of his
experiences after leaving the Soviet Union in 1937, SPSL 244/8. See also
Dethick, *The Liberation of Lake Trasimeno*, 17; and emails to author from
Dethick, 3 January 2017.

49 Secretary, United Kingdom Bureau for German Austrian and Stateless
Persons from Central Europe to Ursell, SPSL, 5 July 1946; Red Cross
Telegram to Luftschift from SPSL, 1 July 1946, SPSL 244/8.

50 Red Cross telegram to Luftschiftz from SPSL, 1 July 1946; Luftschitz to
SPSL, 30 July 1946, SPSL 244/8.

51 Ursell to Luftschitz, 18 November 1940; Ursell to Luftschift, 5 December 1946,
SPSL 244/8.

52 Signor G. Ludovisi-Luftchitz to Ilse Ursel, 20 February 1949, SPSL.

53 Ursell to Signor G. Ludovisi-Luftchitz, 4 March 1949, SPSL 244/8.

54 Semelin, *The Survival*, 264–5.

11. Survival and Triumph

1 Ling, "Ernst Emsheimer," 426.

2 For more on the situation in Sweden, see Gilmour, *Sweden, the Swastika, and
Stalin*.

3 Tagg, "The Goteborg Connection," 225.

4 Ling, "Ernst Emsheimer," 426.

5 Others who supported Muntz included Harald Bohr, Alexander Ostrowski,
Vito Volterra, Tullio Levi-Civita, and Gábor Szegő.

6 Muntz's SPSL file traces his desperate quest for a position outside of
Sweden. SPSL 282/7.

7 Ortiz and Pinkus, "Herman Muntz," 12.

8 Ortiz and Pinkus, "Herman Muntz," 13.

9 Ortiz and Pinkus, "Herman Muntz."

10 Adams to Under Secretary State, Home Office (Aliens Department), 1 July
1938, SPSL 444.

11 Zinnemann and Zinnemann, "Incidence and Significance."

12 This story is recounted by Pam Zinnemann-Hope in her poem "Oma Leah
Arrives in England with Opa Lazar," in *On Cigarette Papers*, Kindle, loc.
560–70 of 1167.

13 Frowiska Crowther to Adams, SPSL, 21 October 1938. See also J.G.
Crowther to Adams, 29 October 1938, SPSL 343.

14 David Cleghorn Thomson, SPSL, to Mr Drew, Under Secretary of State Home Office (Aliens Department), 15 December 1938, SPSL 444.

15 Sadowsky to SPSL, 4 September 1937, SPSL 284/5.

16 Recollections of Hilda Sadowsky, "Nothing New under the Sun," December 1981, private archives of George Sadowsky; See also Schmiedeshoff, "In Memory of Dr Michael Sadowsky," 126.

17 Guy Stanton Ford, 23 November 1937; Oswald Veblen to American Consul General, Jerusalem, 22 November 1937; W.J. Luyten, University of Minnesota, to American Consul General, Jerusalem, 22 November 1937, in the private archives of George Sadowsky. The letter from Einstein does not survive but was referred to in the author's interview with George Sadowsky, 9 April 2021.

18 Zimmerman, "Competitive Cooperation," 172.

19 Betty Drury, AEC, to Dr Duggan, 26 January 1938, AEC 111/37.

20 Interview of Betty Compton by Charles Weiner on 11 April 1968, Niels Bohr Library and Archives, American Institute of Physics, College Park, Maryland, www.aip.org/history-programs/niels-bohr-library/oral-histories/4560-1.

21 For details on Schein trip to the United States, see SPSL 339 and AEC 111/37.

22 Simonson, curriculum vitae, 19 July 1938, with handwritten addendum September; Simonson to A.V. Hill, 9 November 1939, SPSL 419/4.

23 Kurt Zinnemann to SPSL, 3 January 1939, SPSL 429/4.

24 Lawson to G.P. Crowden, London School of Hygiene and Tropical Medicine, 6 March 1939, SPSL 429/4.

25 Simonson to SPSL, 6 April 1939, SPSL 429/4.

26 Simonson to Thompson, SPSL, 18 April 1940; Simonson to Ursell, SPSL, 20 May 1946, SPSL 429/4.

27 Rhode-Juchtern, "Charlotte Schlesinger"; Schlesinger, "Тетрадь." Many of the dates found in Schlesinger's account, dictated in the last few weeks of her life, are inaccurate and have been altered to conform to the official records. See Schlesinger's entry, List or Manifest of Alien Passenger Entry, SS *Volendam*, 13 August 1938, New York, Arriving Passenger and Crew Lists (including Castle Garden and Ellis Island), 1820–1957, US National Archives and Record Service, accessed through ancestry.ca.

28 Shifman, *Standing Together*, 88–91.

29 A list of Baerwald's patents can be found in Google Patents, https://patents .google.com/?inventor=Hans+Baerwald&oq=Hans+Baerwald.

30 Ley to Polanyi, 23 September 1937; Polanyi to Adams, SPSL, 14 October 1937, SPSL 334/1.

31 Levy, "A Correlation Method." In this article Levy is listed as a former member of the Physical Institute of the University of Moscow.

32 Emails from Dan Flanagan, Archives Assistant, JW England Library, to author, 26 and 27 February 2018.

33 In a 1955 article Simons states that he was employed and funded by the "Roche" Study Foundation (Basel). For three years Simons is listed as being associated with the Swiss Serum and Vaccine Institute in Bern. The institute, despite its name, was a private pharmaceutical company. Simons, "Nachweis von Borrelien" and "Is Disseminated Sclerosis Caused by Spirochaetes."

34 Reference Letter for Zinnemann by McLeod, 8 September 1951, private collection of Pamela Zinnemann-Hope.

35 Abel et al., "Tribute to Ernst Simonson (1898–1974)."

36 Mott, "Herbert Frohlich," 161.

37 Rossi, "Prof. Marcel Schein."

38 Marcel Schein obituary, *New York Times*, 21 February 1960.

39 Yodh, "Early History."

40 Tisza, "Adventures, Part II: America"; Friedman, Greytak, and Kleppner, "Laszlo Tisza"; Griffin, "Laszlo Tisza."

41 Khriplovich, "The Eventful Life," 36–7.

42 Amaldi, *The Adventurous Life*, 111–24.

43 O'Connor and Robertson, "Werner Romberg."

44 Schmiedeshoff, "In Memory of Dr Michael Sadowsky"; Illinois, US Federal Naturalization Records, 1856–1991, for Michael Sadowsky, District Court, Northern District, Illinois Petitions, V. 1132–1134, No. 283451–284050, 1943, NARA, accessed through Ancestry.ca.

45 R.G.D. Richardson to H. Wriston, n.d. but late December 1945, in Siegmund-Schultz, *Mathematicians*, ch. 9-D-5.

46 O'Connor and Robertson, "Stefan Bergman."

47 Pierre Potier, "Biography of Edgar Lederer," 1995.

48 Schneider, "In memoriam Ernst Emsheimer."

49 A brief biography of Zeisel is on the website of the Cooper Hewitt Museum in New York, http://collection.cooperhewitt.org/people/18043295/bio. While it is an accurate summary of her ceramic designs, its account of her time in the Soviet Union is mainly inaccurate. No mention is made of Alexander Weissberg.

12. The Ensnared in the Cold War

1 Barbara Ruhemann to SPSL, 21 March 1939, SPSL 154.

2 Simpson to Dr B. Ruhemann, 23 March 1938, SPSL 154. The *Nature* article appeared 17 December 1938.

3 Ruhemann to Simpson, 27 March 1938, SPSL 154.

4 Initially Simpson was not even aware that Ruhemann, who signed her first letter to the SPSL only with the initial B, was a woman.

5 Barbara's sister, Franzisca, married Crowther on 21 February 1934.

6 Brown, *J.D. Bernal*, 106–9.

7   Brown, *J.D. Bernal*, 161.
8   Bernal, *The Social Function of Science*, 212–21, 393–7.
9   Martin Ruhemann, Appendix VII: Notes on Science in the USSR," in Bernal, *The Social Function of Science*, 443–9. Bernal in his foreword to the book lists Ruhemann as one of his principal sources of information (xvi).
10 Brown, *J. D. Bernal*, 162.
11 Polanyi, "Right and Duties of Science."
12 Extract from C.C. report, Taunton, 16 March 1940, NAUK, KV2/4438.
13 Note of Drs M. and B. Ruhemannn contributed by Sir Graham Greene, former secretary of the Admiralty, 7 June 1940, NAUK, KV2/4438.
14 M. Ruhemann, "Science and the Soviet Citizen."
15 "Science and Marxist Philosphy," *Nature*, 6 September 1941, 280–1. Barbara's paper "discussed the economic origins of universal totemism that accompanies the clan family organization."
16 Special Branch Report, 4 November 1941, NAUK, KV2/4438.
17 Weissberg, *The Accused*, Kindle, loc. 1738–42 of 10573.
18 John Close, "Physics By Design: Obituary of Martin Ruhemann," *The Guardian*, 30 December 1993. See also "Dr Martin Ruhemann Retires," *Cryogenic News*, 26 December 1986, 696.
19 Extract of a report made by Ivor Cunnar Martin, "Communist 1949," 22 August 1950, NAUK KV2/4438.
20 Extract of conversation, 31 January 1955; extract on Ruhemann, 3 March 1957, NAUK KV2/4340.
21 *Karl Marx–Frederick Engels: Collective Works*, vols. 4–9 (Moscow: Progress, 1975).
22 Bernhardt, "Gerhard Harig," 9.
23 Chitralla, "Das Franz-Mehring-Institut."
24 Bernhardt, "Gerhard Harig," 20.
25 Connelly, *The Sovietization*, 148–9.
26 Prokop, "Gerhard Harig," 47.
27 Harig as qtd in Nothnagle, "From Buchenwald to Bismark," 104.
28 Prokop, "Gerhard Harig," 81.
29 Connelly, "Ulbricht and the Intellectuals," 355. See also Prokop, "Gerhard Harig."
30 "Harich, Wolfgang, 9.12.1923–15.3.1995, Philosoph, Publizist," Biographical information from *Who Was Who in the GDR?*, https://www.bundesstiftung-aufarbeitung.de/de/recherche/kataloge-datenbanken/biographische-datenbanken/wolfgang-harich?ID=1247.
31 Howell, "The Hagiography of a Secular Saint," 67.
32 W. Krohn, "Wissenschaftliche Revolution und gesellschaftlicher Wandel," *N.T.M.* 18 (2010), 329–35, https://doi.org/10.1007/s00048-010-0020-7.
33 Brinkel, "Institutionalizing *Volkskunde*."
34 Lang, "Wolfgang Steinitz' Russian Textbook."

35 Eckert, "Wolfgang Steinitz und die Vermittlung des Russischen," *Zeitschrift für Slawistik* 30, no. 6 (1985): 916–18. Quote is author's translation.
36 Brinkel, "Institutionalizing *Volkskunde*," 152.
37 Brinkel, "Institutionalizing *Volkskunde*," 152–7.
38 "Deutsche demokratische Volkslieder. Ehrliche, schlichte Songs voll ruhrender Klarheit. Steinitz als Wegbereiter der deutsch-deutschen Folkszene," *Folker!*, http://archiv.folker.de/200504/01steinitz.htm.
39 Steiner, "Ein Intellektueller?," 92–107.
40 Hoffmann, "Fritz Lange," 414.
41 Lederer, "Adventures in Research," 480.
42 "Report of the Sub-Committee on Chemical Warfare in Vietnam," 338–67; Lederer, "Report on Chemical Warfare in Vietnam," 203–25.
43 Lederer, "Adventures in Research," 481–2.
44 Ling, "Ernst Emsheimer," 428.
45 Tagg, "The Goteborg Connection," 237n31.
46 Fussl, "Pestalozzi in Dewey's Realm?"
47 Fussl, "Pestalozzi in Dewey's Realm?," 85.
48 Rhode-Juchtern, "Charlotte Schlesinger." See also *Black Mountain College Bulletin/Bulletin-Newsletter* 5, no. 1 (November 1946), 1, Black Mountain College North Carolina Digital Collection (BMC).
49 *Black Mountain College Community Bulletin*, November 1, 1946; Bulletin 2, p. 8, BMC.
50 Harris, *The Art of Black Mountain College*, 117–19, 163–4.
51 Rhode-Juchtern, "Charlotte Schlesinger."
52 Timmermann, "Bremen, 16. Mai 2009," 372–3.
53 "Vocal and Chamber Music – Bosmans, H. / Weigl, V. / Schlesinger, C. / Kaprálová, V. (EntArteOpera Festival) (Haselbock, Bartolomey, Zeilinger)," Gramola 99183, 1 February 2019.
54 *Hermann Borchardt Werke Band 1*, 302.
55 Hesse, "Beschrieb er der Menschheit," 530–2.
56 Hesse, "Beschrieb er der Menschheit," 531. Author's translation.
57 F. Borchardt, "Hermann Borchardt."
58 The following account is based on Hesse, but Frank Borchardt believes his father wrote the first two acts of Toller's play and that *The Brethren of Halberstadt* came only later. It is interesting to note that none of the studies of Toller make any mention of Borchardt.
59 Frank Borchardt, "Bourchardt." Author's translation.
60 Orville Prescott, "Book of the Times," *New York Times*, 30 July 1943, 13. See also Maxwell Geismar, "Antichrist, Born of Machines," *NYT*, 25 July 1943, BR1; and Joseph Freeman, "Melodrama and Metaphysics," *The New Republic*, 9 August 1943, 205–6.
61 Mann as qtd in Joachim Kersten, "Porträt."

62 Borchardt, *Communism and Fascism.*

63 Frank Borchardt discusses his father's interactions with Booth Luce.

64 As qtd in Kersten, "Porträt." Author's translation.

65 As qtd by Frank Borchardt.

66 Borchardt, *Die Verschworung der Zimmerleute.*

67 In addition to the collected correspondence between Borchardt and Georg Grosz cited above, as of April 2021 this group has published the first volume of an intended four- volume collection of Borchardt's collected works. Herausgegeben von Hermann Haarmann, Hesse, and Laier, eds., *Hermann Borchardt Werke Band 1.*

68 Koestler, *The Invisible Writing,* 477.

69 Scammell, *Koestler* 158–9, 190.

70 Koestler, *The Invisible Writing,* 480.

71 Poulain, "A Cold War Best-Seller."

72 Kubby, "In the Shadow of The Concentration Camp," 154–5.

73 For the background to the trial see Kuby, "In the Shadow of The Concentration Camp," 165–70.

74 The testimony and events of the trial are from *La Bataille de David Rousset, T.Bernard – G. Rosenthal, pour la vérité sur les camps concentratinnaires* (Paris: Editions Du Pavios, 1950). Author's translation.

75 On 8 February 1951, Koestler, with a group of American intellectuals, including the historian Arthur Schlesinger Jr, the theologian Reinhold Niebuhr, and the socialist-pacifist Norman Thomas, wrote a letter to the *New York Times* complaining about the lack of coverage of a trial they said was more important than the Dreyfus trial.

76 Koestler in Weissberg, *The Accused,* Kindle, loc. 165 of 10573.

77 Judith Szapor wrote: "Despite its merits as one of the earliest eyewitness testimonies and analyses of the Gulag – perhaps because Weissberg refused to toe the anti-Communist line – the book sold poorly and *received disappointing reviews.*" The vast bulk of the reviews, however, were positive. Szapor, "Private Archives," 105. Emphasis added.

78 David Dallin, "Victim of the 'Conveyor': Unscathed," *New York Times,* 30 December 1951.

79 Graubard, Review of *The Accused.*

80 Robert Conquest in *The Great Terror* made extensive use of Weissberg's book. Among many recent works to use both books is Halfin, *Stalinist Confessions.* See also Gorelik, "The Paternity of the H-Bombs."

81 Lasky, "Why the Kremlin Extorts Confessions," 1.

82 Barnhisel, "Cold Warriors of the Book," 206.

83 Nye, *Michael Polanyi,* 200.

84 Nye, *Michael Polanyi,* 211.

85 Berghahn, *America and Intellectual Cold Wars,* 133.

86 For more on Polanyi and Science Studies see Aronova, "The Congress for Cultural Freedom."

87 Alexander Weissberg, "Russia's Three Bears: Maneuvers of Malenkov, Beria, and Bulganin," *The Freeman*, 18 May 1953.

88 Gorelik, "The Paternity of the H-Bombs," 183.

89 Oppenheimer as qtd in Polenberg, ed., *In the Matter of J. Robert Oppenheimer*, 15.

90 Ginzburg as qtd in Gorelik, "The Paternity of the H-Bombs," 183.

13. The Long Ordeal

1 List of Manifest SS *Britannic*, 8 July 1939; Germany, Index of Jews Whose Nationality Was Annulled by Nazi Regime, 1935–1944, ancestory.com; email from Margaret Noether Stevens, 11 June 2019.

2 Henderson, *Sons and Soldiers*.

3 Interview with Art Kharlamov and Mica Nava, 4 December 2018; Art Khalamov to author, 28 December 2018; Nava, *Visceral Cosmopolitanism*, 139–41.

4 The accounts of Victoria and Hans Hellmann Jr's lives are from Schwartz and Sabine, *Hans G.A. Hellmann*, 23; Hellmann, "Lebenslauf von Hans Hellmann," 13–15; Petra Netter to author, 27, 28, and 31 May 2020.

5 Hans G.A. Hellmann and the Hellmann-Award, http://www.tc.chemie.uni-siegen.de/hellmann.

6 Hans Hellmann Memorial Award, University Veterinary Medicine at Hannover website, https://www.tiho-hannover.de/en/university/prizes-and-honors/hans-hellmann-memorial-award.

Conclusion: The Ensnared and History

1 Hollander, *Political Pilgrims*, 109.

2 See, for example, Zimmerman, "Competitive Cooperation"; and Leff, *Well Worth Saving*.

3 Frank and Reinisch, eds., *Refugees in Europe*.

4 Lässig, "Strategies and Mechanisms," 774.

5 For an example of recent studies of the social function of science, see Brown, *Bernal*, 158–65; and Nye, *Michael Polanyi*, 184–222.

6 Adam Kirsch, "The Desperate Plight Behind 'Darkness at Noon,'" *The New Yorker*, 30 September 2019.

7 See, for instance, Zuehlsdorff et al., "Nonlinear Spectroscopy"; and Davies, Kemp, Warner, et al., "Variations in Duschinsky Rotations."

Bibliography

Archival Holdings

American Institute of Physics Oral History Collection, Emilio Segre Visual
Archive, and Digital Collections, College Park Maryland
Arolsen Archives, International Center on Nazi Persecution, Bad Arolsen,
Germany
Bodleian Library, Oxford University (Bod), Papers of the Society for the
Protection and Learning (SPSL) and the Academic Assistance Council (AAC)
Jewish Transmigration Bureau Deposit Cards, 1939–1954 (JDC)
J.G. Crowther Papers, Special Collections, University of Sussex
London School of Economics, Beveridge Papers
National Archives United Kingdom, Kew, MI-5 Files (KV-2)
New York Public Library (NYPL), Records of the (American) Emergency
Committee in Aid of Displaced Foreign (before 1938 German) Scholars (AEC)
Private Archive of George Sadowsky
Private Archive of Mica Nava
Private Archive of Pam Zinnemann-Hope
Royal Society Archives, London, Sir Francis (Franz) Simon Papers
State Archive of Kharviv Province (*Derzhavnyi arkhiv Kharkivskoi oblasti*), NKVD
Interrogation Files
Wiener Library, London
Yad Vashem, Jerusalem, Central Database of Shoah Victims' Names

Published Primary Document Collections

Guido Beck. *Guido Beck: The Career of a Theoretical Physicist Seen through His
Correspondence.* Edited by Antônio Augusto Passos Videira and Carlos F. Puig.
Rio de Janeiro: LF Editoria, 2020.

Hermann Borchardt George Grosz: "Lass uns das Kriegsbeil bergraben!" Der Briefwechsel
(*Hermann Borchardt George Grosz: "Let's dig up the hatchet!" The Correspondence*).
Edited by Hermann Haarman and Christoph Hesse. Gottingen: Wallstein
Verlag, 2019.

Hermann Borchardt Werke Band 1: Autobiographische Schriften. Edited by Herausgegeben
von Hermann Haarmann, Christoph Hesse, and Lukas Laier. Gottingen:
Wallstein Verlag GmbH, 2021.

Oral History Interviews

Guido Beck

Guido Beck, Oral History Interview by John Heilbron, 22 April 1967, American
Institute Physics Oral History Collection, https://www.aip.org/history-programs
/niels-bohr-library/oral-histories/4500.

Siegfried Gilde

Bill Chicurel (Gilde's great-nephew), interview with author, 13 November 2019.

Ruth Marx Chicurel (Gilde's niece), oral history interview, *Shoah: Survivors and
Witnesses in Western North Carolina,* n.d., http://toto.lib.unca.edu
/findingaids/oralhistory/SHOAH/chicurel.htm.

Edgar Lederer

Entretien avec Edgar Lederer, by Jean-François Picard et Elisabeth Pradoura,
19 March 1986, http://www.histcnrs.fr/archives-orales/lederer.html. English
translation by author.

Laszlo Tisa

Interview of Laszlo Tisza by Kostas Gavroglou on 15 November 1987, Niels Bohr
Library and Archives, American Institute of Physics, College Park, Maryland,
www.aip.org/history-programs/niels-bohr-library/oral-histories/4915-1.

Tisza, Laszlo. "Adventures of a Theoretical Physicist, Part I: Europe," *Physics in
Perspective* 11 (2009): 46–97.

– "Adventures of a Theoretical Physicist, Part II: America," *Physics in Perspective*
11 (2009): 120–68.

Konrad Weisselberg

Art Kharlamov and Mica Nava (Weisselberg's grandson and niece), interview
with author, 4 December 2018.

Other Oral History Interviews

Fritz Reiche – Session III, American Institute of Physics Oral History Collection, https://www.aip.org/history-programs/niels-bohr-library/oral-histories/4841-3.
Interview of Betty Compton by Charles Weiner on 11 April 1968, Niels Bohr Library and Archives, American Institute of Physics, College Park, Maryland.
Interview of Victor Weisskopf by Thomas S. Kuhn and John L. Heilbron, 10 July 1963, Niels Bohr Library and Archives, American Institute of Physics, College Park, Maryland, www.aip.org/history-programs/niels-bohr-library /oral-histories/4944.
"A Scientist's Odyssey with Victor F. Weisskopf," *Conversations: A Video Series on International Affairs*, Institute of International Studies, University of California, Berkeley, YouTube, https://www.youtube.com/watch?v=utZY2vlgphc.

Other Sources

Abel, Hubert, et al. "Tribute to Ernst Simonson (1898–1974)." *Electrocardiology Physiological, Pathophysiological and Diagnostical Research, Proceedings of the 1st International Congress on Electrocardiology* (Wiesbaden, 14–17 October 1974), ix–xvii.
Alter, Peter, ed. *Out of the Third Reich: Refugee Historians in Post-War Britain*. London: I.B. Tauris, 1998.
Amaldi, Edoardo. *The Adventurous Life of Friedrich Georg Houtermans, Physicist (1903–1966)*. Heidelberg: Springer, 2007.
Aronova, Elena. "The Congress for Cultural Freedom, *Minerva*, and the Quest for Instituting 'Science Studies' in the Age of Cold War." *Minerva* 50 (2012): 307–37.
Ash, Mitchell, and Alfons Sollner, eds. *Forced Migration and Scientific Change after 1933*. Cambridge: Cambridge University Press, 1996.
Balibar, Sébastien. *Savant cherche refuge* [Scholar seeks refuge]. Paris: Odile Jacob, 2019.
Barnhisel, Greg. "Cold Warriors of the Book: American Book Programs in the 1950s." *Book History* 13 (2010): 185–217.
Beck, F., and W. Godin. *Russian Purge and the Extraction of Confession*. New York: Viking Press, 1951. (This book was actually co-authored by Fritz Houtermans and Russian historian Konstantin Shteppa.)
Beck, Guido. *Guido Beck: The Career of a Theoretical Physicist Seen through His Correspondence*. Edited by Antônio Augusto Passos Videira and Carlos F Puig. Rio de Janeiro: LF Editoria, 2020.
Bederson, Benjamin. "Fritz Reiche and the Emergency Committee in Aid of Displaced Foreign Scholars." *Physics in Perspective* 7 (2005): 453–72.

Bell, Stephen. "Prelude to Brazil: Leo Waibel's American Career as a Displaced Scholar." *Geographical Review* 106, no. 1 (January 2016): 5–27.

Bentwich, Norman. *The Rescue and Achievement of Refugee Scholars: The Story of Displaced Scholars and Scientists, 1933–1952.* The Hague: Martinus Nijhoff, 1953.

Berghahn, Volker. *America and Intellectual Cold Wars in Europe.* Princeton: Princeton University Press, 2001.

Bernal, J.D. *The Social Function of Science.* London: Faber and Faber, 1939.

Bernhardt, Hannelore. "Gerhard Harig (1902–1966): Life and Work in Turbulent Times." In *Werk und Wirken von Gerhard Harig und Walter Hollitscher* (Work and Lives of Gerhard Harig and Walter Hollitscher). Leipzig: Rosa-Luxemburg-Stiftung Sachsen, 2004.

Bertolotti, Mario. *The History of the Laser.* Bristol: Institute of Physics, 1999.

Beveridge, Lord. *A Defence of Free Learning.* London: Oxford University Press, 1959.

"Birlad." *Encyclopedia of Jewish Communities in Romania*, vol. 1 (Romania). Translation of "Birlad" chapter from *Pinkas Hakehillot Romania.* (Jerusalem: Yad Vashem, 1969). http://www.jewishgen.org/Yizkor/pinkas_romania/rom1_00017.html.

Blackburn, Henry, and Pentii Rautaharju. "Obituary: Ernst Simonson, M.D., 1898–1974." *European Journal of Cardiology* 3, no. 1 (1975): 77–9.

Borchardt, Frank. "Hermann Borchardt." In *Deutschsprachige Exilliteratur seit 1933. Band 2 (Studien zur deutschen Exilliteratur).* Edited by John M. Spalek, 120–31. New York: Saur Verlag, 1991.

Borchardt, Hermann. *Communism and Fascism: Two of a Kind.* Pamphlet No. 13 in the Series on Communism. New York: Catholic Information Society, 1947.

– *Die Verschwörung der Zimmerleute. Rechenschaftsbericht einer herrschenden Klasse.* Bonn: Weidle Verlag, 2005.

– *I Was a Teacher in Soviet Russia.* New York: Catholic Information Society, 1947.

– "My Last Days in Soviet Russia." *Catholic World*, April 1945, 41–4.

– *The Soviet Caste System.* New York: Catholic Information Society, 1947.

Breitman, Richard, and Alan Kraut. *American Refugee Policy and European Jewry, 1933–1945.* Bloomington: Indiana University Press, 1987.

Brinkel, Teresa "Institutionalizing *Volkskunde* in Early East Germany." *Journal of Folklore Research* 46, no. 2: 141–72.

Brinson, Charmian. "The Strange Case of Dora Fabian and Mathilde Wurm." *German Life and Letters* 45, no. 4 (October 1992): 323–44.

Brinson, Charmian, and Richard Dov. *A Matter of Intelligence: MI5 and the Surveillance of Anti-Nazi Refugees.* Oxford: Oxford University Press, 2014.

Brown, Andrew. *J.D. Bernal: The Sage of Science.* Oxford: Oxford University Press, 2006.

Caron, Vicki. *Uneasy Asylum: France and the Jewish Refugee Crisis, 1933–42.* Stanford: Stanford University Press, 1999.

Castles, S., H. Crawley, and S. Loughna. "States of Conflict: Causes and Patterns of Forced Migration to the EU and Policy Responses." London: Institute of Public Policy Research, 2003.

Caute, David, *The Fellow-Travellers: Intellectual Friends of Communism.* New Haven: Yale University Press, 1988.

Cesarani, David. *Arthur Koestler: The Homeless Mind.* New York: Free Press, 1998.

Chitralla, Birgit. "Das Franz-Mehring-Institut – zentrale Institution der Weiterbildung für das marxistisch-leninistische Grundlagenstudium" (The Franz-Mehring-Institut – Central Institution for Advanced Training for Basic Marxist-Leninist Studies). In *War der Wissenschaftliche Kommunismus eine Wissenschaft? Vom Wissenschaftlichen Kommunismus zur Politikwissenschaft* (Was Scientific Communism a Science? From Scientific Communism to Political Science), 93–115. Wiesbaden: Springer, 1993.

Cohn, Susan. "In Defence of Academic Women Refugees: The British Federation of University Women." In *In Defence of Learning: Academic Refugees – Their Plight, Persecution and Placement, 1933–1980s.* Edited by Shula Marks, Paul Weindling, and Laura Wintour, 161–76. London: British Academy, 2011.

Connelly, John. *The Sovietization of East German, Czech, and Polish Higher Education.* Chapel Hill: University of North Carolina Press, 2000.

– "Ulbricht and the Intellectuals," *Contemporary European History* 6, no. 3 (1997): 329–59.

Conquest, Robert. *The Great Terror.* Oxford: Oxford University Press, 1990.

Crawford, Sally, Katharina Ulmschneider, and Jas Elsner, eds. *Ark of Civilisation: Refugee Scholars and Oxford University, 1930–45.* Oxford: Oxford University Press, 2017.

Croddy, Eric, C. Perez-Armendariz, and J. Hart. *Chemical and Biological Warfare: A Comprehensive Survey for the Concerned Citizens.* New York: Copernicus Books, 2002.

Crowther, J.G. *Industry and Education in Soviet Russia.* London: W. Heinemann, 1932.

– *Soviet Science.* London: Keagan Paul, 1936.

Daum, Andreas W., Hartmut Lehmann, and James J. Sheehan, eds. *The Second Generation: Emigrés from Nazi Germany as Historians.* New York: Berghahn Books, 2016.

Davies, Alexander R., David J. Kemp, Lewis G. Warner, et al. "Variations in Duschinsky Rotations in m-Fluorotoluene and m-Chlorotoluene during Excitation and Ionization." *Journal of Chemical Physics* 152, no. 21 (June 2020). doi:10.1063/5.0009391.

de Hass, E. "Migration Theory – Quo Vadis?" *International Migration Institute Working Papers* (November 2014).

Demidov, S.S., and B.V. Levshin. *The Case of Academician Nikolai Nikolaevich Luzin.* Providence: American Mathematical Society, 2016.

Dethick, Janet Kinrade. *The Liberation of Lake Trasimeno*. Lulu.com, n.d.

Diamond, H. *Fleeing Hitler: France 1940*. Oxford: Oxford University Press, 2007.

Displaced German Scholars: A Guide to Academics in Peril in Nazi Germany during the 1930s [1936]. San Bernardino: Borgo Press, 1993.

Duggan, Stephen, and Betty Drury. *The Rescue of Science and Learning: The Story of the American Emergency Committee in Aid of Displaced Foreign Scholars*. New York: Macmillan, 1948.

Duschinsky, F. "The Importance of the Electron Spectrum in Multi-Atomic Molecules: Concerning the Franck-Condon Principle." *Acta Physicochimica URSS* 7, no. 4 (1937): 551–66.

Dwork, Deborah, and Robert Jan Van Pelt. *Flight from the Reich: Refugee Jews, 1933–1946*. New York: W.W. Norton, 2009.

Eckert, R. "Wolfgang Steinitz und die Vermittlung des Russischen." *Zeitschrift für Slawistik* 30, no. 6 (January 1985): 915–20.

Edgcomb, Gabrielle Simon. *From Swastika to Jim Crow: Refugee Scholars at Black Colleges*. New York: Krieger, 1993.

Ege, Ragip, and Harald Hagemann. "The Modernisation of the Turkish University after 1933: The Contributions of Refugees from Nazism." *European Journal of the History of Economic Thought* 19, no. 6 (2012): 944–75.

Elie, J. "Histories of Refugees and Forced Migration Studies." In *The Oxford Handbook of Refugee and Forced Migration Studies*. Edited by E. Fiddian-Qasmiyeh, G. Loescher, K. Long, and N. Sigona, 22–35. Oxford: Oxford University Press: 2014.

"Exclusion until Today? Musicology and Exile Research: Symposium on the 100th Birthday of the Composer and Music Teacher Charlotte Schlesinger (1909 Berlin–1976 London)." *Die Musikforschung*, 62, Jahrg., H. 4 (Oktober–Dezember 2009): 372–3.

Falk, Barbara. *Caught in a Snare: Hitler's Refugee Academics*. Melbourne: University of Melbourne Press, 1998.

Fiddian-Qasmiyeh, Elena, Gil Loescher, Katy Long, and Nando Sigona. "Introduction: Refugee and Forced Migration Studies in Transition." In *Oxford Handbook of Refugee and Forced Migration Studies*, 1–15. Oxford: Oxford University Press, 2014.

Firsov, Fridrikh. "Dimitrov, the Comintern and Stalinist Repression." In *Stalin's Terror High Politics and Mass Repression in the Soviet Union*. New York: Palgrave Macmillan, 2003.

Fisher, Harold Henry. *The Communist Revolution: An Outline of Strategy and Tactics*. Stanford: Stanford University Press, 1955.

Fitas, Augusto, and Antonio Videira. *Cartas entre Guido Beck e Cientistas Portugueses* (Letters between Guido Beck and Portuguese Scientists). Lisbon: Instituto Piaget, 2004.

Fitzpatrick, Sheila. *Everyday Stalinism: Ordinary Life in Extraordinary Times – Soviet Russia in the 1930s.* New York: Oxford University Press, 1999.

Fleck, Christian. "Austrian Refugee Social Scientists." In *In Defence of Learning: Academic Refugees – Their Plight, Persecution, and Placement, 1933–1980s.* Edited by Shula Marks, Paul Weindling, and Laura Wintour, 127–42. London: The British Academy, 2011.

– "Emigration of Social Scientists Schools from Austria." In *Forced Migration and Scientific Change: Emigre German-Speaking Scientists and Scholars after 1933.* Edited by Mitchell G. Ash and Alfons Söllner. Washington, D.C.: German Historical Institute, 1996.

– *Etablierung in der Fremde. Vertriebene Wissenschaftler in den USA nach 1933* (Establishing in a Foreign Country: Displaced Scientists in the USA after 1933). Frankfurt am Main: Campus Verlag, 2015.

Frank, M., and J. Reinisch. *Refugees in Europe: A Forty Years' Crisis?* London: Bloomsbury, 2017.

Frank, Tibor. "Organizing Rescue Operation in Europe and the United States." In *In Defence of Learning: Academic Refugees – Their Plight, Persecution, and Placement, 1933–1980s.* Edited by Shula Marks, Paul Weindling, and Laura Wintour, 143–60. London: The British Academy, 2011.

Franks, Norman, and Frank Baily. Above the Lines: The Aces and Fighter Units of the German Air Service, Naval Air Service, and Flanders Marine Corps, 1914–1918. London: Grubb Street, 2008.

Franzén, Johan. "Communism versus Zionism: The Comintern, Yishuvism, and the Palestine Communist Party." *Journal of Palestinian Studies* 36, no. 2 (2007): 6–24.

Frenkel, Victor Ya. "Professor Friedrich Houtermans: Works, Life, Fate." In *Physics in a Mad World.* Edited by Shifman Misha. Singapore: World Scientific, 2116. Kindle Edition.

Friedman, Jerome, Thomas J. Greytak, and Daniel Kleppner. "Laszlo Tisza." *Physics Today* 62, no. 7 (2009).

Füssl, Karl-Heinz. "Pestalozzi in Dewey's Realm? Bauhaus Master Josef Albers among the German-Speaking Emigrés' Colony at Black Mountain College (1933–1949)." *Paedagogica Historica* 42, nos. 1–2 (2006): 78–80.

Gatrell, Peter. *The Making of the Modern Refugee.* Oxford: Oxford University Press, 2013.

– "Refugees – What's Wrong with History?" *Journal of Refugee Studies* 30, no. 2 (2017).

– *A Whole Empire Walking: Refugees in Russia during World War 1.* Bloomington: Indiana University Press, 1999.

Gerstengarbe, Sybille. *"Die erste Entlassungswelle von Hochschullehrern deutscher Hochschulen aufgrund des Gesetzes zur Wiederherstellung des Berufsbeamtentums vom 7.4.1933"* (The First Wave of Dismissals of Professors from German

Universities Due to the Law for the Restoration of the Civil Service of 7.4.1933). *Berichte zur Wissenschaftsgeschichte* 17: 17–40.

Gilde, Siegfried. "Vergleichende Versuche über die Kreislaufwirkung der Adenylsäure und des 'Eutonon'" (Comparative Experiments on the Circulatory Effect of Adenylic Acid and "Eutonon"). Berlin: Friedrich-Wilhelms-Universität, 1932.

Gilmour, John. *Sweden, the Swastika, and Stalin: The Swedish Experience in the Second World War.* Edinburgh: Edinburgh UP, 2011.

Ginzburg, Vitaly L. "Notes of an Amateur Astrophysicist." *Annual Review of Astronomy and Astrophysics* 28, no. 1 (1990): 1–37.

Glad, John. "Hermann J. Muller's 1936 Letter to Stalin." *Mankind Quarterly* 43, no. 3 (Spring 2003): 305–19.

Goldberg, Stanley. "Inventing a Climate of Opinion: Vannevar Bush and the Decision to Build the Bomb." *Isis* 83 (1992): 429–52.

Gorelik, Gennady. "The Paternity of the H-Bombs: Soviet–American Perspectives." *Physics in Perspective* 11 (2009): 169–97.

Gottvald, Ales, and Mikhail Shifman. *George Placzek: A Nuclear Physicist's Odyssey.* Singapore: World Scientific, 2018.

Gowing, Margaret. *Britain and Atomic Energy, 1939–1945.* London: St Martin's Press, 1964.

Grant, Andrew. "The Tragic Story of Hans Hellmann." *Physics Today*, 28 September 2018.

Graubard, Mark. "Review of *The Accused*." *Isis* 44, nos. 1–2 (June 1953): 93–4.

Graziosi, Andrea. "Foreign Workers in Soviet Russia, 1920–40: Their Experience and Their Legacy." *International Labor and Working-Class History* 33 (Spring 1988): 39.

Greytak, Thomas J., and Daniel Kleppner. "Laszlo Tisza." *Physics Today* 62, no. 7 (2009): 65.

Griffin, Allan. "Laszlo Tisza (1907–2009): An Appreciation." *Journal of Low Temperature Physics* 157 (2009): 1–5.

Haarmann, Herausgegeben von Hermann, Christoph Hesse, and Lukas Laier, eds. *Hermann Borchardt Werke Band 1: Autobiographische Schriften.* Gottingen: Wallstein Verlag, 2021.

Hagemann, Harald, and Claus-Dieter Krohn, eds. *Biographisches Handbuch der deutschsprachigen wirtschaftswissenschaftlichen Emigration nach 1933* (Biographical Handbook of German-speaking Economic Emigration after 1933), 2 vols. Munich, 1999.

Halfin, Igal. *Stalinist Confessions: Messianism and Terror at the Leningrad Communist University.* Pittsburgh: University of Pittsburgh Press, 2009.

Hall, Karl. "The Schooling of Lev Landau: The European Context of Postrevolutionary Soviet Theoretical Physics." *Osiris* 23, no. 1 (2008): 230–59.

Harris, James. *The Great Fear: Stalin's Terror of the 1930s.* Oxford: Oxford University Press, 2016.

Harris, Mary Emma. *The Art of Black Mountain College.* Cambridge, MA: MIT Press, 1987.

Harris, Sarah Miller. *The CIA and the Congress for Cultural Freedom in the Early Cold War: The Limits of Making Common Cause.* London: Routledge, 2016.

Havas, Peter. "The Life and Work of Guido Beck: The European Years: 1903–43." In *Guido Beck Symposium: Rio de Janeiro, August 29–31, 1994.* Edited by H.M. Nussenzveig and Antonio Augusto Passos Videira, 11–17. Rio de Janeiro: Academia Brasileira de Ciência, 1996.

Helmann, Hans Jr. "Lebenslauf von Hans Hellmann" (Hans Hellmann's CV). In *Einführung in die Quantenchemie* (Introduction to Quantum Chemistry). Heidelberg: Springer Spektrum, 2015.

Henderson, Bruce. *Sons and Soldiers.* New York: William Morrow, 2017.

Hesse, Christoph. "'Beschrieb er der Menschheit ganzen Jammer': Zu Leben und Werk Hermann Borchardts" ("He Described All of Misery to Mankind": The Life and Work of Hermann Borchardt). In *Hermann Borchardt George Grosz: "Lass uns das Kriegsbeil bergraben!" Der Briefwechsel* (Hermann Borchardt George Grosz: "Let's dig up the hatchet!": The Correspondence). Edited by Hermann Haarman and Christoph Hesse, 520–60. Gottingen: Wallstein Verlag, 2019.

Hildebrandt, Sabine. "Anatomy in the Third Reich: Careers Disrupted by National Socialist Policies." *Annals of Anatomy* 194 (2012): 251–66.

Hoffmann, David. *The Stalinist Era.* Cambridge: Cambridge University Press, 2018.

Hoffmann, Dieter. "Fritz Lange, Klaus Fuchs, and the Remigration of Scientists to East Germany." *Physics in Perspective* 11, no. 4 (December 2009): 409–10.

Hollander, Paul. *Political Pilgrims: Western Intellectuals in Search of the Good Society.* London: Routledge, 1998.

Hopwood, Nick. "Biology between University and Proletariat: The Making of a Red Professor." *History of Science* 35 (1997): 367–424.

Hoßfeld, Uwe. "Schaxel, Julius." In *Neue Deutsche Biographie* 22 (2005), S. 597–8.

Houtermans, Friedrich Georg. "Chronological Report of My Life in Russian Prisons." In Shifman, *Physics in a Mad World,* Kindle Edition.

Howell, James F. "The Hagiography of a Secular Saint: Alexander Von Humboldt and the Scientism of the German Democratic Republic." *German Quarterly* 91, no. 1 (Winter 2018): 67–82.

Hugh-Jones, Martin. "Wickham Steed and German Biological Warfare Research." *Intelligence and National Security* 7, no. 4 (1992): 379–402.

Ilic, Melanie. *Soviet Women – Everyday Lives.* London: Routledge, 2020.

Ings, Simon. *Stalin and the Scientists: A History of Triumph and Tragedy, 1905–1953.* New York: Atlantic Monthly Press, 2017.

Johnson, Alvin. "The New School of Social Research." In *The Rescue of Science and Learning: The Story of the American Emergency Committee in Aid of Displaced*

Foreign Scholars. Edited by Stephen Duggan and Betty Drury. New York: Macmillan, 1948.

Johnston, Timothy. *Being Soviet: Identity, Rumour, and Everyday Life under Stalin 1939–1953.* Oxford: Oxford University Press, 2011.

Jones, Nigel. "Silence, Exile, and Cunning." *History Today* 59, no. 4 (April 2009): 25–31.

Josephson, Paul, and Aleksandr Sorokin. "Physics Moves to the Provinces: The Siberian Physics Community and Soviet Power, 1917–1940." *British Journal for the History of Science* 50, no. 2 (June 2017): 297–327.

Jug, Karl, Wolfgang Ertmer, Joachim Heidberg, Manfred Heinemann, and W.H. Eugen Schwarz. "Hans Hellmann: Pionier der modernen Quantenchemie" (Hans Hellmann: Pioneer of Modern Quantum Chemistry). *Chem. Unserer Zeit* 37 (2004): 412–21.

Kersfeld, Daniel. *La migración judía en Ecuador. Ciencia, cultura y exilio 1933–1945* (Jewish Migration in Ecuador: Science, Culture, and Exile 1933–1945). Quito: Academia Nacional de Historia, 2018.

Kersten, Joachim. "Porträt: Der Verlorene: Er war Jude, Protestant, Kommunist, Katholik)" (Portrait: The Lost One: He Was a Jew, a Protestant, a Communist, a Catholic.) *Die Zeit*, 26 February 2004.

Khriplovich, Iosif. "The Eventful Life of Fritz Houtermans." *Physics Today*, July 1992, 29–37.

Kleist, J. Olaf. "The History of Refugee Protection: Conceptual and Methodological Challenges." *Journal of Refugee Studies* 30, no. 2 (2017): 161–9.

Knox, K., and T. Kushner. *Refugees in an Age of Genocide: Global, National, and Local Perspectives during the Twentieth Century.* London: Frank Cass, 1999.

Koestler, Arthur. *The Invisible Writing.* New York: Vintage Classics, 1952.

Kojevnikov, Alexie. *Stalin's Great Science: The Times and Adventures of Soviet Physicists.* London: Imperial College Press, 2004.

Kovner, Michail A. "Hans Hellmann of the Hellmann–Feynman Theorem." In *Culture of Chemistry: The Best Articles on the Human Side of 20th-Century Chemistry from the Archives of the Chemical Intelligencer.* Edited by B. Hargittai and I. Hargittai, 31. New York: Springer Science+Business Media, 2015.

Kreft, Gerald. "'Dedicated to Representing the True Spirit of the German Nation in the World': Phillip Schwartz (1894–1977), Founder of the *Notgemeinschaft*." In *In Defence of Learning: Academic Refugees – Their Plight, Persecution, and Placement, 1933–1980s.* Edited by Shula Marks, Paul Weindling, and Laura Wintour, 127–42. London: The British Academy, 2011.

Krohn, Claus-Dieter. *Intellectuals in Exile: Refugee Scholars and the New School for Social Research.* Amherst: University of Massachusetts Press, 1993.

Kubby, Emma. "In the Shadow of the Concentration Camp: David Rousset and the Limits of Apoliticism in Postwar French Thought." *Modern Intellectual History* 11, no. 1 (2014): 153–75.

Kuenzli, Andreas. "Dem Stalinismus hoffnungslos ausgeliefertEine KGB-Opferlistefür die sowjetischen Esperantisten: Die Tragödie der„bolschewistischen Sprache" in Sowjetrussland," February 2013, http://www.plansprachen.ch/Esperanto_Stalinismus_Sowjetunion_1920-30er.pdf.

Kushner, Tony. *Journeys from the Abyss: The Holocaust and Forced Migration from the 1880s to the Present.* Liverpool: Liverpool University Press, 2017.

Lamberti, Marjorie. "The Reception of Refugee Scholars from Nazi Germany in America: Philanthropy and Social Change in Higher Education." *Jewish Social Studies* 12, no. 3 (Spring–Summer 2006) (New Series): 160–84.

Lang, Ewald. "Wolfgang Steinitz' Russian Textbook: A Reminiscence Summary." *Zeitschrift für Slawistik* 50, no. 2 (June 2005), https://doi.org/10.1524/slaw.2005.50.2.199.

Lasky, Melvin J. "Why the Kremlin Extorts Confessions." *Commentary* 13 (January 1952), 1.

Lässig, Simone. "Strategies and Mechanisms of Scholar Rescue: The Intellectual Migration of the 1930s Reconsidered." *Social Research: An International Quarterly* 84, no. 4 (Winter 2017): 769–807.

Lederer, Edgar. "Adventures in Research." *Selected Topics in the History of Biochemistry: Personal Recollections.* Edited by G. Semenza, special issue of *Comprehensive Biochemistry* 36 (1985): 442–3.

– "Report of the Sub-Committee on Chemical Warfare in Vietnam." In *Against the Crime of Silence*, compiled by the Biochemists and Nutritionists of the Region of Paris, under the Direction of Edgar Lederer, 338–67. London: Bertrand Russell Peace Foundation, 1968.

– "Report on Chemical Warfare in Vietnam." In *Prevent the Crime of Silence*, 203–25. London: Alan Lane, 1971.

Leff, Laurel. *Well Worth Saving.* New Haven: Yale University Press, 2019.

Lehmann, Hartmut, and James Sheehan, eds. *An Interrupted Past: German-Speaking Refugee Historians in the United States after 1933.* Washington, D.C.: German Historical Institute, 1991.

Leitenberg, Milton, Raymond A. Zilinskas, and Jens H. Kuhn. *The Soviet Biological Weapons Program: A History* Cambridge, MA: Harvard University Press, 2012.

Leo, Annette. *Leben als Balance-Akt: Wolfgang Steinitz, Kommunist, Jude, Wissenschaftler* (Life as a Balancing Act: Wolfgang Steinitz, Communist, Jew, Scientist). Berlin: Metropol Verlag, 2005.

Levin, Aleksey E. "Anatomy of a Public Campaign: Academician Luzin's Case in Soviet Political History." *Slavic Review* 49, no. 1 (Spring 1990): 90–108.

Levy, Saul. "A Correlation Method for the Elimination of Errors Due to Unstable Excitation Conditions in Quantitative Spectrum Analysis." *Journal of Applied Physics* 11 (1940): 480. doi:10.1063/1.1712798.

Liepman, Heinz. *Death from the Skies: A Study of Gas and Microbal Warfare*, with the scientific assistance of H.C.R. Simons. Translated from the German by Eden and Cedar Paul. London: M. Secker and Warburg, 1937.

Ling, Jan. "Ernst Emsheimer." *Ethnomusicology* 34, no. 3 (Autumn 1990): 425–8.

Lins, U. *Dangerous Language – Esperanto and the Decline of Stalinism.* Berlin: Springer, 2017.

Löhr, Isabella. "Solidarity and the Academic Community: The Support Networks for Refugee Scholars in the 1930s." *Journal of Modern European History* 12, no. 2: 231–46.

London, Louise. *Whitehall and the Jews: 1933–1948: British Immigration Policy and the Holocaust.* Cambridge: Cambridge University Press, 2000.

Lucas, W. Scott. "Revealing the Parameters of Opinion: An Interview with Frances Stonor Saunders." *Intelligence and National Security* 18 (2003): 15–40.

Marfleet, Philip. "Refugees and History: Why We Must Address the Past." *Refugee Survey Quarterly* 26, no. 3: 136–48.

Marrus, M.R. *The Unwanted: European Refugees from the First World War through the Cold War.* Philadelphia: Temple University Press, 2002.

Marx, E. "The Social World of Refugees: A Conceptual Framework." *Journal of Refugee Studies* 3, no. 3: 189–203.

McLoughlin, Barry, and Kevin McDermott. "Rethinking Stalinist Terror." In *Stalin's Terror: High Politics and Mass Repression in the Soviet Union.* New York: Palgrave Macmillan, 2003.

McRae, Kenneth. *Nuclear Dawn.* Oxford: Oxford University Press, 2014.

Medawar, Jean, and David Pyke. *Hitler's Gift: The True Story of the Scientists Expelled by the Nazi Regime.* New York: Arcade, 2000.

Mills, Eric L. *Biological Oceanography: An Early History 1870–1960.* Toronto: University of Toronto Press, 2014.

Mizelle, William. "More about Peron's Atom Plans." *New Republic,* 31 March 1947, 20–1.

– "Peron's Atomic Plans." *New Republic,* 24 February 1947, 22–3.

Moser, Petra, Alessandra Voena, and Fabia Waldinger. "German Jewish Émigrés and US Invention." *American Economic Review* 104, no. 10 (October 2014): 3222–55.

Mott, Nevill. "Herbert Frohlich. 9 December 1905–23 January 1991." *Biographical Memoirs of the Fellows of the Royal Society* (1992): 140–55.

Motulsky, Arno. "A German-Jewish refugee in Vichy France 1939–1941: Arno Motulsky's Memoir of Life in the Internment Camps at St Cyprien and Gurs." *American Journal of Medical Genetics* 176, no. 6 (June 2018): 1289–95.

Murawkin, Herbert. "Beiträge zur Theorie und Konstruktion des Kreismassenspektrographen." *Annalen der Physik* 400, no. 2 (1931).

– "Massenspektra von Gläsern, Salzen und Metallen nebst Konstruktion eines Kreismassenspektrographen." *Annalen der Physik* 400, nos. 3–4 (1931).

– *Massenspektra von Gläsern, Salzen und Metallen nebst Theorie und Konstruktion eines Kreismassenspektrographen.* Berlin, 1931.

– "Nachtrag zu: Massenspektra von Gläsern, Salzen und Metallen nebst Konstruktion eines Kreismassenspektrographen" and "Nachtrag zu: Beiträge zur Theorie und Konstruktion eines Kreismassenspektrographen." *Annalen der Physik* 401, no. 8 (1931).

Murphy, William T. "First Decade of Soviet Espionage in America: 1924 to 1933." *International Journal of Intelligence and CounterIntelligence* 34, no. 1 (2021): 45–69. Doi:10.1080/08850607.2020.1781442.

– "Soviet Espionage in France between the Wars." *International Journal of Intelligence and CounterIntelligence* (2021). Doi:10.1080/08850607.2020.1861581.

Müssener, Helmut. *Exil in Schweden: politische und kulturelle Emigration nach 1933* (Exile in Sweden: Political and Cultural Emigration after 1933). Munich: C. Hanser, 1974.

Nava, Mica. *Visceral Cosmopolitanism: Gender, Culture and the Normalisation of Difference.* Oxford: Berg, 2007.

Němysova, E.A., and Z.S. Riàbchikova. Вольфганг Штейниц–финно-угровед, исследователь хантыйского языка, *28* февраля *1905–21* апреля *1967* : к *100*-летию со дня рождения (Wolfgang Steinitz – Finno-Ugric scholar, researcher of the Khanty language, 28 February 1905–21 April 1967: To the 100th Anniversary of His Birth). Khanty-Mansiĭsk : Informatsionno-izdatel'skiĭ tsentr: 2005.

Neumark, Fritz. *Zuflucht am Bosporus: dt. Gelehrte, Politiker und Künstler in d. Emigration : 1933–1953* (Refuge on the Bosporus: German Scholars, Politicians, and Artists in Emigration: 1933–1953). Frankfurt am Main: Knecht, 1980.

Nguyen, Anzo, and Frank W. Stahnisch. "From Interned Refugee to Neuropathologist and Psychiatrist." *Journal of the History of the Neurosciences* (April 2019).

"Noether, Fritz Alexander Ernst." *Neue Deutsche Biographie* 19 (1999): 321–2.

Noether, Gottfried E. "Fritz Noether (1884–194?)." *Integral Equations and Operator Theory* 8 (1985): 573–4.

Norwood, Stephen H. "Legitimating Nazism: Harvard University and the Hitler Regime, 1933–1937." *American Jewish History* 92, no. 2 (June 2004): 200–31.

Nothnagle, Alan. "From Buchenwald to Bismark: Historical Myth-Building in the German Democratic Republic, 1945–1989." *Central European History* 26, no. 1 (1993): 91–113.

Nussenzveig, H.M. "Guido Beck." *Physics Today* 43, no. 12 (1990): 89–90.

Nussenzveig, H.M., and Antonio Augusto Passos Videira, eds. *Guido Beck Symposium: Rio de Janeiro, August 29–31, 1994.* Rio de Janeiro: Academia Brasileira de Ciência, 1996.

Nye, Mary Jo. *Michael Polanyi and His Generation: Origins of the Social Construction of Science.* Chicago: University of Chicago Press, 2011.

O'Connor, J.J., and E.F. Robertson. "Fritz Alexander Ernst Noether." MacTutor History of Mathematics archive website, 2016. https://mathshistory.st-andrews.ac.uk/Biographies/Romberg/.

– "Stefan Emmanuilovich Cohn-Vossen." MacTutor Website, 2021, https://mathshistory.st-andrews.ac.uk/Biographies/Cohn-Vossen.

– "Werner Romberg." MacTutor History of Mathematics archive website, 2013, https://mathshistory.st-andrews.ac.uk/Biographies/Romberg.

Oltmer, J. *Migration und Politik in der Weimarer Republik* (Migration and Politics in the Weimar Republic). Gottingen: Vandenhoeck & Ruprecht, 2005.

Orth, Karin. *Die NS-Vertreibung der jüdischen Gelehrten die Politik der Deutschen Forschungsgemeinschaft und die Reaktionen der Betroffenen* (The Nazi Expulsion of Jewish Scholars, the Policy of the German Research Foundation, and the Reactions of Those Affected). Göttingen: Niedersachs Wallstein Verlag, 2016.

Ortiz, Eduardo L., and Allan Pinkus. "Herman Müntz: A Mathematician's Odyssey." *Mathematical Intelligencer* 27 (December 2005): 1–16.

Pantelides, Sokrates, and Massimiliano Di Ventra. "Hellmann-Feynman Theorem and the Definition of Forces in Quantum Time-Dependent and Transport Problems." *Physical Review B* 61, no. 21 (15 June 2000): 16207.

Passos Videira, Antônio Augusto, and Carlos F. Puig, eds. *Guido Beck: The Career of a Theoretical Physicist Seen through His Correspondence.* Rio de Janeiro: LF Editoria, 2020.

Pavlenko, V., N. Raniuk, and A. Khramov. *"Delo" UFTI 1935–1938* (The UFTI Case 1935–38). Kiev: Fenicks, 1998.

Petersen, Neal H., ed. *From Hitler's Doorstep: The Wartime Intelligence Reports of Allen Dulles.* University Park: Pennsylvania State University Press, 1996.

Polanyi, Michael. "Right and Duties of Science" (1939). In *The Comtempt of Freedom.* London: Watts, 1940.

Polenberg, Richard, ed. *In the Matter of J. Robert Oppenheimer: The Security Clearance Hearing.* Ithaca: Cornell University Press, 2018.

Potier, Pierre. "Biography of Edgar Lederer." 1995. http://historique.icsn.cnrs-gif.fr/spip.php?article5.

Poulain, Martine. "A Cold War Best-Seller: The Reaction to Arthur Koestler's 'Darkness at Noon' in France from 1945 to 1950." *Libraries and Culture* 36, no. 1 (January 2001): 172–84.

Powers, Thomas. *Heisenberg's War.* Boston: Little, Brown, 1993.

Primakov, E. Очерки истории российской внешней разведки. Том 3 Примаков (Survey of the History of Russian Foreign Intelligence, vol. 3, ch. 42). https://military.wikireading.ru/21198.

Pringle, Peter. *The Murder of Nikolai Vavilov: The Story of Stalin's Persecution of One of the Great Scientists of the Twentieth Century.* New York: Simon and Schuster, 2008.

Prokop, Siegfried. "Gerhard Harig – First State Secretary for Higher Education in the GDR (1951–1957)." In *Werk und Wirken von Gerhard Harig und Walter Hollitscher* (Work and Lives of Gerhard Harig and Walter Hollitscher). Leipzig: Rosa-Luxemburg-Stiftung Sachsen, 2004.

Qian Peng, Yuanping Yi, and Zhigang Shuai. "Excited State Radiationless Decay Process with Duschinsky Rotation Effect: Formalism and

Implementation." *Journal of Chemical Physics* 126, 114302 (2007).; https://doi
.org/10.1063/1.2710274.

Rechenberg, Helmut. "Ladensburg, Rudolf." *New German Biography* 13 (1982):
391. https://www.deutsche-biographie.de/pnd116643803.html#ndbcontent.

Reichenbach, Maria Cecilia von. "Richard Gans: The First Quantum Physicist in
Latin America." *Physics Perspective* 11 (2009): 302–17.

Reinders, L.J. *The Life, Science, and Times of Lev Vasilevich Shubnikov.* Cham:
Springer, 2018.

Reiß, Christian. "No Evolution, No Heredity, Just Development – Julius Schaxel
and the End of the Evo–Devo Agenda in Jena, 1906–1933: A Case Study."
Theory Bioscience 126 (2007): 158, 162–3.

Reiß, Christian, Susan Springer, Uwe Hoßfeld, Lennart Olsson, and Georgy S.
Levit. "Introduction to the Autobiography of Julius Schaxel." *Theory Biosci* 126
(2007): 165–75.

Rhode-Jüchtern, Anna-Christine. "Charlotte Schlesinger." In *Lexikon verfolgter
Musiker und Musikerinnen der NS-Zeit* (Lexicon of Persecuted Musicians
from the Nazi Era). Edited by Claudia Maurer Zenck and Peter Petersen.
Hamburg: Universität Hamburg, 2009. See also Institute for the Historical
Musicology website, 2009. http://www.lexm.uni-hamburg.de/object/lexm
_lexmperson_00003104.

Rhodes, Richard. *The Making of the Atomic Bomb.* New York: Simon and Schuster,
1986.

Rimmington, Anthony. *Stalin's Secret Weapon: The Origins of Soviet Biological
Warfare.* London: C. Hurst, 2018.

Ritter, Gerhard A, Friedrich Meinecke, and Alex Skinn. *German Refugee Historians
and Friedrich Meinecke, 1910–1977: Letters and Documents.* Boston: Brill, 2010.

Rogacheva, Maria. *The Private World of Soviet Scientists from Stalin to Gorbachev.*
Cambridge: Cambridge University Press, 2017.

Rossi, Bruno. "Prof. Marcel Schein." *Nature* 186, 30 April 1960, 355–6.

Ruhemann, Barbara. "Die Kristallstrukturen von Krypton, Xenon,
Jodwasserstoff, und Bromwasserstoff in ihrer Abhängigkeit von der
Temperatur." PhD diss., Friedrich-Wilhelms-Universität zu Berlin, 1932.

Ruhemann, M. "Appendix VII: Notes on Science in the USSR." In *Social Function
of Science,* 443–9.

– "Kharkov in the Thirties (from Recollections)." *Fizika Nizkikh Temperatur* 18,
no. 1 (1992): 74–5.

– "Science and the Soviet Citizen." *Anglo-Soviet Journal* 1, no. 3 (1940): 220–8.

Rutkoff, Peter, and William Scott. *New School: A History of the New School of Social
Research.* New York: Free Press, 1986.

Rykov, Alexey. *Tesla protiv Eynshteyna* (Tesla vs Einstein). Moscow: Exmo, 2010.

Sadowsky, Michael. *Die räumlich-periodischen Lösungen der Elastizitätstheorie.* (The
Spatial-Periodic Solutions of the Theory of Elasticity). Berlin: A.W. Schade, 1927.

Saltzman, Martin. "Is Science a Brotherhood? The Case of Siegried Ruhemann." *Bulletin of the History of Chemistry Society* 25, no. 2 (2000): 116–21.

Samuelson, Lennart. *Tankograd: The Formation of a Soviet Company Town, Cheliabinsk, 1900s–1950s.* London: Palgrave Macmillan, 2011.

Saunders, Frances Stonor. *Who Paid the Piper? The CIA and the Cultural Cold War.* London: Granta Books, 1999.

Scammell, Michael. *Koestler: The Literary and Political Odyssey of a Twentieth-Century Skeptic.* New York: Random House, 2009.

Schein, Edgar. *Becoming American.* Bloomington: iUniverse, 2016.

Schiffer, M.M. "Stefan Bergman." In *Dictionary of Scientific Biography*, vol. 17, 71–3. New York: Charles Scribner's Sons, 1990.

Schiffer, Menahem, and Hans Samelson. "Dedicated to the Memory of Stefan Bergman." *Applicable Analysis* 8 (1979): 195–6.

Schlesinger, Charlotte. "Тетрадь, найденная на чердаке" (Notebook found in the attic), translated and edited by Misha Shifman, Номер 3(72)март 2016 года. https://7iskusstv.com/2016/Nomer3/MShifman1.php.

Schlügel, Karl. *Moscow 1937.* Cambridge: Polity Press, 2012.

Schmid, Daniel C. "A Learning Journey to Trstena According to Ed Schein's 'Organizational Cultural Model.'" In *Edgar H. Schein – The Spirit of Inquiry.* Innsbruck: Innsbruck University Press, 2019.

Schmidt, W.F. "High Voltages for Particle Acceleration from Thunderstorms: A. Brasch and F. Lange." *IEEE Transactions on Electrical Insulation* 24, no. 2 (April 1989): 222. doi:10.1109/14.90277.

Schmiedeshoff, F.W. "In Memory of Dr Michael Sadowsky." *Journal of Composite Materials* 2 (1968): 126–7.

Schneider, Albrecht. "In memoriam Ernst Emsheimer (1904–1989)." *Jahrbuch für Volksliedforschung* 35 (1990): 110–13.

Schwarz, W.H.E., D. Andraea, S.R. Arnold, J. Heidberg, J., H. Hellmann jr., J. Hinze, A. Karachaliose, M.A. Kovner, P.C. Schmidt, and L. Zülicke. "Hans G.A. Hellmann (1903–1938)," trans. Mark Smith and Schwartz, *Bunsen-Magazin* 1999 (1) 10–21, (2) 60–70, http://www.tc.chemie.uni-siegen.de/hellmann/hh-engl_with_figs.pdf.

Seabrook, Jeremy. *The Refuge and the Fortress: Britain and the Flight from Tyranny,* New York: Palgrave Macmillan, 2009.

Semelin, Jacques. *The Survival of the Jews in France, 1940–44.* Oxford: Oxford University Press, 2019.

Shifman, M., ed. *Physics in a Mad World.* Singapore: World Scientific, 2016.

– *Standing Together in Troubled Times.* Singapore: World Scientific, 2017.

Siegmund-Schultze, Reinhard. *Mathematics Fleeing from Nazi Germany: Individual Fates and Global Impact.* Princeton: Princeton University Press, 2009.

Silberklang, David. "Jewish Politics and Rescue: The Founding of the Council for German Jewry." *Holocaust and Genocide Studies* 7, no. 3 (1993): 333–71.

Simons, Dr. H.C.R. "Escape from Death." Wiener Library, 1656/3/9/60.

Simons, Hellmuth. *Beiträge zur Kenntnis der experimentellen Nagana.* Stuttgart: Buchdruckerei A. Bonz' Erben, 1918.

– "Is Disseminated Sclerosis Caused by Spirochaetes." *Deutsche Medizinische Wochenschrift* 83, no. 28 (1958): 1196–1200

– "Nachweis von Borrelien im Zentralnervensystem durch Desintegration Mittels einer Thedanblau-Kaliumchloratmethode (Tkm), Zugleich ein Technischer Beitrag zur Nachprufung der Spirochatenatiologie der Multiplen Sklerose." *Zeitschrift für Hygiene und Infektionskrankheiten* 141, no. 3 (1955): 197–217.

Simonson, Ernst. "Die wissenschaftliche arbeit in Russland: Vortrag gehalten in der Schlesischen gesellschaft für vaterländische kultur am, 1. 2. 1932" (Scientific work in Russia: Lecture given in the Silesian Society for Patriotic Culture, 1 February 1932). *Abhandlungen der Schlesischen gesellschaft für vaterländische cultur. Geisteswissenschaftliche reihe,* 8.

Skran, C. *Refugees in Inter-War Europe: The Emergence of a Regime.* Oxford: Oxford University Press, 1995.

Sollner, Alfons. "From Public Law to Political Science? The Emigration of German Scholars after 1933 and Their Influence on the Transformation of a Discipline." *Studies in History and Philosophy of Science Part C: Studies in History and Philosophy of Biological and Biomedical Sciences* 31, no. 3 (September 2000): 477–89.

Solomon, Susan Gross, ed. *Doing Medicine Together: Germany and Russia between the Wars.* Toronto: University of Toronto Press, 2006.

Sorokina, Marina Yu. "Within Two Tyrannies: The Soviet Academic Refugees of the Second World War." In *Defence of Learning: Academic Refugees – Their Plight, Persecution, and Placement, 1933–1980s,* 227–45. London: The British Academy, 2011.

SPSL (Society for the Protection of Science and Learning) Annual Reports, 1934–38, SPSL Papers 1/1–4, Bodleian Library, Oxford.

Stahnisch, Frank W. "Karl T. Neubuerger (1890–1972) – Pioneer in Neurology." *Journal of Neurology* (January 2018).

Steiner, Helmut. "Ein Intellektueller im Widerstreit mit der Macht?" (An Intellectual at Odds with Power?). In *Wolfgang Steinitz: Ich hatte unwahrscheinliches Glück. Ein Leben zwischen Wissenschaft und Politik* (Wolfgang Steinitz: I was incredibly lucky. A Life between Science and Politics). Edited by Klaus Steinitz and Wolfgang Kaschuba, 92–107. Berlin: Karl Dietz Verlag, 2006.

Steinitz, Wolfgang. *Der Parallelismus in der finnisch-karelischen Volksdichtung: Untersucht an den Liedern des karelischen Sängers Arhippa Perttunen* (The Parallelism in Finnish-Karelian Folk Poetry: Examined using the Songs of the Karelian Singer Arhippa Perttunen). Helsingfors : Druckerei der Finnischen Literatur-Gesellschaft and Tartu: K. Mattiesens Buchdruckerei, Ant.-Ges., 1934.

– *Verfolgung und Auswanderung deutschsprachiger Sprachforscher 1933–1945.*
http://zflprojekte.de/sprachforscher-im-exil/index.php/catalog/s/451
-steinitz-wolfgang.

Steinmetz, George. "Ideas in Exile: Refugees from Nazi Germany and the Failure
to Transplant Historical Sociology into the United States." *International Journal
of Politics, Culture, and Society* 23, no. 1: 1–27.

Stokes, Lawrence D. "Canada and an Academic Refugee from Nazi Germany:
The Case of Gerhard Hertzberg." *Canadian Historical Review* 52, no. 2 (June
1976): 150–70.

Stortz, Paul. "Rescue Our Family from a Living Death: Refugee Professors
and the Canadian Society for the Protection of Science and Learning, at
the University of Toronto, 1935–1946." *Journal of the Canadian Historical
Association* 14, no. 1 (2003): 231–61.

Strauss, Herbert A. "The Migration of the Academic Intellectuals." In
International Biographical Dictionary of Central European Emigres 1933–1945,
vol. 2: *The Arts, Sciences and Literature*. Edited by Herbert Strauss and Werner
Röder. Munich: K.G. Saur, 1983.

Szajkowski, Zosa. *Jews and the French Foreign Legion.* New York: Ktav, 1975.

Szapor, Judith. "Private Archives and Public Lives: The Migrations of Alexander
Weissberg and the Polanyi Archives." *Jewish Culture and History* 15, nos. 1–2
(2014): 93–109.

Tagg, Philip. "The Goteborg Connection: Lessons in the History and Politics of
Popular Music Education and Research." *Popular Music* 17, no. 2: 219–42.

Tantscher, Svetlana. "Ernst Emsheimer: Ein wissenschaftlicher Wanderer im
Schatten politischer Ideologien." https://musikwissenschaft.univie
.ac.at/fileadmin/user_upload/i_musikwissenschaft/Ueber_uns
/Institutsgeschichte/Tantscher_finale.pdf.

Timmermann, Volker. "Bremen, 16. Mai 2009: Ausgrenzung bis heute?
Musikwissenschaft und Exilforschung. Symposium zum 100. Geburtstag der
Komponistin und Musikpadagogin Charlotte Schlesinger" ("Bremen, 16 May
2009").

Tischler, Carola. *Flucht In Die Verfolgung: Deutsche Emigranten im Sowjetischen Exil,
1933 bis 1945.* Berlin: Lit Vlg. Hopf, MÆCEnst, 1996.

Tortel, Emilien. "Marseille, City of Refuge: International Solidarity, American
Humanitarianism, and Vichy France (1940–1942)." *Esboços* 28, no. 48 (August
2021): 364–85.

Tully, J.C. "Diatomics-in-Molecules." In *Semiempirical Methods of Electronic Structure
Calculation, Modern Theoretical Chemistry*, vol. 7. Edited by G.A. Segal. Springer:
Boston, 1977.

Volkov, Shulamit. *Germans, Jews, and Antisemites: Trials in Emancipation.* Cambridge:
Cambridge University Press, 2006.

Wallace, David B. "An Introduction to the Hellmann-Feynman Theory." MA thesis, University of Central Florida, 2005.

Waloschek, Pedro. *Todesstrahlen als Lebensretter – Tatsachenbericht aus dem Dritten Reich* (Death Rays as a Lifesaver – Factual Report from the Third Reich). Hamburg: Norderstedt, 2004.

Weindling, Paul. *Health, Race, and German Politics between National Unification and Nazism, 1870–1945*. Cambridge: Cambridge University Press, 1989.

– "The Impact of German Medical Scientists on British Medicine: A Case Study of Oxford, 1933–45." *Forced Migration and Scientific Change.*

– "An Overloaded Ark? The Rockefeller Foundation and Refugee Medical Scientists, 1933–45." *Studies in History and Philosophy of Science Part C: Studies in History and Philosophy of Biological and Biomedical Sciences* 31, no. 3 (September 2000): 477–89.

Weissberg, Alexander. *The Accused.* New York: Simon and Schuster: 1951.

– Letter to Marcel and Anna Weiselberg and Gertrude Wagner, n.d. but 1946, original from the collection of Mica Nava. Translated in *Physics in a Mad World.* Singapore: World Scientific, 2016. Kindle Edition.

– *Rußland im Schmelztiegel der Säuberungen Hexensabbat* (Witches Sabbath: Russia in the Melting Pot of the Purges). Frankfurt am Main: Verlag der Frankfurter Hefte, 1951.

Weisselberg, Konrad. *Beiträge zum Nachweis von Magnesium mittels Farbstoffreagentien* (Contributions to the Detection of Magnesium Using Dye Reagents). University of Vienna Library, Wien, 1930.

Wendland, Ulrike. *Biographisches Handbuch deutschsprachiger Kunsthistoriker im Exil. Leben und Werk der unter dem Nationalsozialismus verfolgten und vertriebenen Wissenschafiler* (Biographical Handbook of German-Speaking Art Historians in Exile. Life and Work of the Academics Persecuted and Expelled under National Socialism), 2 vols. Munich, 1999.

Wilford, Hugh. *The CIA, the British Left, and the Cold War: Calling the Tune?* London: Frank Cass, 2003.

Williams, M.M.R. "The Publications of Boris Davison and Some Reminiscences of the Man." *Progress in Nuclear Energy* 8 (1981): 53–75.

Yodh, Gaurang B. "Early History of Cosmic Rays at Chicago." *Centenary Symposium 2012: Discovery of Cosmic Rays* 1516 (2013): 37–45. doi:10.1063/1.4792537.

Zahradník, Rudolf. "Hans Hellmann: Zvotni Príbeh Vedce Ve 20. Století" (Hans Hellmann: A Leader's Story in the 20th Century). *Chem. Listy* 98 (2004): 98–101.

Zamfira, Andreea. "The Enthusiasm of Intellectuals for Communism at the End of First World War in France." *History of Communism in Europe* 2 (2011): 11–28.

Zegenhagen, Evelyn. "Buchenwald Main Camp." *The United States Holocaust Memorial Museum Encyclopedia of Camps and Ghettos, 1933–1945*. Bloomington: Indiana University Press, 2009.

Zeisel, Eva. *Eva Zeisel: A Soviet Prison Memoir*. Edited by Jean Richards. 2011. (Privately published and available as a Kindle.)

Zimmerman, David. "Competitive Cooperation: The Society for the Protection of Science and Learning, the American Emergency Committee, and the Placement of Refugee Scholars in North America. *Yad Vashem Studies* 46, no. 2 (December 2018): 151–78.

– "'A Narrow Minded People': Canadian Academics and the Academic Refugee Crisis, 1933–40." *Canadian Historical Review* 88, no. 2 (June 2007): 291–316.

– "'Protests Butter No Parsnips': Lord Beveridge and the Rescue of Refugee Academics from Europe, 1933–1938." In *In Defence of Learning: Academic Refugees – Their Plight, Persecution and Placement, 1933–1980s*. Edited by Shula Marks, Paul Weindling, and Laura Wintour, 29–44. London: The British Academy, 2011.

– "Review of the Second Generation: Emigrés from Nazi Germany as Historian." *History of Intellectual Culture* (Spring 2019).

– "The Society for the Protection of Science and Learning and the Politicalization of British Science in the 1930s." *Minerva* 1 (2006).

– "The Tizard Mission and the Development of the Atomic Bomb." *War in History* 2, no. 3 (1995): 259–273.

Zinnemann, Kurt. "Ordered to the Donetz Basin." n.d. but likely 1942, from Pam Zinneman-Hope's personal archive.

Zinnemann, K., and I. Zinnemann. "Incidence and Significance of the Types of Diphtheria Bacilli in the Ukraine." *Journal of Pathology and Bacteriology* 48, no. 1 (January 1939): 155–68.

"Zinnemann, K.S., DSC, MD, FRCPATH." *BMJ: British Medical Journal* 297, no. 6644 (30 July 1988): 355.

Zinnemann-Hope, Pam. *On Cigarette Papers*. London: Ward Wood, 2012.

Zolberg, A.R. "The Formation of New States as a Refugee-Generating Process." *Annals of the American Academy of Political and Social Science* 467 (1983): 24–38.

Zuehlsdorff, Tim J., Hanbo Hong, Liang Shi, et al. "Nonlinear Spectroscopy in the Condensed Phase: The Role of Duschinsky Rotations and Third Order Cumulant Contributions," *Journal of Chemical Physics* 153, no. 4 (2020), https://doi.org/10.1063/5.0013739.

Index

Aachen, 19; Technical University of, 57–8, 129

Academic Assistance Council (AAC). *See* Society for the Protection of Science and Learning

academic refugee crisis, 5–7, 10, 14, 34, 68–70, 85–6, 293

Académie des Sciences, 246

Academy of Sciences, USSR, 56, 79, 84, 127, 133, 142, 155

Accused, The, 41, 160, 276–7

Adams, Walter, 136, 202–3, 229, 232

Albers, Josef, 264

Alexandrov, Pavel, 84

All-Union Electro-Technical Institute, Moscow, 124

American Emergency Committee in Aid of Displaced Foreign (German) Scholars (AEC), 5, 34, 297; and academic refugees, 83, 88, 210, 214, 233, 235, 237, 248, 293; policies of, 12, 20, 68–9, 85–6, 216; and the SPSL, 68, 206, 232, 294

Anders, Rudolf or Erwin, 173

antisemitism, 5–6, 27–8, 30, 39, 48, 60, 64, 86, 89, 211–12, 286, 295

Ardenne, Manfred von, 194, 196–7

Argentina, 211

Atkinson, Robert, 23–4, 32

Atomic Energy Commission, 279–81

Aufricht, Ernst Josef, 270

Auschwitz-Birkenau, 190–2, 219–20

Austria, 5, 74, 76, 102, 140, 147, 172, 173, 182, 191, 192, 199, 215, 234; academic refugees, 10, 15, 26, 27, 34, 41, 48, 50, 60, 68, 73, 81, 116, 175, 177, 194, 202–3, 211, 213, 231, 274–5, 286, 288; annexation by Germany, 178, 201, 203; and antisemitism, 27–8, 30, 48, 295; and Communist Party, 7, 41, 51, 60–1, 148

Austria-Hungary, 22, 23, 27–8, 41, 48, 84, 86, 93, 221, 290

Bacteriological Institute, Kharkiv, 65

Baerwald, Hans, 15, 71, 93, 237–8

Bartz, Ilse (Houtermans), 197–8

Batumi, Georgia, 33

Bauman and Berson Children's Hospital, Warsaw, 187

Beck, Guido, 238, 295–6; background, 15, 22–3, 27, 87; migration to USSR, 27, 87–8, 149; life in USSR, 100, 106; at Universities of Kiev and Odessa, 130–1, 132, 146, 149; departure from USSR, 150; in France, 208–11, 223, 293, 294;

Beck, Guido (*continued*)
in Portugal, 211; in Argentina, 211–12; migration to Brazil, 212
Belgian Université Libre, 83
Belgium, 83, 89, 132
Bergman, Stephan: background, 63; in USSR, 55, 63, 92, 143; in France, 208, 210, 223; in USA, 244–5
Beria, Lavrenty, 174, 178
Berlad, Austro-Hungary, 28
Berlin, 19, 23–4, 26, 28, 29, 30, 31, 33, 34, 44, 49, 50, 56, 58–9, 63, 65, 71, 73, 74, 82, 88, 89, 93, 95, 116, 122, 132, 160, 172, 185, 186–7, 205, 228, 230, 235, 237, 253, 263; Berlin Museum of Ethnology, 56; during the Cold War, 255, 257, 259–62, 277, 383; and Houtermanns, 193–6; and left-wing intellectual gatherings, 34, 41–3, 52, 68, 112, 150, 177, 252, 270
Bernal, J.D., 48, 179, 251, 254; and *The Social Function of Science*, 251–2, 254, 295
Bethe, Hans, 53, 195, 220
Beveridge, Lord, 12
Black Mountain College, 263–5
Blackett, Patrick, 32, 176, 179
Bloch, Felix, 177–8, 233
Bohr, Harald, 176, 251
Bohr, Niels, 32, 54, 88, 108, 123, 125, 146, 210, 251, 279
Borchardt, Hermann: background, 15, 30, 52, 295; and antisemitism, 30; and Weimer politics, 52; flight from Germany, 79; migration to USSR, 79–81; in Minsk, 92, 97–8, 99, 137; life in the USSR, 106–7; expulsion from USSR, 137–9, 144; return to Germany, 204–5, 294; arrest and imprisonment, 205, 208; flight to USA, 205–6, 266; in the USA, 266, 269–70, 296; plays, 267;

The Conspiracy of the Carpenters, 268–9, 270; rediscovery of his work, 270, 297
Born, Max, 23, 32, 121, 198, 203
Bracey, Bertha, 185
Bradford University, 255
Brasch, Arno, 24
Brecht, Bertolt, 52, 93, 205, 270
Breslau, 19, 22, 47, 56; Technical University of, 21
Brest-Litovsk, Poland, 183
Briggs, Lyman, 195–6
Brown University, 244
Brussels, 83, 89
Buchenwald concentration camp, 106, 207, 256
Budapest, 42, 53, 120, 149, 209, 293
Building Research Institute, Sverdlovsk-Urals, 85
Butyrka Prison, 140, 182

California Institute of Technology, 177
Cambridge, Massachusetts, 210
Cambridge University, 24, 29, 37, 54, 71, 73, 77, 108, 110, 120, 253
Cammann, Theodor, 199
Canada, 35, 47, 219
Carnegie Foundation, 6, 48
Cartan, Henri, 263
Castello Guglielmi, 221
Caucasus, 94, 101, 116, 283
Centro Brasileiro de Pesquisas Físicas, Rio de Janeiro, 212
Chernovtsy, Austria-Hungary, 28–9
China, 38, 68, 87, 269, 297
CNRS. *See* National Center for Scientific Research
Cockcroft, John, 24
Cohn-Peters, Hans-Jurgen, 93, 225–6, 259, 261
Cohn-Vossen, Margot Marie Elfriede (Friede), 84, 150, 157–8, 180, 283
Cohn-Vossen, Richard, 283

Cohn-Vossen, Stefan: background, 22; flight to USSR, 84; in USSR, 55, 92, 103, 136, 180, 283, 294

Collège de France, 209

Cologne, 19, 29

Comintern (Communist International), 43, 52, 57, 61, 78, 105, 135, 154

Committee for the Reception and the Organization of the Work of Foreign Scientists (Comité français pour l'accueil et l'organisation du travail des savants étrangers), 12, 208–9, 211

Compton, Arthur Holly, 86, 232–3

Congress for Cultural Freedom (CCF), 277–8

Cooper, Edna, 108

Copenhagen, 54, 88, 108, 176, 184, 210–11

Council for At-Risk Academics (CARA), 297

Crowther, J.G., 48, 117, 230, 251

Cybulski, Countess Zofia, 192

Czechoslovakia, 5, 10, 22, 27, 73, 76, 79, 86–7, 132, 201, 211, 215, 219, 221, 232–4

Czestochowa, 63

Czyste Hospital, Warsaw, 184, 186, 188, 190

Dachau concentration camp, 191, 205

Daix, Pierre, 272–3

Darkness at Noon, 270, 272, 276, 296

Davidovich, Semen Abramovich, 110–13, 119, 123, 140, 160, 164

Davison, Boris, 35

Der Knuppel, 52

Dirac, Paul, 25

Dolfuss, Engelbert, 60–1

Dorpat, Russia (Tartu, Estonia), 28

Drancy concentration camp, 190, 219

Dresden, 19, 21, 84, 222

Drösing, Austria, 60–1

Duke University, 32

Dulles, Allen, 218–19

Duschinsky, Erich, 219–20

Duschinsky, Fritz, 238; background, 26; migration to USSR, 83, in USSR, 92, 132–3, 137; and the Rotation Effect, 133, 296; in France, 208; death, 219–20, 223

East Germany. *See* German Democratic Republic

Einstein, Albert, 21, 22, 39, 49; letters of support for refugee scientists, 80, 176–7, 226–7, 231

electric trains, 100

Elsasser, Walter, 38–9

Emergency Committee in Aid of Displaced Foreign (German) Scholars. *See* American Emergency Committee in Aid of Displaced Foreign (German) Scholars

EMI (Electric and Musical Industries), 71, 73, 122

Emsheimer, Ernst: background, 46; in USSR, 46, 92, 116; departure from USSR, 105–6, 144; in Sweden, 106, 156, 225–6, 247; at Museum of Music History, 246–7; political activism, 263, 296

Emsheimer, Mia *née* Wilhelmine, 46, 144, 263

Engineering College of the Industrial Institute, Novocherkassk, 92, 131, 149, 150, 206

Esperanto, 59, 143–4, 172

Esterwegen concentration camp, 205

Estonia, 28, 57, 105, 226

Experimental Zoology and Morphology Laboratory, Leningrad, 92

Fabian, Dora, 76
Fenigstein, Henry (Henryk), 187, 190
Finland, 28, 56–7, 88, 100, 105
First World War, 6, 8, 27, 28–9, 34, 37,
    39, 41, 54, 61, 77, 83, 93, 94, 253
Fleischmann, Anneliese, 264
Fomin, Valentin, 122, 153, 173
France, 13, 29, 75, 76, 100, 194, 251;
    as an initial place of refuge, 15,
    52, 64, 79–83; as a refuge after
    expulsion from USSR, 120, 125,
    154, 190, 208–9, 210–12, 214, 242,
    292; in the Second World War,
    209–20, 223, 245, 247; Vichy, 10,
    213, 216–17; in the Cold War, 272–6
Franck, James, 23, 32, 88, 176
Frankfurt, 19, 46, 64
Franz Mehring Institute, 256
French Committee for the Reception
    and Organization of the Work of
    Foreign Scientists (Comité français
    pour l'accueil et l'organisation du
    travail des savants étrangers), 78,
    208–9
French Communist Party, 79, 272–6
French Foreign Legion, 214–17
Frenkel, Victor, 123
Frenkel, Yakov, 64, 88, 123
Frick, Wilhelm, 56
Friends (Quakers) Committee for
    Refugees and Aliens, 71, 184–5
Fröhlich, Herbert, 230; background,
    63–4; migration to USSR, 64;
    expulsion from USSR, 135–6,
    201–2; in Bristol, 202, 240, 247;
    postwar career, 241, 293
Fromageot, Clause, 212–13
Frumkin, A.N., 63
Fuchs, Klaus, 202

Gablonz, Czechoslovakia, 219
Gamow, George, 23, 32–3, 54, 242

Gans, Richard, 198–9
Gardiner, Margaret, 179
Gengenbach, Germany, 3
Georgia, USSR, 92, 101, 116, 170, 283
German Academy of Sciences, 259,
    261–2
German Communist Party
    (Kommunistische Partei
    Deutschlands; KPD), 9, 256; and
    Esperanto, 59; espionage for, 52,
    73, 154; and the Great Terror, 9,
    150, 154, 174, 207; member of, 19,
    43, 46, 51, 52, 55–6, 58, 67, 150, 266
German Democratic Republic (GDR),
    10, 226, 256, 263, 283, 296; and
    Fritz Lange, 261–2; and Harig,
    256–9; and Steinitz, 259–61
German universities, 24, 37, 39, 133;
    antisemitism in, 27–8, 56, 61–2, 115
Germany, 9, 30, 47, 48, 50, 57, 59, 64,
    67, 71, 74, 76, 81, 88, 100, 106, 115,
    116, 119, 131, 155, 161, 174, 179,
    212, 219, 221–3, 225, 227, 228–9,
    242, 253, 281, 283–5, 289, 291; and
    academic refugees, 3–7, 14–15, 19,
    41, 45, 48, 51–2, 63, 69, 70, 73, 74,
    80, 90, 92, 138, 150, 201–8, 230,
    234–5, 237, 244, 264–70, 297; and
    antisemitism, 5, 27–8, 84, 89; and
    Harig, 156; interwar culture, 93;
    and relations with USSR, 6, 54–5,
    70, 118, 135, 254, 280; and transfer
    of Soviet Prisoners, 181–200;
    treatment of Jews, 64–6; Weimar
    Republic, 29, 32, 58, 295
Gestapo, 9, 19, 23, 57, 71, 73, 76, 79,
    106, 152, 154, 155, 156, 173, 197–9,
    221; and exchange with USSR, 181,
    183–5, 190–4, 272; and return of
    refugees to Germany, 205–8, 231, 266
Gilde, Marina, 99, 102, 158, 184, 190,
    224, 294, 296

Gilde, Ruth *née* Gelinek, 26, 83, 99, 102, 158, 190, 224, 294, 296

Gilde, Siegfried, 13, 193, 224, 238, 292, 296; background, 26; flight from Germany to France, 82–3; in USSR, 91, 99; arrest and imprisonment, 157–9; transfer to Germans, 179, 181, 183–4, 199; in the Warsaw Ghetto, 184–5, 186–9, 200; research for the Hunger Disease project, 187, 189–90

Gilde-Berent, Mascha, 184

Ginzburg, Vitaly, 117–18, 281

Goldstein, Bernhard and Jakob, 191

Gordon, Maurice, 229

Göring, Hermann, 64, 66, 152

Great Britain, 4, 7, 11–12, 14, 15, 19, 26, 29, 35, 36, 37, 47, 68, 69, 84, 114, 124, 133, 176, 179, 194, 219, 220, 251, 254–5, 267, 272, 276, 288, 295; as a place of refuge, 71–8, 202–3, 213, 228–30, 232–7, 292–3; security services, 75–7, 106, 253–4

Great Depression, 6, 37, 41, 44, 48, 53, 75, 86, 88

Great Terror. *See* Purges and Great Terror

Green, Graham, 253–4

Grosz, George, 52, 205, 266, 270

Gugenheim, Franz, 65, 228

Guild for German Cultural Freedom, 267

Hahn, Otto, 194, 198

Haldane, J.B.S., 251, 254

Hanover, 19; Institute of Technology, 25, 290

Harich, Wolfgang, 258

Harig, Georg, 256

Harig, Gerhard: background, 19, 57; and KPD, 19, 51, 58; migration to USSR, 55, 57–8, 61; as historian

of science, 57–8, 129, 258–9, 278; in USSR, 92, 119, 129, 155–6; and NKVD, 105–6, 155–6, 207; at Buchenwald, 106, 207–8, 256, 294; in the GDR, 256–9, 262, 296

Harig, Katharina (Kathe) *née* Heizmann, 58, 102, 107, 155–6, 256–7

Harvard University, 60, 210, 234, 245, 284

Haslund-Christensen, H., 226

Hebrew University, 40, 204

Heidelberg, 19; Kaiser Wilhelm Institute for Medicine, 15, 26, 81

Heisenberg, Werner, 21, 23, 87, 129, 195–6, 211

Heitler, Walter, 202

Hellman, Hans, 13, background, 25; migration to USSR, 63; life in the USSR, 91, 99, 100, 102; science in the USSR, 119, 125–7; arrest and execution, 157, 159, 161, 173–4, 175, 180; scientific achievements and remembrance, 238, 290, 297

Hellmann, Hans Jr., 91, 174, 285, 288–90

Hellmann, Viktoria *née* Bernstein, 25, 63, 91, 99, 159, 285, 288–301

Hellmann–Feynman Theory, 25

Hermitage Museum, 46, 92, 116

Hertz, Gustav, 23, 32

Herzberg, Gerhard, 70

Hesse, Christophe, 267

Hilbert, David, 22, 32, 84

Hill, Archibald Vivian, 47, 151, 233

Hitler, Adolph, 4, 6, 11, 14, 22, 26, 35, 37, 44, 46, 48, 51, 58, 64, 65, 68, 73, 76, 80, 89, 90, 106, 152, 190, 193, 204, 220, 225, 248, 252, 257, 268, 275, 285, 290–1, 294, 296–7

Hofmann, August Wilhelm, 37

Hogben, Lancelot, 251

Holocaust, 10, 190, 221–2, 223, 248, 272–5, 294; and academic refugees, 10, 91, 133, 201; medical research and experiments in, 186–9; in France, 190, 200, 208, 211, 212–13, 215–17, 219–20, 223; in Poland, 186–93

Hopf, Heinz, 84

Hoselitz, Kurt, 202

Houtermans, Charlotte née Riefenstahl, 15, 185–6; background 31, 33, 54; and Oppenheimer, 32, 179, 236, 280; romance with Fritz, 23, 32–3; in Berlin, 42–4, 160; and the KDP, 51–2, 291; in Britain, 71–2, 74; migration to USSR, 72–3, 100; at UFTI, 93, 99, 111, 102, 108, 112, 118; and the Purges, 114, 149, 153; flight from USSR, 157–8, 235, 283; efforts to free Fritz, 176, 236, 251; ending of her marriage, 197; migration to USA, 236; career in USA, 236–7; remarriage to Fritz, 243

Houtermans, Elsa, 23, 176, 193, 236

Houtermans, Fritz: background, 15, 23–4; in Berlin, 34, 42–4, 160; and the KPD, 51–2, 291; migrations to USSR, 55, 72–3, 100; in Britain, 71–2, 73, 74; at UFTI, 93, 99, 108, 111–12, 119; scientific research in USSR, 122–3; and the Purges, 141, 149, 153, 157; arrest and imprisonment, 157, 162, 164–5, 167, 169–71, 173, 252; Russian Purges, 161–2, 277, 296; efforts to free him, 176–9, 271, 275; transfer to Germans, 181–4; in wartime Germany, 193–5, 200; German atomic research, 194–6; divorce from Charlotte, 196–7; mission to USSR, 197–9; rescue of Jews, 198–9; postwar, 242–3

Houtermans, Giovanna, 33

Hungary, 15, 27, 53, 56, 120, 191, 209

Hunger Disease, 187, 189–90

Huxleyy, Julian, 251

Hylleraas, Egil Andersen, 221

Illinois Institute of Technology, 244

Imperial Chemical Industries (ICI), 74–7

Imperial College, Royal College of Science, 74

Imperial Physical Technical Institute (Physikalische-Technische Reichsanstalt), 196

Institut Henri Poincaré, Paris, 210

Institute for German Folklore, 260

Institute for the History of Science and Technology, Leningrad, 92, 129, 155–6

Institute of Chemical Physics, Leningrad, 70

Institute of Communications and Electro-Mechanics, Moscow, 124

Institute of Physical and Chemical Biology, Paris, 212

Institute of Physical Problems, Moscow, 146

Institute of the Peoples of the North (INS), Leningrad, 57, 92, 98, 105, 127, 154

International War Crimes Tribunal, 262–3

Ioffe, Abraham, 58–9, 118

Isle of Man, 230, 239–40

Italy, 5, 34, 84, 217, 221, 223, 247, 252, 255

Ivanov, F.I., 115

Jacobsohn-Lask, Louis, 35

Japan, 87–8, 247

Jena, University of, 21, 55

Jewish Labour Committee (JLC), 216

Jewish Professional Committee (JPC), 74, 77, 213
Jewish Transmigration Bureau, 219
Jews, 25, 26, 39, 41, 45, 47, 49, 50, 58, 63, 83, 85, 86, 95, 107, 150, 194, 195, 210, 226, 227, 264, 269, 285, 289, 296; as academic refugees, 6, 10, 14, 15, 19, 22, 82, 83–4, 87, 88–90, 116, 144, 203–4, 207, 229, 234; in Austria, 27–8; and German racial policies, 19, 21, 23, 30, 56, 57, 61, 64–5, 70, 79, 154, 183, 193, 197, 200, 204, 206, 211, 228; in German universities, 4, 6, 19, 21, 27–8, 294; and the Holocaust, 10, 184–93, 198–200, 205–6, 207–8, 212–24, 274–5, 294; persecution of, 28–9, 30, 34, 295; and Zionism, 40, 64, 77
Johns Hopkins University, 176
Joliot-Curie, Frédéric, 178, 271, 275
Joliot-Curie, Irène, 178, 208, 271, 275
Jost, W., 25
Jourdan, Henri, 205

Kaiser-Wilhelm Institute, 26, 42, 49, 116
Kapitsa, Pyotr, 54, 145, 157, 178, 288, 290
Karl Marx University, 258
Karlstadt (Karlovac), Croatia, 221
Karpov Institute, 63, 125, 127, 159, 174
Kastler, Alfred, 263
Katowice, Poland, 26
Keilin, David, 77–8
Kettering, Karen, 175
Khanty (Ostyak), 128, 139, 154–5
Kharkiv, Ukraine, 8, 39, 44, 47, 50, 53, 54, 59, 60, 61, 65, 66, 69, 71–4, 91, 93, 94, 95, 98–102, 103, 106, 107, 108, 112, 113, 115–16, 120, 122–3, 125, 130–3, 140–2, 144–6,

151, 157, 160, 171, 172, 178, 185, 194, 225, 229, 250, 254, 279–80, 287–9; NKVD at, 108–9, 111, 144, 147–53, 160–1, 165–6, 168, 173–4; in Second World War, 123, 197–8, 285–6
Kharlamov, Alex. See Weisselberg, Alex
Kharlamov, Art (Artjom), 285–8
Kharlamov, Nadia, 286, 288
Kholodnaya Gora Prison, 167, 168, 177
Kiev, 104, 131–2, 144, 150, 167, 197, 228
Kirov assassination, 73, 92, 135–6
Klinger, Rudolf, 191–2
Knoll, Max, 24
Koestler, Arthur: in Berlin 43, 270; in Ukraine, 94–5, 140, 270; and efforts to rescue Weissberg, 178, 251, 271–2, 296; in the Cold War, 273, 275, 276–8
Kopfermann, H., 49
Korets, Moisej, 112–13, 140
Kraków, Poland, 41, 184, 190–2, 228
Kreditel Mainer oil refinery, Drösing, 60
Kuhn, Richard, 26
Kurella, Alfred, 150, 158, 283
Kurella, Heinrich, 150

Ladenburg, Rudolf, 49, 195–6, 237
Lanczos, Cornelius, 39
Landau, Lev: and Tisza, 53–4, 108, 120–1, 149, 209, 279; at UFTI, 108–13, 122, 140, 145–7, 164; in Moscow, 157, 178
Landsberg, Grigory, 49, 116–17
Lange, Fritz, 43, 93, 225; background, 24, 51, 59; migration to USSR, 55, 73; in Britain, 71, 73; at UFTI, 93, 120, 123; survival in USSR, 159–61, 180; in the Cold War, 256, 261–2, 283, 296
Langevin, Paul, 79, 80–1, 209

Lassalle, Ferdinand, 52
Laue, Max von, 193–5
Law for the Restoration of the Professional Civil Service, 3, 22, 61
Lederer, Edgar, 15; background, 25–6; and antisemitism in Austria, 27–8; flight from Germany, 52; route to USSR, 81; in USSR, 81–2, 92, 98, 100, 132, 136; flight from USSR, 154; in France, 154, 208, 212–13, 223, 245, 247; postwar scientific career, 245–6; political activism, 262–3, 265, 296
Lederer, Hélène née Fréchet, 81102
Leff, Laurel, 12
Lehmann-Russbueldt, Otto, 75–7
Leipunski, Alexander "Sasha," 71–3, 110–11, 118, 122, 148, 153, 160
Leipzig, 22, 57–8, 87, 207, 256
Lenin, 79, 129, 139
Leningrad, 94, 102, 104, 105, 117, 130, 139, 140, 260; and academic refugees, 46, 48, 56, 57, 58, 77, 81–4, 89–92, 98, 105, 116, 124–5, 127–8, 132, 212, 214, 240; Kirov assassination and expulsions, 92, 135–6, 202; and the Purges, 142, 144, 147, 150, 154–5, 219, 225–6
Leningrad Electrical Physical Institute (LEFTI), 92, 124
Leningrad Institute of Physical Chemistry, 70
Leningrad Physical Technical Institute (LFTI), 58, 64, 88, 92, 122, 129, 136
Leningrad State University, 40, 70, 84, 92, 115, 136, 154
Les Milles detention camp, 216–17, 219
Levy, Saul: background, 49, 50; in USSR, 117–19
Lewis, Cecil Day, 179
Lewy, Ernst, 56–7

Liberec (formerly Reichenberg), Czechia, 22
Liepman, Heinz, 78
Ling, Hans, 225–6, 246, 263
Lingens, Helene and Pepi, 191
Lithuania, 49, 57, 157–8, 184, 237
Łódź, Poland, 27, 39, 202
London, Fritz, 209
London, Heinz, 202
Lowinsky, Edward, 264
Lubyanka Prison, 55, 157
Luce, Claire Boothe, 269–70
Luftschitz, Heinrich: background, 20–1; Flight from Germany, 84–5; in Second World War, 221–2, 294; in the postwar, 222–3
Luzin, Nikolai, 142–3
Lvov (Lviv), Poland, 48

Maisky, Ivan, 177, 179, 251
Makeevka (Makiivka), Ukraine, 103
Makower, A.J., 213
Manhattan Project, 196, 266
Mansi (Vogul), 128
Margerie, M. Roland de, 82, 158
Margulies, Marek, 228
Marseille, 210, 213, 215–17
Massachusetts Institute of Technology, 53, 210, 221, 242
McLeod, J.W., 103, 132, 229, 238
Meitner, Lise, 194
Melnikov, G.B., 65
Mezhlauk, V.I., 152
MI5, 75, 77, 106, 253, 255
Military Medical Academy, Leningrad, 77, 124, 214
Millikan, Robert, 177–8
Mineralogical-Geological Institute, Moscow, 118
Minsk, 79–81, 117, 137–8, 266
Mises, Richard von, 63, 210, 245
Morgan, Claude, 273–5

Moscow, 52, 65, 81, 128; academic refugees and families in, 34, 49, 55, 59–60, 63, 78, 83, 85, 91–2, 99–100, 103, 117–18, 124–5, 127, 138–9, 155–6, 161, 173–4, 179, 204, 237, 283, 286, 288–9, 294; prisons, 3, 55; and prisoner exchanges, 181–3; in the Purges, 9, 134, 140, 140–2, 145–6, 149–50, 152, 154, 157–9, 177–8, 235, 271; and science, 94, 130, 198, 290; as seat of government, 80, 105, 109–11, 113, 118–19, 140, 273, 276

Moscow State University, 49, 55, 59, 80, 83–5, 116

Mount Sinai, Hospital, Milwaukee, 234–5, 239

Muller, Hermann, 35

Munich, 19, 44–5, 64, 240; Munich Accord, 221, 233, 272

Munro, James, 74–5

Muntz, Herman (Chaim): background, 39–40, 295; and Einstein, 39–40, 226; in the USSR, 40–1, 100, 115, 136; assistance to refugees, 55, 92, 136; expulsion from USSR, 154; in Sweden, 226–7

Murawkin, Herbert: background, 26, 58; and KPD, 51, 55, 291; flight to USSR, 57, 59, 61; and Esperanto, 59, 143–4, 172; and espionage, 59, 291; in USSR, 92, 123–4; arrest and execution, 143–4, 157, 171–2, 174; rehabilitation, 290

Museum of Ethnic History, Sweden, 226

Museum of Music History (Musikmuseet), Sweden, 246

Mykalo, (Anna) Galia, 101–2, 146, 166, 172, 180, 285–6

Nathansohn, Alexander, 34

National Center for Scientific Research (CNRS), 209, 211–12, 245–6

National Committee to Free Germany, 179

Nava, Mica, 13, 288

Nazis, 4, 5, 6, 10, 11, 14, 15, 19, 21, 28, 30, 33, 34, 35, 46, 49, 51, 56–61, 66–7, 91, 153, 154, 194, 198, 214, 220–1, 222, 225, 229, 244, 260, 269, 284, 292, 297; in the Holocaust, 190, 204, 205, 206, 208, 215, 218, 220, 223, 266–7, 273–6, 294; medical experiments by, 186; opposition to, 64, 65, 72, 75–8, 79, 179, 182, 191, 193, 198–9; Soviet conspiracy theories about, 152, 174, 183

Nazi–Soviet Non-Aggression Pact, 97, 253, 271, 280

Needham, Joseph, 251, 254

Neher, Carola, 182

Nernst, Walter, 24

Neumann, John von, 195

New School of Social Research, New York, 86

Niederwürschnitz, Germany, 57

NKVD, 3, 9, 50, 59, 94, 102, 158, 160–1, 176, 178, 271; and academic refugees, 50, 63, 97, 134, 141, 146–7, 149–53; arrests by, 33, 100, 140, 144–5, 148, 150, 151–2, 154, 156, 157, 159, 174, 186, 288–9; and espionage for, 105–8, 139, 154–6; executions by, 292; and expulsions from USSR, 138; interrogations by, 13, 60, 122, 135, 162–6, 172–4, 288; treatment of prisoners by, 162–71, 175, 228, 272, 277, 296; and transfer of prisoners to Germany, 181–5, 199; at UFTI, 8n11, 109–13, 146–9, 207

Noether, Emmy, 21

Noether, Fritz Alexander, 3, 179, 283; background, 21; expulsion from Germany, 61–2; at Tomsk, 62, 92,

Noether, Fritz Alexander (*continued*)
100, 119, 143; arrest, imprisonment,
and execution of, 159, 174–5,
180–3, 284, 292
Noether, Gottfried, 3, 284–5, 295
Noether, Herman(n), 3, 284–5, 295
Noether, Max, 21
Noether, Regina, 3, 92
Norway, 221, 224
Norwegian Institute of Technology,
243–4
Notgemeinschaft Deutscher
Wissenschaftler im Ausland, 85
Nuremberg Laws, 5
Nutrition Institute, Moscow, 91

Obremov, Ivan Vassiliyevitch, 41
Observatorio Astronómico,
Córdoba, 211
Odessa; 100, 117, 149, 151;
international physics congress, 45, 54
Office of Strategic Services, 218–19
OGPU. *See* NKVD
Oppenheimer, Robert, 23, 32, 179,
236, 279–81
Ordzhonikidze, Sergo, 109, 145
Orel, USSR, 3, 183
OSGO, Ukraine, 119–20
Ostyak. *See* Khanty
Otto, Karin, 12
Oxford Brookes University, 186
Oxford University, 11, 232

Palacz, Michal, 186
Palestine, 15, 77; immigration to, 40,
64, 85, 204, 206–7; migration from,
231, 244
Paris-Sud University, 245
Pasteur Institute, 81, 214
Pauli, Wolfgang, 23, 33, 69, 72, 195
Pauling, Linus, 23
Pawlowsky, E.N., 77, 124–5

Peierls, Rudolf, 33, 53, 220, 230, 232
People's Commissariat of Heavy
Industry (Narkomtjazhprom), 109,
111, 117, 119, 145, 152
Perón, Juan, 212
Perrin, Jean, 178, 209, 271, 275
Perttunen, Arhippa, 57
Physical Technical Institute in
Dnipopetrowsk, 64
Physico-Technical Institute of the
Urals, 48, 136
Physics Journal of the Soviet Union,
118–19, 121, 123, 129
Piccard, Auguste, 83
Placzek, George, 44, 280; at UFTI,
146, 172, 173; and Weissberg, 178,
185–6, 271, 275
Polanyi, Michael, 32, 42, 49, 177, 237,
252; in the Cold War, 277–8
Prague, 26, 28, 33, 87, 211; German
University of, 15, 49, 88; as a place
of refuge, 151, 158, 220–1,
232–5, 244
Princeton University, 49, 84, 177,
231, 237
Prinsgsheim, Peter, 26
Proletarian Esperantist International
(IPE), 59, 143
Purges and Great Terror, 9–11, 33, 50,
60, 91, 100, 134–73, 178, 186, 292,
288–9

Rapkine, Louis, 208–9, 211, 213
Red Cross, 197, 222, 228, 289
Regener, Erich, 25
Reiche, Friederich, 195
Rensselaer Polytechnic Institute, 244
Research Institute of Communications
and Electro-Mechanics, 59
Riga, Lithuania, 57, 157–8, 176
Ritchie Boys, 285
Rivesaltes detention camp, 217

Rockefeller Foundation, 6, 85, 213, 278; fellowships to scholars, 24, 48, 54, 69, 84, 86, 108

Roger, Max, 213

Romberg, Werner, 14; background, 26, 64; migration to USSR, 64; flight from USSR, 220–1; in Norway, 221, 223; postwar career, 243–4

Rosenfeld, Ludwig, 34–5

Rousset, David, 272–5

Royal College of Chemistry, 37

Royal Society, 29, 233, 241, 293

Rozenkevich, Lev, 152

Ruhemann, Barbara, 44, 92, 99, 102, 164; background, 31; *Low Temperature Physics*, 120; political views, 45, 96–7, 106, 146, 249–50; support of Communism, 249–51, 253–6, 262, 281, 296; at UFTI, 93, 95–7, 111, 118–19, 141, 156

Ruhemann, Martin, 13, 92, 99, 164, 238; background, 26, 29, 37; views of Soviet Science, 7, 252, 295; and Weissberg, 43; recruitment to UFTI, 44–5; political views, 44, 96, 106; at UFTI, 93, 95–7, 110, 119–20, 141, 146, 148, 156; *Low Temperature Physics*, 120; in Britain, 253–5

Ruhemann, Siegfried, 37

Ruhemann, Stephen, 253, 255

Russian Purge and the Extraction of Confession, 162, 277

Russian Revolution and Civil War, 6, 8, 28, 38, 79, 109, 202, 276

Rust, Bernhard, 89

Sachsenhausen concentration camp, 205

Sadowsky, Hilda *née* Bein, 88, 102, 231

Sadowsky, Michael, 13, 295; background, 28, 37, 88–9; flight to USSR, 89–90, 92; in the USSR, 92,

131–2, 136; expulsion from, 149–50; in Germany, 206; in Palestine, 206–7, 208; in USA, 231–2, 244

Saint-Cyprien detention camp, 219

Salazar, Oliveira, 211

Schaxel, Julius: and KPD, 19, 21, 55–6; background, 21, 55–6; migration to USSR, 51, 55–6, 57, 61; arrest and imprisonment, 55, 157, 159, 175, 170; in USSR, 77, 92, 127, 159, 179

Schein, Edgar, 86, 100–1

Schein, Hilde *née* Schoenbeck, 86

Schein, Marcel: background, 15, 86; in the USA, 86; at the University of Zurich, 86; expulsion from Switzerland, 87; migration to USSR, 87–8; life in USSR, 100–1, 130; scientific work in Odessa, 119, 131; flight from USSR, 131, 151; migration to USA, 232–3, 247–8, 294; scientific work in USA, 241–2; meeting with Oppenheimer, 280

Schlesinger, Charlotte "Lotte," 265, 270; background, 33, 42; and the Houtermans, 34, 99, 157; flight to USSR, 34, 55, 73–4; in USSR, 94, 103–4, 144, 150; flight from USSR, 150, 158, 283; migration to USA, 235–6; at Black Mountain College, 263–5; and her music, 265, 297

Schmellenmeier, Heinz, 199

Schneider, Georg, 35

Scholar Rescue Fund, 297

Scholars at Risk Network (SAR), 297

Schreker, Franz, 33

Schwartz, Laurent, 263

Schwarzkopf, Rudolph, 33, 235–6

Second World War, 13, 26, 85, 106, 115, 284–5; academic refugees in, 10, 186, 201, 214–24, 226, 237, 242, 245, 265, 291, 293–4, 131

Shifman, Misha, 13, 43, 161

Shoenberg, Isaac, 71
Shteppa, Konstantin F., 162, 167, 170, 198, 270
Shubnikov, Lev, 108, 110, 119, 152
Siberia, 3, 70, 88, 91–2, 104, 128–9, 137, 139, 143, 160, 166, 286
Siberian Physical Technical Institute (SFTI), 143
Siemens and Halske, 15
Simons, Helmuth: background, 15, 29–30, 295; in Britain, 74–8; and espionage, 75–7, 218–19; in USSR, 124–5, 214; in France, 13, 208, 213–17, 223, 294; escape to Switzerland, 217–18; postwar, 238
Simons, Sir Francis (Franz), 11, 31, 45, 232
Simonson, Ernst: background, 46; migration to USSR, 47; and Communism, 47–8; flight from USSR, 48, 151; in USSR, 93, 115–16; migration to USA, 232–5, 247–8, 292, 294; in USA, 235, 239–40
Simonson, Sophie, 102
Simpson, Esther, 184, 203, 206, 250
Sinelnikov, Kirill, 108, 110
Skarowa, Janka, 192
Skepper, C.M., 82–3
Social Function of Science, 251–3, 295
Social Hygiene Investigation Office, 47
Socialist Democratic Party (SDP), 52, 56
Society for Intellectual Liberty, 179
Society for the Protection of Science and Learning (SPSL), 3, 5, 12, 34–5, 38, 68, 70, 71, 77, 151, 163, 209, 240, 247–8, 249–50, 292–4; and Duschinsky, 83–4, 219–20; and Fröhlich, 202; and Gilde, 82–3, 184; and the Houtermans, 176, 178–9, 198–9, 236; and Luftschitz, 84–5,

222–3; policies, 69, 237; post-war tracing of missing scholars, 12, 219–20; role in academic rescue, 12, 21, 71, 74, 220, 222, 226; and Sadowsky, 89, 206, 231; and Schien, 222–3; and Simons, 74, 213; and Simonson, 233–5; and the United States, 232; and Wasser, 136, 202–4; and Zinnemann, 228–9
Sommerfeld, Arnold, 63
Soviet Esperanto Union (SEU), 143
Soviet Union (USSR), 4, 5, 13, 21, 24, 28, 42, 144, 155, 190, 193–4, 197, 209, 210, 212–14, 219–20, 225, 230, 233, 237, 242, 246, 247, 256–7, 259, 293–4; accusations against Houtermans, 198–200; arrests and executions in, 140, 144, 148, 156, 157–9, 161, 171–4, 182–3, 292; campaigns to release prisoners from, 174–9, 236; in the Cold War, 249–50, 257, 260, 261–2, 265–6, 273–81; contributions of scientists to, 114–33; departures from, 10, 154, 223, 232; disease in, 103–4, 229; and espionage, 43, 52, 57, 59, 105–6, 154, 156, 182–3, 193, 226; expulsions from, 135–8, 149–52, 155, 181–4, 156, 175–6, 180, 201–2, 204, 207–8, 226, 228, 231–2, 235, 238–41, 244; families of executed academics in, 283–90; migration of refugee scholars, 6–8, 11, 19, 20–2, 26–7, 34–6, 38, 40–1, 43–6, 48–50, 53, 55–8, 60–7, 68–81, 83–5, 88–90, 220, 291; prisons in, 162–71, 174, 182–3, 296; in the Purges and Great Terror, 9, 11, 60, 134–73, 178, 186, 292; and refugee scholars, 51–3; and rehabilitation of victims, 3, 284, 286, 289, 290; relations with Germany, 6, 54, 118, 181; life in,

8–9, 91–113, 181–3; science in, 7–8, 14–19, 23, 47–8, 54, 70, 79, 93–4, 108–13, 122, 145, 152, 178, 240, 258, 292; views of, 7–8, 10, 47, 53, 68–70, 79, 107, 117, 198, 250–5, 263, 266–7, 271–2, 291, 295–6; women scholars in, 31–4

Sozialistische Einheitspartei Deutschlands (SED), 256, 258, 261

Spain, 5, 10, 147, 212, 217, 263, 271

Special Branch, 76–7, 253

Spooner, Hertha, 32

Squire, Charles, 209–10, 242

SS (*Schutzstaffel*), 65, 189, 199, 207, 256

St Lawrence University, 185–6

Stalin, Joseph, 4, 11, 36, 43, 67, 80, 109, 133, 170, 190, 201, 205, 220, 223, 230, 241, 248–9, 257, 266, 284–6, 291–2, 295, 297; attitudes toward, 68–70, 90, 96–7, 106–7, 111, 113, 146, 160–1, 179, 180, 250–2, 254, 261, 269, 272, 280–1, 296; conspiracy theories against, 60, 141, 172–3, 175–6; Five-Year Plan, 8, 47, 94; and the Great Famine, 94; Purges and Great Terror, 9–10, 13, 41, 105, 114, 134–5, 141, 155, 161, 162, 163, 167, 172, 186, 275–8; personal appeals to, 138, 143–4, 155, 177, 178, 271

Stalinism, 258–62, 263, 265, 273, 273, 283

Stalino (Donetsk), 103

Stammberger, Friedrich, 35

Stark, Moyzis, 33

State Academic College of Music, Berlin, 33–4

State Optical Institute, Leningrad, 83, 92, 132

Steed, Wickham, 75, 124

Stein, Josef, 187, 189

Steinitz, Minna (Inge) *née* Kasten, 56, 98–9, 102, 104–6, 154–5

Steinitz, Wolfgang, 14, 262, 263; espionage, 51, 55–7, 291, 105; background, 56–7; migration to USSR, 57, 61; in Leningrad, 92, 98–9, 100; research in USSR, 127–9, 139–40; in the Purges, 139–40, 154–5; in Sweden, 106, 155–6, 225, 226; in the GDR, 226, 256, 259–61, 278, 296; research in folk music, 260–1

Stettin (now Szczecin, Poland), 207

Strachey, John, 179

Strassman, Fritz, 194

Striker, Eva, 250, 252; in Berlin, 42–3, 68, 150, 177, 270; in Kharkiv, 44, 69, 94, 100; arrest and imprisonment, 140–1, 144, 147, 163, 167, 251; release from prison, 175–6; efforts to rescue Weissberg, 177–8, 185–6, 251; marriage to Zeisel, 177; ceramic designer, 247; and Koestler, 43, 270–2, 296

Striker, Laura, 94, 100, 140–1, 175, 177, 251

Switzerland, 15, 24, 56, 76, 83–5, 100, 127, 196, 232; and expulsion of Schein, 86–7; treatment of Jewish refugees, 217–18; Simons in, 218, 238

Szilard, Leo, 32

Tammann, Gustav, 32

Tartu, Estonia, 28, 57

Technical University, Berlin, 28, 58, 88–9, 122

Teller, Edward, 53–4, 210, 242, 279–81, 293, 296

Tesla, Nikola, 124

Theresienstadt concentration camp, 198, 211

Thomson, George Derwent, 254
Thuringia, Germany, 55
Tisza, Laszlo, 26; background, 15; and Communism, 15, 53–5, 266; migration to USSR, 54–5; at UFTI, 93, 108, 119–22, 125; life in USSR, 98, 112, 132, 149; departure from USSR, 149–50; in France, 208–10; migration to USA, 208, 210, 223, 293; at MIT, 242; in the Cold War, 279
Tisza, Veronoka (Vera) née Benedek, 209
Tjomkin (or Temkin), Michail, 159
Toller, Ernst, 267–8
Tomsk, USSR, 3, 53, 55, 62–3, 91–2, 142–3, 159, 174
Tousckek, Bruno, 199
Traubenberg, Heinrich Rausch von, 198
Traubenberg, Marie Rausch von, 198–9
Treaty of Rapallo, 54
Tristena (Trstená, Slovakia), 86
Turin, Italy, 34
Turkey, 15, 297
Tyndall, A.M., 202

Udmurts, 57
Ukraine, 39, 63, 108, 145, 160–1, 170, 288, 297; and academic refugees, 44–5, 47, 51, 53, 73, 91, 93, 101, 103, 115, 144, 149, 220, 244, 266; famine, 55, 94–6; in Second World War, 197–9, 285–6
Ukrainian Coal Chemical Research Institute, 61, 93, 102, 146–7, 173
Ukrainian Commissariat of Labour, 47
Ukrainian Institute of Physical Culture and Education, 115
Ukrainian Physical Technical Institute (UFTI), 8, 11, 13, 59, 125, 164, 270, 279, 292; foreign scientists at, 38,

41–5, 54–5, 73, 93, 107–8, 118; life at, 93, 95–6, 99, 102, 107–9, 112–18, 132; struggle for control of, 109–11, 112–13, 140; during the Purges, 145–9, 152–3, 157, 160–1, 164, 172–3, 185–6; Houtermans wartime visit to, 197–8
Ulbricht, Walter, 258, 261, 283
Union of Soviet Socialist Republics (USSR). See Soviet Union
United States, 4, 49, 73, 120, 151, 160, 176, 193, 197, 269; and academic refugees, 7, 11–12, 14–15, 19, 35–6, 50, 63, 68, 83, 84, 85–90, 133, 204, 214, 220, 239–45, 247, 263–5, 293; and the atomic bomb, 194–6; and the Cold War, 139, 272, 280–1; and Communists, 139, 177–8, 231–7, 238, 280; and German academics, 33, 37–8; immigration to, 151, 178, 185, 205–6, 208–10, 213, 215–16, 217, 219, 228, 231–7, 248, 294
universities: Bern, 243; Breslau, 57, 61; Bristol, 202, 240, 293; Chicago, 86, 232–3, 241–2; Coimbra, 211; Cologne, 22, 84; Düsseldorf, 29; Frankfurt, 47, 240; Freiburg, 29, 46; Gottingen, 23, 25, 54, 63, 84, 121 (Houtermans at, 23–4, 32, 196, 198, 236, 242–3, 280); Greifswald, 30, 46; Halle, 257; Heidelberg, 86, 244; Istanbul, 210; Jena, 21, 55; Kansas, 88, 100; Kharkiv, 115, 145; Kiev, 69, 130, 146, 162; Liverpool, 241; London, 179; Lyon, 212–13; Minnesota, 13, 89, 206, 231, 239; Minsk, 92, 97, 99, 107, 204; Munich, 63, 64, 220; Nanking, 87; Odessa, 88, 100, 129–32, 241; Oslo, 221, 243–4; Paris, 82, 210, 212, 245; Porto, 211; Rochester, 69; Saskatchewan, 70; Stuttgart, 25, 31,

44–5; Tbilisi, 55, 92, 143; Tomsk, 55; Uppsala, 221; Würzburg, 29; Zurich, 86–8
University of Berlin (aka Friedrich-Wilhelms-Universität), 20, 24, 26, 29, 39, 49, 56, 58, 63, 73, 82–3; as Humboldt University, 259
Urbain, George, 82, 209
Urban Curt, 24
Ursell, Ilse, 219–20, 222–3

Vassar, 33, 236
Viborg, Finland, 28
Vienna, 23, 29, 41, 44, 46, 60, 73–4, 82, 102, 125, 144, 176–8, 191, 202–4, 236, 286; University of, 15, 22, 26, 27–9, 41, 48, 57, 60, 63
Villefranche Zoological Station, 29
Vishinsky, Andrey, 141, 173–4, 178
Vitamin Institute of Leningrad, 81, 92, 132

Wallace, Henry, 211–12
Warburg, Max, 205
Warburg, Otto, 205
Warsaw, 133, 184, 186–8, 190, 192–3, 220, 228, 274; Ghetto, 187, 296
Warschavsky, Adam, 85
Wasser, Elizabeth, 294
Wasser, Emanuel, 226; background, 27, 48; migration to USSR, 48; expulsion, 92, 135–6, 201; search for refuge, 201–4, 293; in France, 208, 219, 223
Wasser, Sara, 294
Webb, Beatrice, 179
Webb, Sydney, 179
Weimer Republic. See Germany
Weissberg, Alexander, 3, 34, 291, 294; background, 41–2; in Berlin, 41–4, 68; and Striker, 42–3, 94, 140–1, 175, 177–8, 185–6, 247; and

Koestler, 43, 95, 178, 270–1, 277; at UFTI, 43–4, 50, 55, 69, 73–4, 92–3, 98, 107–9, 111–13, 118–19, 122, 141–2, 160; and the Ruhemanns, 44–5, 95–7, 148, 156, 250, 254; and Weisselberg, 60–1, 93, 102, 147, 147, 172–3, 174; and Communism, 41, 43, 68, 95–7; arrest of, 144, 146–50, 153, 157, 160; and The Accused, 41, 160, 276–7; in prison, 162–71, 180–2, 252; international campaign for release, 175–9, 251, 271; transfer to the Germans, 181–4; in Poland, 184–6, 190–3, 199–200; Rousset trial, 272–6; in the Cold War, 277–9, 296
Weisselberg, Alex, 102, 172, 180, 285–90, 295
Weisselberg, Anna. See Mykalo, (Anna) Galia
Weisselberg, Kiffer, 288
Weisselberg, Konrad, 13, 15, 175; background, 27–9, 51; migration to USSR, 60–1; in Kharkiv, 93, 102, 146–7; arrest, imprisonment, and execution, 148–9, 157, 159, 160, 163, 166, 173–4; family of, 180, 283–8, 290, 295
Weissenberg, Karl, 18, 82, 158, 184
Weisskopf, Victor, 68; in the USSR, 69–70, 95, 151, 280; in USA, 70; and rescue of Weissburg, 185, 280
Weizmann, Chaim, 77–8
Weller, Grigory Maksimovich, 144
Wigner, Eugene, 195–6
Wilhelmshaven, 25
Winthrop College, 33
Wodak, Walter, 286
Wurm, Mathilde, 76

Yad Vashem, 191
Yeshiva College, 244

Yezhov, Nikolia, 173–4
Yugoslavia, 221–2

Zehden, Walter, 237; migration to USSR, 49, 116; in USSR, 102, 116–17; in Britain, 161, 230, 238, 248, 292; family, 102, 161, 230
Zeisel, Eva. *See* Striker, Eva
Zeisel, Hans, 177–8, 185–6
Zhukhovitskij, A.A., 159
Zinnemann, Irene (Ena) *née* Loesch: romance with Kurt, 64–6, 132; arrest and imprisonment, 149, 151–2, 157, 162, 167–70; release and flight from USSR, 176, 228

Zinnemann, Kurt, 234, 240, 250; flight from Germany, 64–6; in Kharkiv, 93, 103, 132; arrest and imprisonment of, 13, 60, 149, 151–2, 157, 162–3, 165; release and flight from USSR, 176, 228; in Britain, 228–30; at the University of Leeds, 238–9
Zinnemann, Lazar, 65, 229–30
Zinnemann, Leah, 65, 229–30
Zinnemann-Hope, Pamela, 65, 151, 163, 170
Zinoviev–Kamenev show trial, 141–2
Zionism, 40, 64, 77–8
Zoppot, Germany, 23
Zurich, 69, 84, 86, 100, 115, 218

Milton Keynes UK
Ingram Content Group UK Ltd.
UKHW011449150124
436070UK00014B/122/J

9 781487 543655